Starch and Starchy Food Products

Starch is one of the main staples in human food, its consumption having both positive and negative aspects. The exploration and exploitation of starches from alternative botanical sources has been increasing recently due to interest in the economic and social development of tropical and sub-tropical regional economies and in support of sustainability. The book reviews existing research on various aspects of starch, including physicochemical, nutritional and functional properties, plus applications in addition to foods. Emphasis is on the various physical and chemical modifications, which are aimed at improving the properties and applicability of starch.

Key Features

- Analyzes the state of the art of the scientific and technological problems associated with starch
- Describes various applications of starch in foods
- Provides a broad view on the field of starch and starchy foods

T0290897

Food Biotechnology and Engineering

Series Editor
Octavio Paredes-López

Volatile Compounds Formation in Specialty Beverages
Edited by Felipe Richter Reis and Caroline Mongruel Eleutério dos Santos

Native Crops in Latin America: Biochemical, Processing, and Nutraceutical Aspects
Edited by Ritva Repo-Carrasco-Valencia and Mabel C. Tomás

Starch and Starchy Food Products: Improving Human Health
Edited by Luis A. Bello-Pérez, José Alvarez-Ramírez, and Sushil Dhital

Bioenhancement and Fortification of Foods for a Healthy Diet
Edited by Octavio Paredes-López, Oleksandr Shevchenko, Viktor Stabnikov, and Volodymyr Ivanov

Starch and Starchy Food Products

Improving Human Health

Edited by
Luis A. Bello-Pérez, PhD,
José Alvarez-Ramírez, PhD,
and Sushil Dhital, PhD

CRC Press
Taylor & Francis Group
Boca Raton London New York

CRC Press is an imprint of the
Taylor & Francis Group, an **informa** business

First edition published 2023
by CRC Press
6000 Broken Sound Parkway NW, Suite 300, Boca Raton, FL 33487-2742

and by CRC Press
4 Park Square, Milton Park, Abingdon, Oxon, OX14 4RN

CRC Press is an imprint of Taylor & Francis Group, LLC

Library of Congress Cataloging in Publication Data

Names: Bello Pérez, Luis Arturo, editor. | Alvarez-Ramirez, Jose, editor. | Dhital, Sushil, editor.
Title: Starch and starchy food products : improving human health / edited by Luis Arturo Bello-Perez, Jose Alvarez-Ramirez, and Sushil Dhital.
Description: First edition. | Boca Raton : CRC Press, 2023. | Includes bibliographical references and index. | Summary: "Starch is one of the main staples in human food, its consumption having both positive and negative aspects. The exploration and exploitation of starches from alternative botanical sources recently has been increasing, due to interest in the economical and social development of tropical and sub-tropical regional economies, and in support of sustainability. The book reviews existing research on various aspects of starch, including physicochemical, nutritional and functional properties, plus applications in addition to foods. Emphasis is on the various physical and chemical modifications, which are aimed at improving the properties and applicability of starch"-- Provided by publisher.
Identifiers: LCCN 2022009145 (print) | LCCN 2022009146 (ebook) | ISBN 9780367543433 (hardback) | ISBN 9781032346243 (paperback) | ISBN 9781003088929 (ebook)
Subjects: LCSH: Starch. | Food--Composition.
Classification: LCC TP415 .S724 2023 (print) | LCC TP415 (ebook) | DDC 664/.2--dc23/eng/20220520
LC record available at https://lccn.loc.gov/2022009145
LC ebook record available at https://lccn.loc.gov/2022009146

ISBN: 9780367543433 (hbk)
ISBN: 9781032346243 (pbk)
ISBN: 9781003088929 (ebk)

DOI: 10.1201/9781003088929

Typeset in Kepler Std
by Deanta Global Publishing Services, Chennai, India

Contents

Series Preface

BIOTECHNOLOGY – OUTSTANDING FACTS

The beginning of agriculture started about 12,000 years ago and ever since has played a key role in food production. We look to farmers to provide the food we need but at the same time, now more than ever, to farm in a manner compatible with the preservation of the essential natural resources of the earth. Additionally, besides the remarkable positive aspects that farming has had throughout history, several undesirable consequences have been generated. The diversity of plants and animal species that inhabit the Earth is decreasing. Intensified crop production has had undesirable effects on the environment (e.g., chemical contamination of groundwater, soil erosion, exhaustion of water reserves). If we do not improve the efficiency of crop production in the short term, we are likely to destroy the very resource base on which this production relies. Thus, the role of so-called sustainable agriculture in the developed and underdeveloped world, where farming practices are to be modified so that food production takes place in stable ecosystems, is expected to be of strategic importance in the future; but the future has already arrived.

Biotechnology of plants is a key player in these scenarios of the 21st century. Nowadays especially, molecular biotechnology is receiving increased attention because it has the tools of innovation for agriculture, food, chemical and pharmaceutical industries. It provides the means to translate an understanding of, and ability to modify, plant development and reproduction into enhanced productivity of traditional and new products. Plant products, including seeds, fruits, and plant components and extracts, are being produced with better functional properties and longer shelf life, and they need to be assimilated into commercial agriculture to offer new options to small industries and finally to consumers. Within these strategies it is imperative to select crops with larger proportions of edible parts as well, thus generating less waste; it is also imperative to consider the selection and development of a more environmentally friendly agriculture.

The development of research innovations for products is progressing, but the constraints of relatively long times to reach the marketplace, intellectual property rights, uncertain profitability of the products, consumer acceptance and even caution and fear with which the public may view biotechnology are tempering the momentum of all but the most determined efforts. Nevertheless, it appears uncontestable that food biotechnology of plants and microbials is intensifying and will emerge as a strategic component to provide food crops and other products required for human well-being.

FOOD BIOTECHNOLOGY AND ENGINEERING SERIES

Series Editor: Octavio Paredes-López, PhD

The Food Biotechnology and Engineering Series focusses on addressing a range of topics around the edge of food biotechnology and the food microbial world. In the case of foods, it includes molecular biology, genetic engineering and metabolic aspects, science, chemistry, nutrition, medical foods and health ingredients, and processing and engineering with traditional- and innovative-based approaches. Environmental aspects to produce green foods cannot be left aside. At the world level there are foods and beverages produced by different types of microbial technologies on which this series will give attention. It will also consider genetic modifications of microbial cells to produce nutraceutical ingredients, and advances of investigations on the human microbiome as related to diets.

STARCH AND STARCHY FOOD PRODUCTS: IMPROVING HUMAN HEALTH

Editors: Luis A. Bello-Pérez, PhD, José Alvarez-Ramírez, PhD, and Sushil Dhital, PhD

As the editors of this book indicate, starch and starchy materials are commonly the major components of many food products and versatile ingredients of industrial materials. Research studies in the last century, and especially in the last decade, have led to formulation of numerous and attractive food products with outstanding healthy properties for human consumption; additionally, new industrial uses, based in the native and modified structure of starch have reached the commercial market. As it is clearly stated in the Series

Preface, the intention of this book is to be a meeting point for diverse visions related to research and development on starch.

This book contains 11 chapters with authors from 19 different universities, centers of research and other organizations who are doing basic and applied research and innovations in 8 countries located in 4 continents. In other words, this publication contains very rich viewpoints and experiences in a good number of starch and starchy components in traditional and innovative fields. In brief, this book provides a meeting space to induce the reader into the exciting world of this macromolecule, which is playing a key role in the scientific and technological challenges of the 21st century.

Thanks are due to the editors and to all authors for their excellent works in this book. Acknowledgments are also due to the editorial group of CRC Press, especially to Mr. Stephen Zollo and to Ms. Laura Piedrahita.

Preface

Starch and starchy materials are the major ingredients of many food products and versatile materials for various industrial applications. An accurate understanding of the molecular and granular structure of starch and the physicochemical properties have been the focus of intensive academic and industrial research since the early years of the 20th century. The results obtained from the research efforts have led to the formulation of more food products with improved health properties for human consumption. In addition, new and more diverse industrial applications have been derived from innovative modifications of the native structure of starch. A plethora of papers and books documenting diverse features of starch biology, characteristics, modifications and applications have been published in the last 100 years. Despite the vast amount of knowledge gained on the subject, starch offers new questions and innovative applications that renew the interest of academic communities. However, research problems turn out to be increasingly difficult and complex, which have required the collaboration of multiple fields of science and technology.

The meeting of diverse disciplines, as well as the unsuspected contribution of various fields of knowledge, have triggered the quantity and quality of research and application of starch. The intention of this book is to provide a meeting point for diverse visions related to research and development on starch. Chapter 1 by José Alvarez-Ramírez describes generalities about the structure, composition and modification of starch. Chapter 2 by Ming Li, Yu Tian and Sushil Dhital provides a stimulating and subtle description of unconventional sources of starch. Chapter 3 by Kai Wang, Bin Zhang and Sushil Dhital reviews innovative techniques to measure structure from the molecular to the granular level. Gelling, pasting, viscosity and rheological properties are critically described by Guantian Li and Fan Zhu in Chapter 4. Maribel Ovando-Martínez and Senay Simsek, in Chapter 5, offer a fine framing of innovative methods for starch modification with an emphasis on starch digestibility. Luis A. Bello-Pérez, Daniel E. Garcia-Valle and Roselis Carmona-Garcia contributed Chapter 6 to provide an exciting view of the nutraceutical properties of starch focusing on food and pharmaceutical applications. The great expertise of Darrell W. Cockburn is reflected in Chapter 7 by revealing the main results and research issues on the dietary fiber characteristics of starch. The important issue of starchy flour modifications for altering digestibility is contributed by Edith Agama-Acevedo and her coworkers Madai Lopez-Silva and Andres Aguirre-Cruz in Chapter 8. Pulses, pseudocereal, tubers and roots as unconventional starch sources for traditional and functional food applications are discussed in Chapter 9 by Pablo Martín Palavecino et al. Chapter 10 by José A. Gallegos-Infante et al. provides a motivating discussion on legume starches, their application in pasta foods and associated health benefits. The emerging concept of the food matrix, its influence on starch digestibility and subsequent metabolic responses is discussed in depth by Yolanda Elizabeth Pérez-Beltrán, Juscelino Tovar and Sonia G. Sáyago-Ayerdi in Chapter 11. The collection of descriptions, critical opinions and proposals in the 11 chapters of the book should form a meeting space that motivates the reader to enter the exciting world of starch research and applications.

Series Editor

Octavio Paredes-López, PhD, a biochemical engineer and food scientist, earned bachelor's and master's degrees in biochemical engineering at the National Polytechnic Institute in Mexico City, and later earned a master's degree in biochemical engineering at Czechoslovak Academy of Sciences, Prague. He earned a PhD (1980) at the University of Manitoba, Canada, which awarded him an honorary Doctor of Science in 2005. He is a past president of the Mexican Academy of Sciences, a founding member of the International Academy of Food Science and Technology, and an ex-member of the Governing Board of the National Autonomous University of Mexico. He received the National Prize for Arts and Sciences in 1991, the highest scientific recognition in Mexico, and the Academy of Sciences of the Developing World Award (previously Third World Academy of Sciences) in 1998, Trieste, Italy. Paredes-López has conducted research and post-doctoral stays in the United States, Canada, Britain, France, Germany, Switzerland, and Brazil. He has served as general Editor of the journal *Plant Foods for Human Nutrition*, and editorial board member of *Critical Reviews in Food Science and Nutrition*, *Frontiers in Food Science and Technology*, and 10 other international journals. He has authored over 300 publications in indexed journals, book chapters, and books and over 150 articles in newspapers in Mexico, USA, and France. His h-index is 55.

Editors

Luis A. Bello-Pérez, PhD, is a Distinguished Professor at the Department of Technological Development, Center for Development of Biotic Products at the National Polytechnic Institute, Mexico. He supervises the research of graduate students and visiting scholars and works closely with local and global research and education. He earned a Bachelor in Biochemical Engineering from Acapulco Institute of Technology, Mexico, Master in Bioengineering and PhD in Plant Biotechnology from CINVESTAV. He also worked as a postdoctoral research fellow at the National Institute of Agronomic Research and took a sabbatical year at Whistler Center for Carbohydrate Research, Purdue University. He received the National Prize for Food Science & Technology in 2012, the highest scientific recognition in Mexico in this area. His research interest is on starch: chemistry, technology and digestibility, with an emphasis on starch from non-conventional sources. His research includes the development of healthy foods, mainly bakery products, snacks and pasta with high dietary fiber content, and gluten-free products. Bello-Pérez has published 395 peer-reviewed research papers, 22 book chapters, and many other reviews and abstracts. He is a member of the Institute of Food Technologists and also of the Cereals & Grains Association. He is Associate Editor of *Journal of the Science of Food and Agriculture*, *Cereal Chemistry* and *LWT*.

José Alvarez-Ramírez, PhD, is a full-time Professor of Chemical Engineering at the Universidad Autonoma Metropolita, Campus Iztapalapa, Mexico. He supervises the research of graduate students and visiting scholars and works closely with local and global research and education. He earned a Bachelor in Chemical Engineering at Universidad Autonoma de San Luis Potosi, Mexico, and PhD in Chemical Engineering at Universidad Autonoma Metropolitana, Campus Iztapalapa. He was a visiting researcher at Exxon Research and Engineering Company, Annandale, New Jersey, in 1992, and spent a sabbatical year at Instituto Mexicano del Petroleo, Mexico City, in 2002. His major research interests are chemical process dynamics and control and food science. His special research interest is on starch properties, functionality and applications in food technology. The main focus of his research group is development of healthy foods, including bakery products, snacks and pasta with high dietary fiber content, and gluten-free products. He has published about 400 peer-reviewed research papers.

Sushil Dhital, PhD, is a Senior Lecturer at the Department of Chemical and Biological Engineering at the Monash University, Australia. Sushil completed his PhD at the University of Queensland (UQ), Australia, in 2011, elucidating the structure–property relations of starch granules, and after which he accepted a research fellow position at UQ, examining the starch–plant cell wall interactions. Sushil joined Monash University in July 2019. He currently serves as Editor for the journal *Carbohydrate Polymers* and is on the Editorial Board of journals *Food Hydrocolloids*, *Food Chemistry*, and *Bioactive Carbohydrates and Dietary Fibre*. He also serves as the Vice President of The Australasian Grain Science Association (AGSA). Sushil carries out fundamental and applied research on elucidating the structure–property–function–health relationships of food and food ingredients. His stronghold and research interest are in relating the plant molecular structures to macroscopic properties relevant to food, health, and product development. He uses cross-disciplinary approaches drawing from physics, chemistry, biology, and engineering. He uses various in-vitro and in-vivo models to elucidate the fundamental mechanisms beyond the nutritional and processing functionality of food and food ingredients. Sushil has published more than 100 peer review research and review articles exceeding 4500 citations.

Contributors

Edith Agama-Acevedo
Instituto Politécnico Nacional
CEPROBI
Yautepec, Morelos, México
Orcid.org/0000-0002-8224-0790

Andres Aguirre-Cruz
Universidad del Papaloapan
Tuxtepec, Oaxaca, México
Orcid.org/0000-0001-9232-3330

José Alvarez-Ramírez
Universidad Autónoma
Metropolitana
Iztapalapa, México City, México
Orcid.org/0000-0002-6482-889X

Luis A. Bello-Pérez
Instituto Politécnico Nacional
CEPROBI
Yautepec, Morelos, México
Orcid.org/0000-0001-7286-273X

Mariela Cecilia Bustos
Instituto de Ciencia y Tecnología
de los Alimentos Córdoba
(CONICET-UNC)
Universidad Nacional de Córdoba
(UNC)
Córdoba, Argentina
and
Instituto de Ciencias Básicas y
Aplicadas
Universidad Nacional
de Villa María
Córdoba, Argentina
Orcid.org/0000-0003-4444-065X

Roselis Carmona-Garcia
Tecnológico Nacional de
México-Tuxtepec
Oaxaca, México
Orcid.org/0000-0003-0307-1413

Darrell W. Cockburn
Department of Food Science
Pennsylvania State University
State College, Pennsylvania
Orcid.org/0000-0001-6791-6599

Sushil Dhital
Department of Chemical and
Biological Engineering
Monash University
Clayton, Victoria, Australia
Orcid.org/0000-0001-9405-1244

Claudia G. Félix-Villalobos
Tecnológico Nacional de
México-Durango
Durango, México
Orcid.org/0000-0003-0922-1304

José A. Gallegos-Infante
Tecnológico Nacional de
México-Durango

Durango, México
Orcid.org/0000-0001-6450-0483

Daniel E. Garcia-Valle
Instituto Politécnico Nacional
CEPROBI
Yautepec, Morelos, México
Orcid.org/0000-0002-0670-7707

Rubén F. González-Laredo
Tecnológico Nacional de
México-Durango
Durango, México
Orcid.org/0000-0001-6329-1413

Alberto Edel León
Instituto de Ciencia y Tecnología
de los Alimentos Córdoba
(CONICET-UNC)
and
Facultad de Ciencias Agropecuarias
Universidad Nacional de Córdoba
(UNC)
Córdoba, Argentina
Orcid.org/0000-0002-2260-3086

Guantian Li
School of Chemical Sciences
University of Auckland
Auckland, New Zealand
Orcid.org/0000-0002-3294-2153

Ming Li
Ministry of Agriculture and Rural
Affairs
Beijing, China
Orcid.org/0000-0003-2204-3944

Madai Lopez-Silva
Instituto Politécnico Nacional
CEPROBI

Yautepec, Morelos, México
Orcid.org/0000-0002-1890-779X

Martha R. Moreno-Jiménez
Tecnológico Nacional de
México-Durango
Durango, México
Orcid.org/0000-0002-5865-9583

Maribel Ovando-Martínez
Departamento de Investigaciones
Científicas y Tecnológicas
Universidad de Sonora
Sonora, México
Orcid.org/0000-0002-3282-9636

Pablo Martín Palavecino
Instituto de Ciencia y Tecnología
de los Alimentos Córdoba
(CONICET-UNC)
Universidad Nacional de Córdoba
(UNC)
Córdoba, Argentina
and
Instituto de Ciencias Básicas y
Aplicadas
Universidad Nacional de Villa María
Córdoba, Argentina
Orcid.org/0000-0001-6318-6234

Yolanda Elizabeth Pérez-Beltrán
Tecnológico Nacional de México/
Instituto Tecnológico de Tepic
Tepic, Mexico
Orcid.org/0000-0001-9638-6048

Pablo Daniel Ribotta
Instituto de Ciencia y Tecnología
de los Alimentos Córdoba
(CONICET-UNC)
and

Facultad de Ciencias Exactas, Físicas
y Naturales
Universidad Nacional de Córdoba
(UNC)
Córdoba, Argentina
Orcid.org/0000-0001-7883-8856

Pedro Losano Richard
Instituto de Ciencia y Tecnología
de los Alimentos Córdoba
(CONICET-UNC)
Universidad Nacional de Córdoba
(UNC)
Córdoba, Argentina
Orcid.org/0000-0003-3487-9589

Nuria E. Rocha-Guzmán
Tecnológico Nacional de
México-Durango
Durango, México
Orcid.org/0000-0002-5715-8939

Walfred Rosas-Flores
Tecnológico Nacional de
México-Durango
Durango, México
Orcid.org/0000-0002-0783-3019

Sonia G. Sáyago-Ayerdi
Tecnológico Nacional de México/
Instituto Tecnológico de Tepic
Tepic, Mexico
Orcid.org/0000-0002-4430-1273

Senay Simsek
Department of Food Science
Purdue University

West Lafayette, Indiana
Orcid.org/0000-0003-0238-5947

Yu Tian
Department of Plant and
Environmental Sciences
University of Copenhagen
Frederiksberg, Denmark

Juscelino Tovar
Department of Food Technology,
Engineering and Nutrition
Lund University
Lund, Sweden
Orcid.org/0000-0002-6944-4991

Kai Wang
College of Food Science
South China Agricultural University
Guangzhou, China
Orcid.org/0000-0002-8072-7220

Bin Zhang
School of Food Science and
Engineering
South China University of Technology
Guangzhou, China
Orcid.org/0000-0003-4465-4826

Fan Zhu
School of Chemical Sciences
University of Auckland
Auckland, New Zealand
Orcid.org/0000-0003-3344-7741

Chapter 1

Starch

An Overview

José Alvarez-Ramírez

CONTENTS

1.1 INTRODUCTION

Starch is the most abundant fraction in food products and, together with cellulose, is the most used worldwide material for a large diversity of purposes. By 2020, the US starch economic output represented about 47.50 billion USD, supporting 168,000 direct jobs (https://internationalstarch.org/, retrieved 20 November 2021). Starch-related products included sweeteners, dextrin, and modified and pregelatinized starch. Worldwide, China, India, Germany and

DOI: 10.1201/9781003088929-1

the US are the leading starch exporters, accounting for about 1 million tons per year (https://wits.worldbank.org/trade, retrieved 20 November 2021).

In recent decades, the application of starch has extended to many economic and industrial sectors, including textile, paper, petroleum extraction, health and medicine (Wertz & Goffin, 2020). Starch is used in the food industry as a main compound (e.g., bread and pasta), additive (e.g., soups) and texture modifier (e.g., sausages). Its application in the pharmaceutical industry is mainly in drug delivery systems (Builders et al., 2016). Low cost, nontoxicity and no irritability characteristics make starch quite suitable and with many advantages over most biopolymers as a drug excipient.

Maize, wheat, rice and potato are the traditional botanical sources of starch. However, alternative sources have emerged in recent years, driven mainly by regional economic and ecological (e.g., sustainability) aims. Cassava, also called tapioca or yuca, is a subtropical plant cultivated under low nutrient availability and relatively frequent droughts (Burrell, 2003; Tonukari, 2004). Starch is extracted from the swollen cassava root, with a yield that is about 40% higher than rice and 25% more than maize. By the early 2000s, about 75% of cassava production was intended for human consumption (Nwokoro, 2002) and the remaining fraction for animal feed and minor industrial applications. However, these figures are changing as cassava starch is increasingly used in diverse modified forms for industrial applications. The recent two decades have witnessed a huge increase in reports on starch extracted from nontraditional botanical sources. Unripe banana is highly available in tropical and subtropical regions, and has been explored as a starch source for improving the socioeconomic activity in low-income regions (Ovando-Martinez et al., 2009). Yam tubers from different species offer sustainable starch production and at the same time provide native and modified materials with attractable properties for the food industry (Zhu, 2015). Taro starch is increasingly used as an additive given its amylose content, granule size and crystallinity degree characteristics, making this starch affordable for food structuring (Andrade et al., 2017).

In a broad sense, starch is still the most consumed staple for human feeding. Although traditional starch sources have been the basis of the historical development of human beings, the recent decades have seen the flowering of research efforts and industrial applications for exploring new botanical sources and developing alternative forms of starch structures with marked advantages and tailored characteristics.

1.2 MOLECULAR COMPOSITION

Starch is a homopolysaccharide formed by glucose units. Linear chains or chains with few branches are identified as the amylose fraction and consist of approximately 500–2000 glucose units. On the other hand, long branched chains become the amylopectin fraction, containing about 10^6 glucose units. The amylose and amylopectin fractions are approximately 98%–99% db of the composition of most native starches (Schmiele et al., 2019). The linear backbone is an α-(1,4)-linked glucan, and branches are linked by an α-(1,6) glycosidic bond. Amylose presents a moderate molecular weight of about 10^6 daltons, while amylopectin has a higher molecular weight of about 10^8 daltons. The amylose/amylopectin ratio is a distinctive feature of the botanical source and is a major determinant of physicochemical (e.g., retrogradation and gelatinization) and functional properties (e.g., swelling, syneresis and viscoelastic behavior of hydrogels). The amylose content varies in the range of 20%–30% for most native sources. In terms of the amylose content, starches commonly are classified as waxy (<15%), normal (20%–35%) and high amylose (>40%). Starches from mutants with higher amylose content (HAS), up to 95%, are commercially available, although their presence in nature is quite rare. Instead, normal botanical sources are modified to shift the enzymatic system to build up amylose over amylopectin chains (Yano et al., 1985; Rolland-Sabaté et al., 2012). It has been postulated that the increase of amylose content can be achieved in three ways: (1) increase granule-bound starch synthase action to favor amylose synthesis, (2) decrease relative amylopectin levels via inhibition of starch synthases or (3) decrease amylopectin branching by inhibition of starch-branching enzymes (Tetlow, 2011; Wang et al., 2017). In general, modifications to regulate the enzymatic action (e.g., synthetase and branching enzymes) are utilized to obtain crops with desired characteristics, such as amylose-free, high-amylose starch and altered amylopectin structures.

1.3 SUPRAMOLECULAR ORGANIZATION

1.3.1 Granule Structure

The biopolymeric chains of native starch are organized in a compact granular structure. It is commonly accepted that the external granule layers have high amylose contents compared to the internal core of the granule (Jane, 2006). Amylopectin is also contained in the periphery, but it consists of shorter branch chains. Amylopectin is preponderantly located in the inner regions of the granule, with amylose chains interspersed to form a complex structure. The α-1,4-glucans with some α-1,6-branches placed in parallel allow the formation of double helices, which are very densely packed with a high degree of regularity. Branched chains are located at specific sites in a brushlike manner to lead to the formation of amorphous lamellae (about 4.5 nm thick) and regions of linear chains having a crystalline lamella configuration, almost equal in thickness.

The lamellar repeats are highly conserved (9–9.5 nm) based on the botanical sources. Each lamella is made up of about 100 double-stranded helices, each with about 20 glucose units. On a larger microscopic scale, lamellas are packed to form well-defined growth rings (Gallant et al., 1997; Coultate, 2009), resulting in alternating shells of amorphous and *semicrystalline* regions, between 100 and 800 nm thick. Atomic force microscopy studies have revealed the existence of blocklets within the growth rings, which are more or less spherical with a size of 20–100 nm (Baker et al., 2001; Tang et al., 2006). Although the double-helices configurations forming lamellae and the upscaled growth ring structure have been characterized to some detail, the presence of intermediate-scale structures like superhelices and blocklets is still controversial (Wang et al., 2020). Even more, the link between the structures at different scales is still an issue of detailed research (Perez & Bertoft, 2010).

The branched nature of amylopectin poses some interesting problems on its spatial organization into the starch granule. Nikuni (1969, 1978) proposed that amylopectin is organized as periodic clusters. French (1972) postulated that the existing experimental evidence pointed out that a homogeneous structure of inner branched clusters was not possible and proposed two potential scenarios: a modified trichitic structure and a cluster (racemose) hypothesis. It was apparent that the cluster scenario was consistent with the structures of Nägeli amylodextrins. Hizukuri (1986) proposed that the periodicity in the size distribution of constituent 1,4-linked chains might be explained by proposing that linear chains were either located within a single cluster or served to connect two or more clusters. The aforementioned studies triggered a plethora of studies oriented to elucidate the branched organization of amylopectin. An idea that has an increasing acceptance is that the branches do not follow a random pattern, but rather they follow an organized although complex geometry (Thompson, 2000). More recent studies have abounded on the complexity of the amylopectin structure. For instance, Bertoft (2013) postulated a two-directional backbone model, where the direction of the clustered chains is perpendicular to the direction of an amorphous backbone of long B-chains. Also, sidechains are attached to the backbone, incorporating additional chains to form building blocks of varying sizes. Current experimental evidence is still controversial, with some results favoring the cluster model and others the backbone configuration (Bertoft, 2017; Ai & Jane, 2018).

1.3.2 Morphology

The distribution of granule size and shape are strongly determined by the germplasm where starch was stored. Climatic conditions and agronomic factors also contribute to the morphology of starch granules (Li et al., 2021). Scanning electronic microscopy (SEM) is the usual approach to determining granule size and morphological characteristics. Carbohydrates predominate (90%–95%) on the granule surface, having a determinant effect on the granule morphology. The shape of granules is quite diverse, with pentagonal and polyhedral geometry for most rice cultivars (Wani et al., 2012), and angular-shaped for corn. Potato exhibits oval-shaped granules, whereas wheat shows spherical and flat circular shapes. Rice granules have a very small size (about 6 μm). In contrast, potato granules are relatively large (about 100 μm). The mean diameter of corn starch granules is about 35 μm with a monomodal distribution. In contrast, wheat starch exhibits bimodal distributions with small granules of 4 μm and large granules averaging 14 μm (Singh et al., 2003). Starch granules are heterogeneous structures containing porous and channels. Commonly, A-polymorphic cereal starches exhibit pores on the surface leading to serpentine channels (Fannon et al., 1993). Pores can penetrate the granule structure to form large cavities and channels as in maize granules (Huber & BeMiller, 1997). The channeled geometry of starch granules affects the susceptibility of starch to enzymatic reactions, chemical actions and dye adsorption (Huber & BeMiller, 2000). The heterogeneous structure of starch granules has potential benefits for ethanol production given the increased surface available for enzymatic hydrolysis (Fannon et al., 2004). The diffusion of dextran in pores of maize starch was studied with fluorescent dextran probes, finding to be either fast with diffusion coefficients in the order of 10^{-6} cm^2.s^{-1} or slow with diffusion coefficients in the order of 10^{-7} cm^2.s^{-1} (Dhital et al., 2013).

1.3.3 Crystallinity and Double-Helical Structure

One assumption generally accepted in the literature is that native starch granules are composed of only a double-helical structure and an amorphous phase. The double-helical structure is responsible for the ordered structures. Amylose is the main fraction of the amorphous regions. Less ordered amylopectin is also found in amorphous forms (Morrison, 1995). The packing of adjacent double helices leads to regular spatial geometrical patterns. It was also found that amylose can be a significant contributor to the crystallinity for high-amylose starches (Tester et al., 2000). The crystallinity of starch granules is usually determined by wide-angle X-ray diffractometry, giving indices of absolute and relative crystallinity contents. The former refers to crystalline components relative to the amorphous components of the diffractogram. The type of starch crystallinity is designated as A, B or C depending on the X-ray diffraction peaks. The crystallinity differences are due to the bonded water content and the pattern of crystallite packing (Imberty et al., 1991). B-type crystallinity exhibits large voids in the lattice where water molecules can be attached, leading to a hexagonal configuration. In contrast, A-type crystallinity displays reduced voids that limit the bonding of water molecules, resulting in an orthorhombic structure (Rodriguez-Garcia et al., 2021). C-type crystallinity is a combination of A-type and B-type crystallinities.

It has been reported that the granule crystallinity is correlated with the double-helical content (Lopez-Rubio et al., 2008a). However, the determination of the double-helical structure and content in starch granules is a more subtle problem. Tan et al. (2007) used 13C CP/MAS NMR to estimate and relative proportions of amorphous, single and double-helical components in starch granules. In contrast to previous results, it was shown that starch from different sources contains amorphous phases with different 13C NMR spectral characteristics. Fan et al. (2013) used the 13C CP/MAS NMR approach to characterize the impact of different heating methods on the ordered structures of starch granules. The results showed that microwave, oil bath and conventional slow heating have similar effects in double helices, the V-type single helix and amorphous structures with rising temperature. The characterization of the double-helical structures and the potential links with structural and functional characteristics of starch granules are issues that still have many open questions.

1.4 MINOR COMPONENTS

Lipids and proteins are found in starch granules in very low fractions (commonly <1%). Starch granules from cereals contain about 0.25% proteins and 1% lipids (Baldwin, 2001). The lipids in starch granules are found as starch–lipid inclusion complexes (Morrison, 1998). Normal rice starches contain about 0.9%–1.3% lipids. Waxy rice starches contain negligible amounts of lipids (Azudin & Morrison, 1986). However, starches from nontraditional botanical sources like amaranth (Tapia-Blácido et al., 2007) and taro (Gonzalez-Soto et al., 2011) have higher contents of proteins. It has been reported that a high number of proteins are located at the starch granule surface, and their association with lipids appears to influence the properties of starch granules and food products (Baldwin, 2001).

The role of minor starch granule components in the molecular organization of starch granules has been a matter of debate in the recent two decades. It has been postulated that proteins maintain the integrity of the granule shells via the formation of complexes (Han & Hamaker, 2002). It has been also suggested that proteins, lipids and amylose are involved in swelling mechanisms of starch granules (Debet & Gidley, 2006). Recent studies have revealed that deproteinization and defatting treatments affected the morphology of starch granules as aggregation disappears and the starch granule structure was disrupted (Zhang et al., 2021). Besides, it was found that the role of endogenous proteins in physicochemical changes is more crucial than that of endogenous lipids (Ding et al., 2021).

1.5 PROPERTIES

1.5.1 Gel Viscoelasticity

The rheological behavior of starch gels is an important issue for a diversity of applications, including sausages, doughs, pasta extrusion and adhesives. An accurate characterization of the rheological response of starchy food provides valuable insights on the viscoelasticity and flowability of starchy food matrices. The effect of starch concentration on the viscoelastic properties of starch gels depends on the botanical source, chemical modifications and amylose content (Evans & Lips, 1992; Campo & Tovar, 2008). Although amylose is the main determinant of gel viscoelasticity, it has been reported that amylopectin chains also impact the pasting and rheological patterns (Li et al., 2019a, b). Some results have focused on the effects of salts (Ahmad & Williams, 1999), ionic liquids (Devi & Das, 2020), chemical modifications (Yousefi & Razavi, 2015) and dietary fiber (Liu et al., 2016). Most of the rheology studies reported in the specialized literature focused on the behavior in the linear regions. The use of large amplitude oscillatory shear (LAOS) tests has been increasingly considered in the recent few years (Precha-Atsawanan et al., 2018). This technique is more realistic for processing since starchy food matrices are usually subjected to large elongational and oscillatory deformations. In particular, information on the nonlinear dynamical characteristics of the starch gels can be drawn from LAOS analysis.

1.5.2 Digestibility

The degradation of starch provides much of the energy needed by humans. However, the process is also responsible for undesirable body responses or metabolic disorders, such as excessive energy intake and high blood glucose levels. These consequences can become major risk factors resulting in increasingly common diseases such as obesity and type 2 diabetes.

The enzymatic degradation of starch chains provides most of the energy required for human activity. The postprandial metabolism depends strongly on the mechanisms involved in starch hydrolysis and absorption in the digestion tract. Salivary α-amylase starts starch degradation, followed by hydrolysis by gastric amylase in the stomach and subsequently by pancreatic α-amylase in the small intestine. The degradation products are small oligomers and α-limit dextrins. These products are further degraded by glucoamylases into glucose as the final product (Nichols et al., 2003). An accurate understanding of the mechanisms involved in the degradation of starch chains should guide the design and fabrication of food products for human consumption. In particular, the impact of starchy products on human health is of primordial importance. The pioneering work by Englyst et al. (1992) and the simple hydrolysis model by Goñi et al. (1997) triggered a plethora of studies oriented to classify the digestible starch fractions and their relationship with the multiscale starch structure. In general, it is accepted that the amylose content is negatively correlated with lower starch digestibility (Witt et al., 2010). Barriers to enzyme binding and structural characteristics controlling the transport of enzymes have been mentioned as the mechanisms dominating the enzymatic starch degradation (Dhital et al., 2017). In particular, the fraction that escapes the attack of amylolytic enzymes in the small intestine and is thus excursed to the colon is called resistant starch, which has been considered of prime importance to regulate the glucose absorbed in the digestive tract.

The potential links between starch structure and digestibility have concentrated the focus of many studies. The results are not conclusive and sometimes controversial (Singh et al., 2010). The granule morphology has an impact on the enzymatic degradability of starch chains (Benmoussa et al., 2006) as observed from the slow hydrolysis rate of B- and C-polymorphic starches compared to A-polymorphic starches with pores and channels on the surface. Apart from the surface structure, structural features such as crystallinity, double helix and average chain length area are also known to affect the enzyme digestion of starches (Dhital et al., 2021). Chemical substitutions such as hydroxypropylation, acetylation and oxidation reduce the susceptibility of starch chains to amylolytic enzymatic attack and increase the resistant starch content (Chung et al., 2008). Hydrothermal treatments induce reorganization of the amylose and amylopectin chains, resulting in modifications of the digestibility properties (Ovando-Martínez et al., 2013). Lopez-Rubio et al. (2008b) postulated that the resistance to enzyme digestion results from the competition between the kinetics of enzyme hydrolysis and the kinetics of amylose retrogradation. The excessive ingestion of starchy foods has been linked to metabolic disorders, including type 2 diabetes (Kulkarni, 2004). The reduction of starch degradability has been proposed as a viable strategy to reduce the incidence of metabolic syndrome. In this way, most of the focus has been oriented to raise the content of resistant starch in food products (Sajilata et al., 2006). Especially, the role of high-amylose starch in postprandial metabolism has been a matter of intense research (Li et al., 2019). In situ or processing-induced interaction of starch with other food components, mainly lipids and proteins, and their impact on starch degradability has been of interest in recent years (Tan & Kong, 2020). The formation of lipid–amylose inclusion complexes is a relatively direct way to modulate the digestibility of starch (Dithal et al., 2019; Guo & Kong, 2021). The addition of phenolic compounds has been postulated as a way to induce the formation of resistant starch (Hernandez et al., 2021). Overall, a large number of reports have focused on the characterization and improvements of starchy foods oriented to reduce the adverse effects on human health.

1.6 MODIFICATIONS

1.6.1 Chemical Modifications

Starch is rarely used in its native form. Starch is subjected to a series of modifications to obtain a structure more affordable for food applications. Chemical modifications are widely used to obtain starch with prescribed properties. The chemical reactivity of starch chains depends strongly on the hydroxyl functions of the α-D-glucan regions. Salient examples of chemical modifications are described as follows. Hydrolytic processes are widely used to obtain derivatives with better functional properties and processing applications in the food industry. Hydrolysis is a common starch modification and involves the addition of a water molecule and cleaving across a bond, resulting in products with hydroxyl group or alcohol functionality. Commonly, acidic or enzymatic methods are used to achieve starch hydrolysis. Starch hydrolysis generates a rich variety of products, including dextrin, maltose and glucose (Nakazawa & Wang, 2003). Starch esterification takes place between the carboxylic acid group of fatty acids or fatty acid chlorides and the alcohol group of the glucose units. This reaction aims to enhance the lipophilic characteristics of starch chains, while weakening the intermolecular bonding responsible for the granule integrity. As a consequence, the granule shape and size and several other properties (e.g., solubility) are modified (Mei et al., 2015). Acetylation with acetic anhydride involves the hydroxyl group of the glucose units. This chemical modification causes steric hindrance to the alignment of the linear chains. In turn, the water percolation is increased, resulting in increased granule swelling power, solubility and reduced lower gelatinization temperature and enthalpy (Chi et al., 2008). Starch can be esterified with octenylsuccinic anhydride (OSA) or octenylsuccinic acid under alkali conditions. The octenyl or dodecyl group increases the level of lipophilicity, favoring the interfacial activity of the modified starches. Commonly, starches modified with OSA are employed to stabilize oil-in-water emulsions with applications in beverage concentrates incorporating flavor and clouding oils (Dokić et al., 2012). Oxidation of starch chains takes place at the primary alcohol group -OH, which is oxidized to give the corresponding carbonyls. Oxidation with strong oxidizing agents can lead to the disruption of inter- and intramolecular bonds and even partial depolymerization (Parovuori et al., 1995). Sodium hypochlorite is the most used oxidizing agent. Chlorine, hydrogen peroxide and potassium permanganate are less often used for food products. Oxidized starches exhibit intermediate viscosity and softer gels, such that they are commonly used for dough and baked products. Cross-linking is intended to form bonds between the hydroxyl groups of the starch chains. Cross-linking reactions improve the structural stability of granules and gels. Carboxylic acids (malic, tartaric, citric, malonic, succinic, glutaric and adipic acid) and sodium trimetaphosphate are commonly used to achieve starch cross-linking (Seidel et al., 2001). Starch cross-linking is a useful modification to improve the stability of starch to processing conditions, including strong acidic/alkali conditions and variations of temperature.

1.6.2 Physical Modifications

Chemical reactions are not involved in the physical modifications of starch chains (Ojogbo et al., 2020). Pregelatinization is a physical treatment that impacts the integrity of the starch granules. This modification involves the cooking of starch to achieve complete gelatinization and subsequent drying to obtain a powder. As a consequence, the swelling power and solubility in water are greatly improved without requiring the application of heat. Pregelatinized starches are intended for applications as thickeners for soups and pasta (Liu et al., 2017). Heat moisture treatment is a technique with low impact on the granule integrity since the modification takes place at low levels of moisture (<35%) and temperatures in the range 80°C–140°C. Crystallinity, solubility and susceptibility to enzymatic hydrolysis are among the characteristics that are affected by heat moisture treatment (da Rosa Zavareze et al., 2011). This approach produces starches with applications in different food products, particularly for foods for babies and the elderly. Annealing is a hydrothermal treatment carried out with a higher moisture content (>65%) at a temperature below the onset of gelatinization. Although annealing can induce starch gelatinization, the granule is not destroyed. Instead, the most ordered structures are retained, such that the swelling temperature of the annealed starch is higher than that of the native starch (Tester & Debon, 2000). This effect is attributed to a rearrangement of the granular structure, mainly the reduction of the amorphous layers and the increase of the crystallinity content. It has been reported that the amylopectin structure and interactions formed during annealing have a significant impact on the digestibility and glycemic index of starches (Chung et al., 2009). Sonication has been explored in the recent two decades to modify the structure and functionality of starches (Falsafi et al., 2019). This method is used alone or in combination with other methods like esterification and hydrolysis (Shabana et al., 2019). High-pressure treatments impact noncovalent chemical linkages in and between starch molecules, increasing starch swelling and solubility (Castro et al., 2020). Cold plasma and pulsed electrical fields are nonthermal treatments that have been explored in recent years to modify certain starch structures for food applications (Grgić et al., 2019).

1.7 CONCLUSIONS

The importance of starch for industrial applications and food production is continuously rising thanks to its versatility, source diversity and biodegradability. Its applicability has extended from an energy source for human and animal feeding to a rich diversity of sophisticated nonfood products. The relevance of starch in technological applications arises from the susceptibility to chemical and physical modifications, converting the starch into a material that can be manufactured to induce desirable properties. The growing interest and applicability of starch in the modern industry have triggered important academic research efforts oriented to characterize the most basic properties of native starch, leading to strategies for obtaining modified starches for a wider group of applications. Nowadays, research is reorienting its focus to address problems linked to new socioeconomic necessities. Health food products to reduce the incidence of metabolic syndrome, economic and ecological sustainability, and innovative products for exigent consumers are instances of issues that should drive the investigation on starch in the immediate future. The topics exposed in the present book should serve as a motivating reading for those readers interested in the research and development of starchy food and nonfood products.

REFERENCES

Ahmad, F. B., & Williams, P. A. (1999). Effect of salts on the gelatinization and rheological properties of sago starch. *Journal of Agricultural and Food Chemistry*, 47, 3359–3366.

Ai, Y., & Jane, J. L. (2018). Understanding starch structure and functionality. In Starch in food (pp. 151–178). Woodhead Publishing.

Andrade, L. A., Barbosa, N. A., & Pereira, J. (2017). Extraction and properties of starches from the non-traditional vegetables Yam and Taro. *Polímeros*, 27, 151–157.

Azudin, M. N., & Morrison, W. R. (1986). Non-starch lipids and starch lipids in milled rice. *Journal of Cereal Science*, 4, 23–31.

Baker, A. A., Miles, M. J., & Helbert, W. (2001). Internal structure of the starch granule revealed by AFM. *Carbohydrate Research*, 330, 249–256.

Baldwin, P. M. (2001). Starch granule-associated proteins and polypeptides: A review. *Starch/Stärke*, 53, 475–503.

Benmoussa, M., Suhendra, B., Aboubacar, A., & Hamaker, B. R. (2006). Distinctive sorghum starch granule morphologies appear to improve raw starch digestibility. *Starch/Stärke*, 58, 92–99.

Bertoft, E. (2013). On the building block and backbone concepts of amylopectin structure. *Cereal Chemistry*, 90, 294–311.

Bertoft, E. (2017). Understanding starch structure: Recent progress. *Agronomy*, 7, 56.

Builders, P. F., & Arhewoh, M. I. (2016). Pharmaceutical applications of native starch in conventional drug delivery. *Starch/Stärke*, 68, 864–873.

Burrell, M. M. (2003). Starch: The need for improved quality or quantity: An overview. *Journal of Experimental Botany*, 54, 451–456.

Campo, L., & Tovar, C. (2008). Influence of the starch content in the viscoelastic properties of surimi gels. *Journal of Food Engineering*, 84, 140–147.

Castro, L. M., Alexandre, E. M., Saraiva, J. A., & Pintado, M. (2020). Impact of high pressure on starch properties: A review. *Food Hydrocolloids*, 106, 105877.

Chi, H., Xu, K., Wu, X., Chen, Q., Xue, D., Song, C., et al. (2008). Effect of acetylation on the properties of corn starch. *Food Chemistry*, 106, 923–928.

Chung, H. J., Shin, D. H., & Lim, S. T. (2008). In vitro starch digestibility and estimated glycemic index of chemically modified corn starches. *Food Research International*, 41, 579–585.

Chung, H. J., Liu, Q., & Hoover, R. (2009). Impact of annealing and heat-moisture treatment on rapidly digestible, slowly digestible and resistant starch levels in native and gelatinized corn, pea and lentil starches. *Carbohydrate Polymers*, 75, 436–447.

Coultate, T. P. (2009). *Food: The Chemistry of its Components*. Royal Society of Chemistry.

Debet, M. R., & Gidley, M. J. (2006). Three classes of starch granule swelling: Influence of surface proteins and lipids. *Carbohydrate Polymers*, 64, 452–465.

Devi, L. S., & Das, A. B. (2020). Effect of ionic liquid on sol-gel phase transition, kinetics and rheological properties of high amylose starch. *International Journal of Biological Macromolecules*, 162, 685–692.

Dhital, S., Shelat, K. J., Shrestha, A. K., & Gidley, M. J. (2013). Heterogeneity in maize starch granule internal architecture deduced from diffusion of fluorescent dextran probes. *Carbohydrate Polymers*, 93, 365–373.

Dhital, S., Warren, F. J., Butterworth, P. J., Ellis, P. R., & Gidley, M. J. (2017). Mechanisms of starch digestion by α-amylase: Structural basis for kinetic properties. *Critical Reviews in Food Science and Nutrition*, 57, 875–892.

Dhital, S., Brennan, C., & Gidley, M. J. (2019). Location and interactions of starches in planta: Effects on food and nutritional functionality. *Trends in Food Science & Technology*, 93, 158–166.

Dhital, S, Warren, F. J., Butterworth, P. J., Ellis, P. R., & Gidley, M. J. (2021). Mechanisms of starch digestion by α-amylase: Structural basis for kinetic properties. *Critical Reviews in Food Science and Nutrition*, 57, 875–892.

Ding, Y., Chen, J., Lin, Q., Wang, Q., Wang, J., & Yu, G. (2021). Effects of endogenous proteins and lipids on structural, thermal, rheological, and pasting properties and digestibility of adlay seed (Coix lacryma-jobi L.) starch. *Food Hydrocolloids*, 111, 106254.

Dokić, L., Krstonošić, V., & Nikolić, I. (2012). Physicochemical characteristics and stability of oil-in-water emulsions stabilized by OSA starch. *Food Hydrocolloids*, 29, 185–192.

Englyst, H. N., Kingman, S. M., & Cummings, J. H. (1992). Classification and measurement of nutritionally important starch fractions. *European Journal of Clinical Nutrition*, 46(Supplement 2), S33–50.

Evans, I. D., & Lips, A. (1992). Viscoelasticity of gelatinized starch dispersions. *Journal of Texture Studies*, 23, 69–86.

Fan, D., Ma, W., Wang, L., Huang, J., Zhang, F., Zhao, J., ... & Chen, W. (2013). Determining the effects of microwave heating on the ordered structures of rice starch by NMR. *Carbohydrate Polymers*, 92, 1395–1401.

Falsafi, S. R., Maghsoudlou, Y, Rostamabadi, H, Rostamabadi, M M, Hamedi, H, Hosseini, S. M. H. (2019). Preparation of physically modified oat starch with different sonication treatments. *Food Hydrocolloids*, 89, 311–320.

Fannon, J. E., Shull, J. M., & BeMiller, J. N. (1993). Interior channels of starch granules. *Cereal Chemistry*, 70, 611–612

Fannon, J. E., Gray, J. A., Gunawan, N, Huber, K. C., & BeMiller, J. N. (2004). Heterogeneity of starch granules and the effect of granule channelization on starch modification. *Cellulose*, 11, 247–254.

French, D. (1972). Fine structure of starch and its relationship to the organization of starch granules. *Journal of the Japanese Society of starch Science*, 19, 8–25.

Gallant, D. J., Bouchet, B., & Baldwin, P. M. (1997). Microscopy of starch: Evidence of a new level of granule organization. *Carbohydrate Polymers*, 32, 177–191.

Gonzalez-Soto, R. A., de la Vega, B, García-Suarez, F. J., Agama-Acevedo, E., & Bello-Pérez, L. A. (2011). Preparation of spherical aggregates of taro starch granules. *LWT-Food Science and Technology*, 44, 2064–2069.

Goñi, I., Garcia-Alonso, A., & Saura-Calixto, F. (1997). A starch hydrolysis procedure to estimate glycemic index. *Nutrition Research*, 17, 427–437.

Grgić, I., Ačkar, Đ., Barišić, V., Vlainić, M., Knežević, N., & Medverec Knežević, Z. (2019). Nonthermal methods for starch modification-A review. *Journal of Food Processing and Preservation*, 43, e14242.

Guo, J., & Kong, L. (2021). Inhibition of in vitro starch digestion by ascorbyl palmitate and its inclusion complex with starch. *Food Hydrocolloids*, 121, 107032.

Han, X. Z., & Hamaker, B. R. (2002). Association of starch granule proteins with starch ghosts and remnants revealed by confocal laser scanning microscopy. *Cereal Chemistry*, 79, 892–896.

Hernández, H. A. R, Gutiérrez, T. J., & Bello-Pérez, L. A. (2021). Can starch-polyphenol V-type complexes be considered as resistant starch? *Food Hydrocolloids*, 107226.

Hizukuri, S. (1986). Polymodal distribution of the chain lengths of amylopectins, and its significance. *Carbohydrate Research*, 147, 342–347.

Huber, K. C., & BeMiller, J. N. (1997). Visualization of channels and cavities of corn and sorghum starch granules. *Cereal Chemistry*, 74, 537–541.

Huber, K. C., & BeMiller, J. N. (2000). Channels of maize and sorghum starch granules. *Carbohydrate Polymers*, 41, 269–276.

Imberty, A, Buléon, A, Tran, V, & Perez, S. (1991). Recent advances in knowledge of starch structure. *Starch/Stärke*, 43, 375–384.

Jane, J. L. (2006). Current understanding on starch granule structures. *Journal of Applied Glycoscience*, 53, 205–213.

Kulkarni, K. D. (2004). Food, culture, and diabetes in the United States. *Clinical Diabetes*, 22, 190–192.

Li, H., Gidley, M. J., & Dhital, S. (2019a). High-amylose starches to bridge the "Fiber Gap": Development, structure, and nutritional functionality. *Comprehensive Reviews in Food Science and Food Safety*, 18, 362–379.

Li, H., Lei, N., Yan, S., et al. (2019b). The importance of amylopectin molecular size in determining the viscoelasticity of rice starch gels. *Carbohydrate Polymers*, 212, 112–118.

Li, M., Daygon, V. D., Solah, V., Dhital, S. (2021). Starch granule size: Does it matter? *Critical Reviews in Food Science and Nutrition*, 1–21.

Liu, C. M., Liang, R. H., Dai, T. T., Ye, J. P., Zeng, Z. C., Luo, S. J., & Chen, J. (2016). Effect of dynamic high pressure microfluidization modified insoluble dietary fiber on gelatinization and rheology of rice starch. *Food Hydrocolloids*, 57, 55–61.

Liu, Y., Chen, J., Luo, S., Li, C., Ye, J., Liu, C., & Gilbert, R. G. (2017). Physicochemical and structural properties of pregelatinized starch prepared by improved extrusion cooking technology. *Carbohydrate Polymers*, 175, 265–272.

Lopez-Rubio, A., Flanagan, B. M., Gilbert, E. P., & Gidley, M. J. (2008a). A novel approach for calculating starch crystallinity and its correlation with double helix content: A combined XRD and NMR study. *Biopolymers: Original Research on Biomolecules*, 89, 761–768.

Lopez-Rubio, A., Flanagan, B. M., Shrestha, A. K., Gidley, M. J., & Gilbert, E. P. (2008b). Molecular rearrangement of starch during in vitro digestion: Toward a better understanding of enzyme resistant starch formation in processed starches. *Biomacromolecules*, 9, 1951–1958.

Mei, J. Q., Zhou, D. N., Jin, Z. Y., Xu, X. M., & Chen, H. Q. (2015). Effects of citric acid esterification on digestibility, structural and physicochemical properties of cassava starch. *Food Chemistry*, 187, 378–384.

Morrison, W. R. (1995). Starch lipids and how they relate to starch granule structure and functionality. *Cereal Foods World*, 40, 437–46.

Morrison, W. R. (1998). Lipids in cereal starches: A review. *Journal of Cereal Science*, 8, 1–15.

Nakazawa, Y, & Wang, Y. J. (2003). Acid hydrolysis of native and annealed starches and branch-structure of their Naegeli dextrins. *Carbohydrate Research*, 338, 2871–2882.

Nichols, B. L., Avery, S, Sen, P, Swallow, D. M., Hahn, D, & Sterchi, E. (2003). The maltase-glucoamylase gene: Common ancestry to sucrase-isomaltase with complementary starch digestion activities. *Proceedings of the National Academy of Sciences*, 100, 1432–1437.

Nikuni, Z. (1969). Starch and cooking. *Science of Cookery*, 2, 6–14.

Nikuni, Z. (1978). Studies on starch granules. *Starch/Stärke*, 30, 105–111.

Nwokoro, S. O., Orheruata, A. M., & Ordiah, P. I. (2002). Replacement of maize with cassava sievates in cockerel starter diets: Effect on performance and carcass characteristics. *Tropical Animal Health and Production*, 34, 163.

Ojogbo, E., Ogunsona, E. O., & Mekonnen, T. H. (2020). Chemical and physical modifications of starch for renewable polymeric materials. *Materials Today Sustainability*, 7, 100028.

Ovando-Martinez, M, Sáyago-Ayerdi, S, Agama-Acevedo, E, Goñi, I, & Bello-Pérez, L. A. (2009). Unripe banana flour as an ingredient to increase the undigestible carbohydrates of pasta. *Food Chemistry*, 113, 121–126.

Ovando-Martínez, M, Whitney, K, Reuhs, B. L., Doehlert, D. C., & Simsek, S. (2013). Effect of hydrothermal treatment on physicochemical and digestibility properties of oat starch. *Food Research International*, 52, 17–25.

Parovuori, P., Hamunen, A., Forssell, P., Autio, K., & Poutanen, K. (1995). Oxidation of potato starch by hydrogen peroxide. *Starch/Stärke*, 47, 19–23.

Pérez, S., & Bertoft, E. (2010). The molecular structures of starch components and their contribution to the architecture of starch granules: A comprehensive review. *Starch/Stärke*, 62, 389–420.

Precha-Atsawanan, S., Uttapap, D., & Sagis, L. M. (2018). Linear and nonlinear rheological behavior of native and debranched waxy rice starch gels. *Food Hydrocolloids*, 85, 1–9.

Rodriguez-Garcia, M. E., Hernandez-Landaverde, M. A., Delgado, J. M., Ramirez-Gutierrez, C F, Ramirez-Cardona, M, Millan-Malo, B. M., & Londoño-Restrepo, S. M. (2021). Crystalline structures of the main components of starch. *Current Opinion in Food Science*, 37, 107–111.

Rolland-Sabaté, A, Sánchez, T, Buléon, A, Colonna, P, Jaillais, B, Ceballos, H, & Dufour, D. (2012). Structural characterization of novel cassava starches with low and high-amylose contents in comparison with other commercial sources. *Food Hydrocolloids*, 27, 161–174.

da Rosa Zavareze, E, & Dias, A. R. G. (2011). Impact of heat-moisture treatment and annealing in starches: A review. *Carbohydrate Polymers*, 83, 317–328.

Sajilata, M. G., Singhal R. S., & Kulkarni P. R. (2006). Resistant starch: A review. *Comprehensive Reviews in Food Science and Food Safety*, 5, 1–17.

Schmiele, M., Sampaio, U. M., & Clerici, M. T. P. S. (2019). Basic principles: Composition and properties of starch. In *Starches for Food Application* (pp. 1–22). Academic Press.

Seidel, C, Kulicke, W. M., Heß, C, Hartmann, B, Lechner, M. D., & Lazik W. (2001). Influence of the cross-linking agent on the gel structure of starch derivatives. *Starch/Stärke*, 53, 305–310.

Shabana, S., Prasansha, R., Kalinina, I., Potoroko, I., Bagale, U., & Shirish, S. H. (2019). Ultrasound assisted acid hydrolyzed structure modification and loading of antioxidants on potato starch nanoparticles. *Ultrasonics Sonochemistry*, 51, 444–450.

Singh, N, Singh, J, Kaur, L, Sodhi, N. S., Gill, B. S. (2003). Morphological, thermal and rheological properties of starches from different botanical sources. *Food Chemistry*, 81, 219–231.

Singh, J., Dartois, A., & Kaur, L. (2010). Starch digestibility in food matrix: A review. *Trends in Food Science & Technology*, 21, 168–180.

Tan, I., Flanagan, B. M., Halley, P. J., Whittaker, A. K., & Gidley, M. J. (2007). A method for estimating the nature and relative proportions of amorphous, single, and double-helical components in starch granules by 13C CP/MAS NMR. *Biomacromolecules*, 8, 885–891.

Tan, L., & Kong, L. (2020). Starch-guest inclusion complexes: Formation, structure, and enzymatic digestion. *Critical Reviews in Food Science and Nutrition*, 60, 780–790.

Tang, H., Mitsunaga, T., & Kawamura, Y. (2006). Molecular arrangement in blocklets and starch granule architecture. *Carbohydrate Polymers*, 63, 555–560.

Tapia-Blácido, D., Mauri A. N., Menegalli, F. C., Sobral, P. J., & Añón, M. C. (2007). Contribution of the starch, protein, and lipid fractions to the physical, thermal, and structural properties of amaranth (Amaranthus caudatus) flour films. *Journal of Food Science*, 72, E293–E300.

Tester, R. F., & Debon, S. J. (2000). Annealing of starch: A review. *International Journal of Biological Macromolecules*, 27, 1–12.

Tester, R. F., Debon, S. J. J., & Sommerville, M. D. (2000). Annealing of maize starch. *Carbohydrate Polymers*, 42, 287–299.

Tetlow, I. J. (2011). Starch biosynthesis in developing seeds. *Seed Science Research*, 21, 5–32.

Thompson, D. B. (2000). On the non-random nature of amylopectin branching. *Carbohydrate Polymers*, 43, 223–239.

Tonukari, N. J. (2004). Cassava and the future of starch. *Electronic Journal of Biotechnology*, 7, 5–8.

Wertz, J. L., & Goffin, B. (2020). *Starch in the Bioeconomy*. CRC Press.

Wang, J., Hu, P., Chen, Z., Liu, Q., & Wei, C. (2017). Progress in high-amylose cereal crops through inactivation of starch branching enzymes. *Frontiers in Plant Science*, 8, 469.

Wang, S., Xu, H., & Luan, H. (2020). Multiscale structures of starch granules. In *Starch Structure, Functionality and Application in Foods* (pp. 41–55). Springer.

Wani, A. A., Singh, P., Shah, M. A., Schweiggert-Weisz, U., Gul, K., & Wani, I. A. (2012). Rice starch diversity: Effects on structural, morphological, thermal, and physico-chemical properties: A review. *Comprehensive Reviews in Food Science and Food Safety*, 11, 417–436.

Witt, T, Gidley, M. J., & Gilbert, R. G. (2010). Starch digestion mechanistic information from the time evolution of molecular size distributions. *Journal of Agricultural and Food Chemistry*, 58, 8444–8452.

Yano, M., Okuno, K., Kawakami, J., Satoh, H., & Omura, T. (1985). High amylose mutants of rice, Oryza sativa L. *Theoretical and Applied Genetics*, 69, 253–257.

Yousefi, A. R., & Razavi, S. M. (2015). Dynamic rheological properties of wheat starch gels as affected by chemical modification and concentration. *Starch/Stärke*, 67, 567–576.

Zhang, K., Zhao, D., Zhang, X., Qu, L., Zhang, Y., & Huang, Q. (2021). Effects of the removal of lipids and surface proteins on the physicochemical and structural properties of green wheat starches. *Starch/Stärke*, 73, 2000046.

Zhu, F. (2015). Isolation, composition, structure, properties, modifications, and uses of yam starch. *Comprehensive Reviews in Food Science and Food Safety*, 14, 357–386.

Chapter 2

Starch Extraction

Principles and Techniques

Ming Li, Yu Tian and Sushil Dhital

CONTENTS

DOI: 10.1201/9781003088929-2

2.1 LOCATION AND MORPHOLOGY OF STARCH IN CEREALS, LEGUMES AND TUBERS

Starch is produced from photosynthesis and stored in different organs of plants, including leaves, fruits, seeds and tubers. Grains and tubers are the most important cultivated staple energy sources that provide carbohydrates. For cereal crops, such as corn, wheat, rice and sorghum, starch granules are stored in endosperm cells, enmeshed in protein matrix (Rooney & Pflugfelder, 1986). The endosperm is heterogeneous in terms of compactness (interaction of starch and proteins), contents (starch, protein, lipid) and cell wall polysaccharides (dietary fiber) structure. Therefore, heterogeneity exists between the botanical sources and the seed of the same botanical source (Shewry et al., 2020), and the properties of the starch (gelatinization and pasting properties) show variance by the gradients in the different pearling fractions (Tosi et al., 2009).

Cereals in the tribe Triticeae (wheat, rye, and barley) have two morphologically distinct types (bimodal distribution) of starch granules: large round or lenticular granules with an average diameter of 25 µm (A-type); and small round granules ranging from 5 to 8 µm (B-type). In wheat, the deposition of large A-type granules is initiated approximately 4–12 days postanthesis (cell division stage), and the granules can grow throughout grain filling. On the other hand, deposition of small B-type granules is initiated between 16 and 22 days postanthesis (cell enlargement stage) in protrusions stemming from amyloplasts containing lenticular granules (Langeveld et al., 2000; Li et al., 2021). The starch granules are distributed differently within the starchy endosperm cells, with only a few small granules present in the protein-rich sub-aleurone cells (Figure. 2.1) (Jackowiak et al., 2005). In maize, endosperm contains a central area composed of powdery endosperm, surrounded by horny endosperm, with an area ratio of about 1:2 (Wang & Wang, 2019). In minor cereals, such as amaranth or quinoa (Figure 2.2), the small starch granules (0.4–2.0 µm) are aggregated in larger numbers (14,000–20,000) in a protein matrix. The treatment of proteolytic enzymes (e.g., pepsin) enhances their separation (Ruales & Nair, 1994).

Legumes have higher protein content (21%–27%) compared with cereals, and the starch accounts for 45%–51% of the total mass (Table 2.1) (Chung et al., 2008; Ren et al., 2021; Reyes-Moreno et al., 1993). The starch granules in legumes are embedded in a matrix of protein bodies and are surrounded by a cellulosic–pectic cell wall (Figure 2.1). Thus the disruption of strong cell walls and protein matrix is essential for isolating starch from legumes. Legume starch is often extracted from the by-product stream after the protein extraction process. Legume starch granules appear in oval, spherical, kidney and

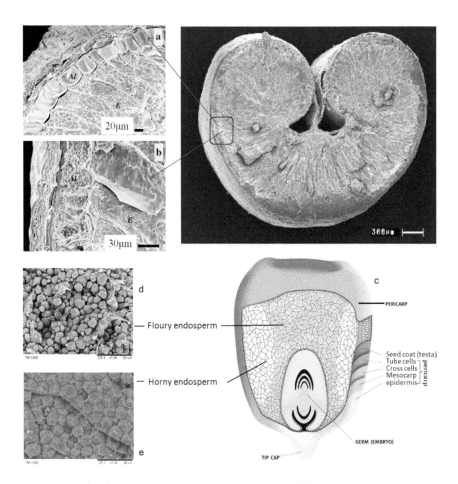

Figure 2.1 (Top) The cross-section anatomy of wheat. (a) represents aleurone cells, and (b) represents endosperm (Jackowiak et al., 2005). (Bottom) In a maize kernel, the different morphology of starch granules is observed in the (c) floury endosperm and (d) horny endosperm (Wang & Wang, 2019).

irregular shapes, with average granule sizes of approximately 15 to 28 μm (Ren et al., 2021).

Starch granules in the roots and tubers are embedded in cellulosic matrices held together by pectin substrates. Potato starch is mostly spherical, with sizes ranging from 5 to 100 μm. Furthermore, the cotyledon cell walls of roots and tubers are fragile, easy to break, releasing starch just by the hydromechanical process without chemical and enzymic pretreatment (Li et al., 2021).

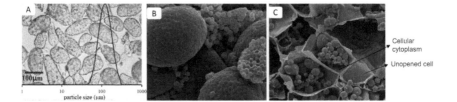

Figure 2.2 Images of starch granules in the endosperm/cotyledon of (A) pulse, (B) quinoa and (C) potato. Magnification of (B) is 3000×, and (C) is 300×. Distinguished morphology of starch granules and granule packing is observed with varying botanical sources (Contreras-Jiménez, Torres-Vargas, & Rodríguez-García, 2019; Singh et al., 2013; Xiong et al., 2018).

TABLE 2.1 COMPOSITION OF TYPICAL CEREALS, BEANS, TUBERS AND ROOTS

	Starch (g/100 g db)	Protein (g/100 g db)	Lipid (g/100 g db)	Fiber (g/100 g db)	Reference
Corn	72.6	9.8	4.2	1.2	(Srichuwong et al., 2017)
Rice*	77.63	7.17	1.51	0.22	(Chang et al., 1986)
Wheat	68.1	13.2	2.5	1.6	(Srichuwong et al., 2017)
White sorghum	72.2	11.4	5.0	5.8	(Srichuwong et al., 2017)
White pea	46.30	21.50	3.09	3.69	(W. Li et al., 2014)
Pea (smooth)	45.7–51.2	21.3–26.8	1.0–2.8	14.2–20.0	(Chung et al., 2008; Ren et al., 2021)
Potato*	8.89–18.83	1.45–2.35	0.25–1.04	0.49–0.80	(Leonel et al., 2017)
Cassava	80.1–86.3	1.2–1.8	0.1–0.8	1.5–3.5	(Charles et al., 2005)

*Based on the whole weight. db = dry matter basis.

2.2 EFFECT OF PROTEIN AND LIPIDS ON THE STRUCTURE AND PHYSICOCHEMICAL PROPERTIES OF ISOLATED STARCHES

The primary purpose of starch extraction is to isolate the starch in its purest form with the least possible damage to the surface structure of isolated granules. However, the extraction process should overcome the strong in-planta interaction of macronutrients (starch, protein, lipids) and the encapsulating cell wall materials (Dhital et al., 2019).

2.2.1 Protein

Protein is the second-largest molecule in cereals and legumes after starch, accounting for 8%–14% and 20%–40%, respectively. In biology, starch and protein are formed simultaneously in endosperm or cotyledon. Storage proteins in cereals form discrete protein bodies during grain filling and bind into a continuous protein matrix around starch granules at the later stage of development (Figure 2.1). Therefore, a robust physical interaction is naturally built between the starch and protein in planta. This interaction plays a vital role in the form and function of starch and flour (Dhital et al., 2019; Pérez et al., 2004).

Wheat with different hardness produces different degrees of damaged starch during milling, and the main determinant of wheat endosperm hardness is the adhesion strength between starch granules and the surrounding protein matrix. Endosperm hardness can affect the properties and final application of isolated starch. Simmonds's early studies showed differences in the cementing layer between starch granules and storage proteins in hard wheat and soft wheat (Simmonds, 1971). Hard wheat granules have more adhesion proteins than soft wheat granules; thus, starch extraction from hard wheat is more complicated than soft wheat varieties.

Starch granules contain only a small amount of protein. The protein associated with the starch granules can be broadly classified as surface and granule-associated proteins (GAPs). Surface proteins are remnants of storage proteins (globulins and prolamins) that adsorbed on the granule surface after extraction. The granule-associated proteins are nonstorage proteins; they are either tightly embedded on the surface (granule-surface proteins, GSPs) or inside the starch granules (e.g., in channels; granule-channel proteins, GCPs) and are mostly biosynthetic or degradative enzymes, typically in the size range of 60–150 kDa (Baldwin, 2001; Han et al., 2005; Zhan et al., 2020). The content of protein in starch depends on the degree of purification in the process of starch extraction (Lowy et al., 1981; Schofield & Greenwell, 1987). A typical well-washed cereal and legume starch sample contains around 0.2%–0.8% protein (Zhan et al., 2020), while a typical root or tuber starch includes a relatively lower amount of protein (<0.16%). Still, it is increasingly recognized that their presence significantly affects both the granule properties and the properties of starch-derived products (Schofield & Greenwell, 1987; Ye et al., 2019; Zhan et al., 2020).

Starch protein significantly affects the starch swelling, water absorption and amylose leaching (Figure 2.3) (Debet & Gidley, 2006; Lim et al., 1999; Wang

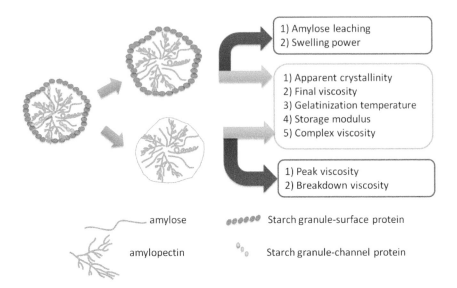

Figure 2.3 Starch granule-associated proteins affecting the physicochemical properties of rice starch (modified from Zhan et al., 2020).

et al., 2014). After protein removal by alkali treatment or surfactant, the peak viscosity, swelling power and solubility of amylose increased significantly; however, the gelatinization temperature decreased (Wang et al., 2014; Zhan et al., 2020), suggesting that the granule-bound proteins mainly provide the physical hindrance of water diffusion, absorption and swelling of the starch granules during processing. It is to be noted that the effect of proteins on starch properties depends upon the type of protein (storage proteins, granule-surface proteins, granule-channel proteins) as studied by selectively removing these proteins (Benmoussa et al., 2010, Tao et al., 2021; Zhan et al., 2020) and was recently reviewed (Dhital et al., 2019). A schematic diagram of the effect of various proteins on the properties of starch is shown in Figure 2.3.

2.2.2 Lipids

Lipids are present in small amounts in native starch on the surface (nonstarch lipids) as well as internal lipids (starch lipids) (Morrison, 1981). Starch isolated from whole grain also contain lipids originally from the aleurone layer and germ due to steeping, milling and other isolation processes. Compared with cereals (wheat, rice and corn) starches, tuber (cassava, potato) and lentil starch contain a lower amount of both total lipids and internal lipids (Vasanthan & Hoover, 1992). For example, the contents of total lipids and internal lipids of potato are 0.54% and 0.21%, which are relatively lower than wheat starch, at 1.19% and 0.88%, respectively (Baszczak et al., 2003). Within the same species, the smaller granules with comparatively less amylose content have more lipids than larger ones (Dhital et al., 2011; Raeker et al., 1998). This may be due to the greater surface-to-volume ratio in small starch granules, which allows adherence of more lipids on the surface (Dhital et al., 2019).

Surface lipids are originally from the amyloplast membrane around the developing starch granules and the endoplasmic reticulum membrane around the starch granules. Surface and internal lipids may exist in a free state or as amylose inclusion complexes, or bind to hydroxyl groups of starch components through ionic or hydrogen bonds (Dhital et al., 2019; Vasanthan & Hoover, 1992). A strong correlation exists between amylose and lipid contents in cereal starches. High-amylose mutants possess a higher proportion of lipids (~1% db) compared to normal maize starches (~0.5%), whereas lipids are almost absent in waxy maize starches (~0.08%) (Morrison, 1988; Morrison et al., 1984). The lipids can reduce the retrogradation of amylose or amylopectin (Eliasson & Ljunger, 1988; Germani et al., 1983), retarding the aging of starchy food and maintaining product quality. Oxidation of lipids also happens during storage, and lipids will take part in further chemical interactions with starch and other nonstarch components. Thus, to some extent, the properties of starch depend upon the storage condition and time (Shi et al., 2021).

The amount of lipids associated with starch depends upon the initial lipid contents in cereals/tubers, particle size, extraction time and choice of solvent. The surface lipids can be removed by soaking the starch/flour in petroleum ether at room temperature, followed by refluxing. Afterward, the internal lipids can be extracted with water-saturated butanol (WSB, n-propanol/water, 3:1 v/v) at 37°C. The removal of internal lipids (starch lipids) results in a higher swelling power but a lower peak viscosity of rice starch than surface lipids (nonstarch lipids) (Zhang et al., 2019), suggesting the starch lipids have a much greater influence on the starch physicochemical properties changes than nonstarch lipids. A higher degree of milling (smaller particle size) and longer extraction time is required for the complete removal of lipids. The extraction of lipids is not commercially viable, as it involves the application of solvents to dissolve the lipids. Thus, lipid-rich grains such as soy and nuts are not suitable for starch extraction. Lipid extraction, however, is needed to obtain pure starch granules for some pharmaceutical and analytical applications.

Defatting reduces the gelatinization temperature and the gel viscosity of starch (Maniñgat & Juliano, 1980). The gelatinization temperature is found to be negatively correlated with the amount of lipid extracted from the starch (Champagne et al., 1990). Once lipids are extracted from starch, the swelling behavior of regular corn and wheat starch is much more rapid, comparable to waxy maize and tapioca starch (Debet & Gidley, 2006). Differently, lipid-bound amylose has a strong inhibitory effect on the starch swelling (Tester & Morrison, 1990). Furthermore, the swelling of high-amylose maize/potato is also affected by protein/lipid extraction (Debet & Gidley, 2006). This suggests that the functional properties of starch are also dependent upon the minor constituent such as protein and lipids.

2.3 MILLING-INDUCED CHANGES IN STARCH STRUCTURE

Wet milling and dry grinding are the most common methods for separating starch from grains, tubers, legumes and other nonconventional products (Leewatchararongjaroen & Anuntagool, 2016). Wet milling starts with the steeping process. Then, the grain constituents are separated with the mechanical splitting of the grains, followed by sieving to obtain pure starch granules (Myers & Fox, 1994). In comparison, dry milling is done using refined or whole grain flour (particle size reduction) with high quality by hammer mills, pin mills, or disk mills under room temperature or cryogenic conditions (Li et al., 2014).

2.3.1 Wet Milling

Wet milling commonly implies the soaking of the grains in an alkaline solution or with added sulfur dioxide (SO_2) (Correia & Beirão-da-Costa, 2012; Han & Hamaker, 2002). This is aimed to loosen grain component interaction as well as inhibit microbial growth. The softened grain is milled, and its components are separated by screening, centrifuging, hydrocloning and washing to produce starch and other components. The wet condition is less damaging, and the starch granules have minimal damage compared to dry milling. However, a significant amount of starch is lost in the protein and bran fractions (Andersson et al., 2001). The starch recovery of starch after wet milling is between 85% for corn and 78% for buckwheat (Dowd, 2003; Haros et al., 2003; Zheng et al., 1997).

2.3.2 Dry Milling

The purity of the starch extracted using drying milling is lower, as dry milling cannot completely separate starch from the protein matrix. Therefore, a water-washing step is required after the air classification to remove part of the attached protein.

Starches or flours obtained from the severe dry milling process have poor processing qualities due to a higher starch damage (Barak et al., 2014; León et al., 2006) and the altered physicochemical properties (Hasjim et al., 2012; Tran et al., 2011). For example, up to 15% damaged starch is obtained after the jet milling of barley, whereas the degree of damage after roller milling is about 5% (Lee et al., 1996). Many factors, not limited to types of starch, particle size and milling strength (Dhital et al., 2010; Hasjim et al., 2012; Roa et al., 2014), can affect the degree of starch damage, leading to further changes in the molecular, crystalline, granular structure and the related physicochemical properties, e.g., swelling and pasting properties (Asmeda et al., 2016; Hasjim et al., 2013).

Compared to high-amylose starches, waxy or starch with a higher amount of amylopectin is much degraded with dry milling. It is hypothesized that the semicrystalline structure of amylopectin is more susceptible to mechanical force than amorphous amylose (Li et al., 2014). Therefore, the higher energy or longer milling time, though it could grind the grain into a smaller size, can have severe consequences, e.g., a higher degree of surface and molecular damage. Thus, the level of damaged starch increases with the reduction of particle sizes (Guan et al., 2020). Detailed analysis of grain milling on starch structures, flour/starch properties and its effect on starch-based food systems are well-reviewed elsewhere (Li et al., 2014; Wang et al., 2020).

2.4 PRINCIPLES TO OBTAIN STARCH IN THE PUREST FORM

The starch isolation methods greatly depend on the plant source, which varies in grain hardness, fiber content, protein solubility, protein and starch interaction, etc. These factors will further affect the starch separation method and are described in the following sections.

2.4.1 Release of Starch from Cellular Matrices

In plant tissues, structures such as aleurone layers and cell walls protect starch from damage. Thus, these higher peripheral structures need to be destroyed to release starch during starch isolation. Extraction of starch granules from hydrated tubers (in potatoes, for example, which are neither embedded in a protein matrix nor closed within a strong cell wall) requires fewer processing steps and less mechanical energy to disintegrate cell walls. Cereals and legumes are mostly stored in dry conditions (mc <14% wb), so hydrating the grains to maximum equilibrium moisture content (e.g., >55% wet basis) is achieved by steeping the grains in water followed by sequential milling to extract the starch-containing endosperm.

The outer layers, e.g., hull, bran and seed coat, are removed to maximize the water diffusion process. The presence of hydrated (swollen) outer layers interfere with the sieving process in subsequent steps. As the by-product, hulls contain mainly phytochemicals and dietary fiber, and are often ground into fine particles and used as supplements in food processing. The micron-sized fibers, insoluble proteins and soluble fiber as impurities are settled with starch during centrifugation steps. In addition, more starch granules are entrapped in solid residuals after filtration of the fibers. Thus the process needs optimization that is capable of separating the fractions into individual pure streams. The steeping time, residual moisture content, particle size and hydrocyclone parameters are important for the extraction of starch concerning both yield and purity.

Steeping can promote the diffusion of water into the germ, endosperm and cellular components. The steeping conditions can be varied by controlling the temperature, time and steeping medium concentration. SO_2, lactic acid and NaOH, with or without enzymes, have been used as the steeping medium (Pérez et al., 2001; Shandera & Jackson, 1996). Sulfur dioxide (SO_2) is frequently used in the steeping of corn and sorghum, which can control the growth of spoilage microorganisms, denature endogenous enzymes and soften protein matrix (encapsulating starch granules) (Eckhoff & Watson, 2009), enhancing the starch release in preceding steps. However, these processing aids also bring about environmental and health problems. Thus proper wastewater treatment facility and washing of starch are required. In addition, high SO_2 concentration and high steeping temperature lead to increased protein disintegration and dispersion, increasing the interaction of starch and proteins (Cox et al., 1944). The hindrance of water diffusion from starch–protein interaction leads to higher peak temperatures and a narrower gelatinization range (Pérez et al., 2001). Wang et al. (2000) reported significant differences in the starch yield, protein content in starch and starch color between SO_2 concentrations of 0.2% and 0.3%. Lactic acid can increase the effectiveness of SO_2 and starch yield (Dailey, 2002; Pérez et al., 2001) mostly due to ionic interactions, facilitating the removal of protein.

2.4.2 Protein Removal

Removal of storage proteins and granule-bound proteins are essential to obtain starch in its purest form. Among different methods used in the lab and industry, chemical extraction (described in the next paragraph) is considered the most economical and efficient, which helps water penetrate into seeds and break the starch–protein network (Singh et al., 2018). Most proteins are removed through massive washing, but some starch granule-associated proteins (SGAPs) that are strongly bounded with starch may remain unremoved (Bancel et al., 2010).

Alkali extraction is the most commonly used chemical method for starch isolation. The grains are submerged in solutions (0.3% sodium sulfite solution, 0.45% sodium pyrosulfite, 0.02%–0.3% sodium hydroxide, sodium bicarbonate, 0.1% sodium bisulfite and SO_2 at 25°C–50°C for 4–24 h) and then disintegrated using mechanical mixers (Halal et al., 2019). Alkali extractions remove most of the proteins, leading to higher yield and purity (lower content of protein and lipid). These methods are primarily applicable for isolating starches with smaller sizes or higher protein content, such as rice and amaranth (Lim et al., 1999). A higher concentration of alkali, though able to remove the surface proteins and lipids of starch granules (Han & Hamaker, 2002), can solubilize the starch (mostly leaching of amylose) and also degrade the double-helical and crystalline structure (Lee et al., 2009; Zhang et al., 2019). This will ultimately affect the physicochemical properties, such as swelling, pasting and gelling (Han & Hamaker, 2002; Palacios-Fonseca et al., 2013), and enzyme susceptivity (Wang & Copeland, 2012; Wang et al., 2014) of starch. For example, a NaOH concentration higher than 0.24% (w/v) leads to significant alteration of the granular organization (Cardoso et al., 2007). The alkali sensitivity of starch is also directly related to its species (Maher, 1983) and variety (Cai et al., 2014); potato, wheat and wild rice are harder to gelatinize by alkali with the same concentration than corn starch or rice with high-amylose content. Thus, an optimum alkali concentration is required to recover starch with the least contaminants (nonstarch polysaccharides, protein and lipids) and damage to the starch surface and internal structure.

To overcome these problems, a sole enzyme or enzyme combinations are used. Enzymes such as protease and cellulase are evaluated in the corn wet milling process to shorten the steeping time and improve starch yield (Correia & Beirão-da-Costa, 2012; Pedersen et al., 2015; Radosavljevic et al., 1998; Wittrock et al., 2008). Several kinds of protease have been used for starch isolation, such as prozyme from *Bacillus amyloliquefaciens*; prozyme from *Aspergillus niger* (Choi et al., 2018); and protease from *Aspergillus oryzae, Bacillus subtilis, Streptomyces griseus*, and *Rhizopus* sp. (Lim et al., 1992). Steinke and Johnson (1991) found that soaking in the presence of a variety of enzymes (cellulase, hemicellulase, β-glucanase, pectinase and bromelain) and sulfur dioxide can significantly reduce the soaking time (from 48 h to 24 h) and produce starch with high purity. Compared with amaranth starch separated by the traditional alkali wet milling procedure, isolation with steeping in low alkaline solution and protease can reduce the amount of NaOH and produce starch of high quality and recovery (Villarreal et al., 2013). Johnston and Singh (2004) and Ramírez et al. (2009) also reported that enzymatic grinding could shorten the soaking time and possibly eliminate the use of SO_2. Johnston and Singh (2004) proposed a two-stage extraction process: first, the grain was soaked in water, coarsely grounded and incubated with enzymes (bromelain) for 6 h. Coarse grinding reduces particle sizes and diffusion barriers for enzymes to penetrate grain kernels and decompose the protein matrix around starch granules. The purity and yield from alkali and enzymic methods are dependent upon many factors, such as concentration and type of alkali and enzymes, soaking time, particle size, etc. The enzyme treatment is more environmentally friendly and less damaging to the starch surface morphology and internal structure, as the enzyme has more substrate selectivity than the alkali. However, the purity of the enzyme is a critical factor, as even small contamination of amylase can degrade the surface and induce pores to starch granules (Correia & Beirão-da-Costa, 2012; Radosavljevic et al., 1998).

A solvent, such as toluene, is also used to solubilize the protein and extract the starch, mainly in the lab and noncommercial practices. In this method, the ground cereals are mixed with toluene–water solution. The solvent dissolves the protein and forms a suspended layer on top of the water layer. Then, the toluene layer is syphoned, and fresh toluene is added until the top toluene layer is clear, indicating a complete removal of protein. However, the toluene–protein layer entraps small starch granules (e.g., B-type wheat starch granules), thus reducing the yield and small granules of starch in the extracted starch (McDonald & Stark, 1988).

2.5 EXAMPLES OF STARCH EXTRACTION

There is no specific method for the extraction of starch. A selection of physical, chemical and enzymatic processes depends on the cost of extraction, environmental regulation and the desired degree of purity. For example, alkaline extraction is more cost-effective than enzyme extraction but less environmentally friendly (Halal et al., 2019). Besides, it is found that the lentil starch extracted at high pH (pH 9.5) can cause >1.0% starch damage (Lee, Htoon, Uthayakumaran, & Paterson, 2007). Therefore, although high extraction pH conditions and temperatures contribute to higher starch yields, this also results in higher starch damage values. Extracting with 0.45% sodium bisulfite for 6 h and repeatedly washing the starch with water is the best extraction method for chickpea starch, which does minor damage to the molecular structure (Zhao et al., 2020). On the other hand, potato or tapioca starch can easily be separated by mechanical disintegration of peeled tubers and recovering the almost pure starch with hydrocyclones (Breuninger et al., 2009), since tuber starches have very low protein and lipid contents, which is different from cereal starches. A generic step to extract the starch from cereals/legumes using chemical and enzymatic methods is shown in Figure 2.4.

There are two commercial starch separation methods, which consist of either wet extraction or dry grinding. Wet extraction is usually used for maize and potato starch, with or without involving the alkali dissolution of protein. The dry grinding method releases starch through needle grinding and air classification, which is usually used for pea and bean starch. The small particle size flours are obtained by multiple crushing with hammer mills or needle mills, and then the low protein starch components are separated by air classification.

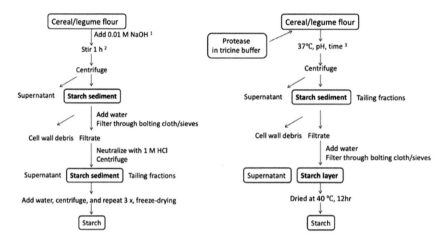

Figure 2.4 Starch isolation using alkaline and enzyme solution. Notes: (1) The alkaline solution can be sodium hydroxide (pH ~8), sodium hydroxide (pH ~12), sodium bisulfite, sodium metabisulfite, etc. (2) Soaking time can vary from 1 h to 24 h. (3) pH and incubation time can also be varied.

However, the purity of starch extracted by the dry grinding method is low, as air classification cannot completely separate the protein from the starch particles. Therefore, after air classification, a water-washing step is used to remove partially attached proteins. In addition, starch after dry milling and air classification contains a higher content of damaged starch than other methods (Wani et al., 2016).

For the purest form of starch, more steps are needed for the removal of SGAPs (Ye et al., 2019). The starch obtained from a wet or dry method is mixed in 4 times the volume of extraction buffer [1.5% SDS, 2% mercaptoethanol 50 mM Tris-HCl (pH 8), 10% glycerol], and stirred for 48 h, and then centrifuged (6000 × g, 15 min). The starch layer is resuspended in ethanol (80%, 10-time volume) and filtered through the system that retains particles above 0.5 µm. After drying at room temperature, the starch is ground and passed through a 100 mesh screen.

2.5.1 Starches with Strong Starch–Protein Interactions: Amaranthus Starches

Some kinds of starches are difficult to isolate due to the small size of the granules and close binding with protein, such as rice starch (3–8 µm), amaranth starch (0.5–3 µm) (Loubes et al., 2012) and oat starch (3–10 µm) (Halal et al., 2019). Compared with other starches, it is difficult to remove the protein, which makes the starch separation process more costly. Alkaline extraction can be used for these starches, and the cost is relatively low. However, the alkaline soaking of these starches requires a high concentration to remove protein, which consequently promotes starch granule damage and results in a reduced starch yield. To overcome this problem, protease is used to facilitate protein removal along with alkali. This will reduce the total amount of alkali as well as promote the separation process, thereby obtaining a higher starch yield and less damaged starch (Lim et al., 1992; Puchongkavarin et al., 2005; Villarreal et al., 2013). Here we have discussed the method of extracting amaranth starch as an example (Villarreal et al., 2013).

Starch was steeped in a NaOH solution (0.05% w/v) at room temperature with shaking for 15 h. The steeped suspension was neutralized using HCl solution, decanted for 30 min and the resulting precipitate washed with distilled water (50 ml) on a 200 mesh stainless steel mesh. The solids were mixed with water, and this mixture was added with the protease (0.1 g/100 g) for 5 h at 37°C (pH 7.5) (stage 2). Starch extraction with the combination of protease and alkaline solution allows obtaining higher yield and recovery and low protein contents (35.32%, 53.53% and 0.266%, respectively) of the extracted starches than using only alkaline solution (31.42%, 47.74% and 0.512%, respectively).

The aforementioned process, however, can be optimized by selecting the alkali type and its concentration and source of protease. There is not a universal recommendation for starch extraction, but combination techniques are more gentle and have lesser damage to the starch surface and internal structure, and hence are preferable to single extractions steps.

2.5.2 Starch Extraction from Nonconventional Sources

The starch used in industry is mainly separated from traditional sources such as cereals, legumes and tubers. However, sourcing starch from nontraditional sources has become increasingly important due to its unique properties and associated benefits (by-product utilization, grown in less fertile land, produced by marginal groups, etc.).

Fruit waste contains a large amount of starch (20%–55%), which has unique characteristics and can expand the possible uses in the industry (Kringel et al., 2020). Pineapple (Kushwaha et al., 2021), banana (Marta et al., 2019) and jackfruit (Kushwaha et al., 2021) starch are extracted through a wet sieving process that consists of washing and cutting of fruits to small pieces followed by soaking in water (1:2) for 0.5–24 h. Then, the hydrated fruit pieces in water are ground mechanically, then passed through the sieves of different mesh sizes. Finally, the starch in the slurry is recovered by decantation or centrifugation. Extraction aids such as citric acid (0.3%–5%; Hernández-Carmona et al., 2017; Jaiswal & Kumar, 2015), NaOH (0.5%; Jaiswal & Kumar, 2015) and sodium metabisulfite (0.16%–0.5%; Guo et al., 2018; Kaur & Bhullar, 2016; Stevenson et al., 2006) have been applied in soaking water to facilitate the hydration, loosening of cell walls and improve the recovery of starch.

Chestnuts and acorns are also starch-rich substances (30%–80%). Starch can be extracted through low shear under alkaline conditions (1 part flour: 2 parts 0.25% NaOH, 24 h, 5°C) and protease treatment under low shear conditions (5U *Aspergillus oryzae* protease per gram of flour, 6 h at 37°C).

2.6 NOVEL EXTRACTION TECHNIQUES

2.6.1 Gluten Washing

A nondestructive patented method with the principle of "gluten washing" was developed by Al-Hakkak and Al-Hakkak (2007) to separate starch with high purity from plant flours that contain a mixture of starch and protein. With the application in the wheat starch isolation on an industrial scale, this method can also be used for other cereals (such as barley, oats and rye), legumes (such as peas, chickpeas and lentils) and seeds (such as amaranths). The basic principle of the method is to add wheat gluten to the desired plant flour in a ratio of 10%–20% (w/w) gluten/flour. Water is then added to the mixture. Upon hydration, wheat protein aggregates, and a gluten network is formed when the mixture is kneaded. The dough is left to proof for 1 h to allow further protein agglomeration. The starch is then washed out of the mixed protein network with water, recovered by centrifugation and drying. Salt is added to the mixture of flour and gluten, which can make the dough firmer and reduce the amount of gluten used.

2.6.2 Ultrasound-Assisted Milling

Compared with the traditional starch extraction technology, ultrasound, as a new, clean and green auxiliary extraction technology, can be used for the separation of starch. Ultrasonic wave is suitable for grain endosperm with a tight combinations of starch and protein, corn and sorghum, for example. The cavitation effect of ultrasonic waves can promote the disruption of the cell membrane and break the cell wall for enhancing the release of starch from the protein matrix. Liu et al. (2019) used ultrasound-assisted wet milling to assist the high-amylose starch extraction. It was found that the starch yield (56.68%) is increased by 11%, SO_2 content in steep water can be reduced by 75% and steeping time is reduced by 25% compared with the conventional wet milling process. In addition, the thermal properties and morphologies of high-amylose starch were not changed by ultrasonic treatment. However, some studies reported that the ultrasonic treatment could induce molecular disruption and cause damage to the granular, crystalline and molecular structure of starch (Hu et al., 2014; Wei et al., 2021). Ultrasonic disruption may preferentially occur to the amorphous regions (amylose part) rather than the crystalline region (Luo et al., 2008). Wei et al. (2021) reported that the breakpoints first occurred at B3 chains (midpoint scission) and then at B2 chains (midpoint scission) with an increase of contents on A chains, resulting in a much narrower molecular size distribution. Enzymatic digestion of protein has also been applied to starch isolation with or without ultrasonic treatment (Cameron & Wang, 2006; Correia & Beirão-da-Costa, 2012; Sit et al., 2014). Considering the changes in the structural and physicochemical properties of ultrasonically treated starch, ultrasonic treatment can be used to improve the hydration process, cooking properties and textural attributes of cereals (H. Kaur & Gill, 2019).

2.6.3 Microwave-Assisted Starch Extraction

Microwave irradiation treatment is used to treat lotus seed flour at different times (Nawaz et al., 2018). The yield of starch is increased with an increase in microwave treatment time; however, microwave-induced variations in structures significantly affects the functional properties of starch, such as the water-holding capacity, oil-holding capacity, swelling capacity and emulsifying stability. Chen et al. (2021) found that microwave irradiation causes more significant changes on the in-kernel starch than isolated starch, and the amylopectin molecules were significantly degraded after the microwave treatment.

2.7 CONCLUSIONS

In plants, starch coexists with protein, lipids, cell walls and other components, and their interactions are dependent on the type of plant. Their interaction increases the difficulty of starch purification and varies the structure and physicochemical properties of the purified starch. Although starch extraction aims to obtain starch materials with higher purity, the structure and physicochemical properties of the purified starch change due to the removal of starch-associated lipids and proteins. The extraction methods and steps depend on the biological variations (cereals, tubers, legumes, etc.), the desired purity and functional properties. Further, in some cases, the starch extraction techniques are carefully selected considering the amount of sample, environmental concerns (regulation), availability of water and appropriate separation techniques. Even with the predetermined methods, the extraction steps need to be optimized, making starch extraction a complicated industrial process.

REFERENCES

Al-Hakkak, J., & Al-Hakkak, F. (2007). New non-destructive method using gluten to isolate starch from plant materials other than wheat. *Starch - Stärke*, 59(3–4), 117–124.

Andersson, A., Andersson, R., & Åman, P. (2001). Starch and by-products from a laboratory-scale barley starch isolation procedure. *Cereal Chemistry*, 78(5), 507–513.

Asmeda, R., Noorlaila, A., & Norziah, M. (2016). Relationships of damaged starch granules and particle size distribution with pasting and thermal profiles of milled MR263 rice flour. *Food Chemistry*, 191, 45–51.

Baldwin, P. M. (2001). Starch granule-associated proteins and polypeptides: A review. *Starch-Stärke*, 53(10), 475–503.

Bancel, E., Rogniaux, H., Debiton, C., Chambon, C., & Branlard, G. (2010). Extraction and proteome analysis of starch granule-associated proteins in mature wheat kernel (*Triticum aestivum* L.). *Journal of Proteome Research*, 9(6), 3299–3310.

Barak, S., Mudgil, D., & Khatkar, B. (2014). Effect of flour particle size and damaged starch on the quality of cookies. *Journal of Food Science and Technology*, 51(7), 1342–1348.

Baszczak, W., Fornal, J., Amarowicz, R., & Pegg, R. (2003). Lipids of wheat, corn and potato starch. *Journal of Food Lipids*, 10(4), 301–312.

Benmoussa, M., Hamaker, B. R., Huang, C. P., Sherman, D. M., Weil, C. F., & BeMiller, J. N. (2010). Elucidation of maize endosperm starch granule channel proteins and evidence for plastoskeletal structures in maize endosperm amyloplasts. *Journal of Cereal Science*, 52(1), 22–29.

Breuninger, W. F., Piyachomkwan, K., & Sriroth, K. (2009). Chapter 12: Tapioca/Cassava Starch: Production and use. In J. BeMiller & R. Whistler (Eds.), *Starch* (Third Edition) (pp. 541–568). San Diego: Academic Press.

Cai, J., Yang, Y., Man, J., Huang, J., Wang, Z., Zhang, C., . . . Wei, C. (2014). Structural and functional properties of alkali-treated high-amylose rice starch. *Food Chemistry*, 145, 245–253.

Cameron, D. K., & Wang, Y. J. (2006). Application of protease and high-intensity ultrasound in corn starch isolation from degermed corn flour. *Cereal Chemistry*, 83(5), 505–509.

Cardoso, M. B., Putaux, J.-L., Samios, D., & da Silveira, N. P. (2007). Influence of alkali concentration on the deproteinization and/or gelatinization of rice starch. *Carbohydrate Polymers*, 70(2), 160–165.

Champagne, E., Marshall, W., & Goynes, W. (1990). Effects of the degree of milling and lipid removal on starch gelatinization in the brown rice kernel. *Cereal Chemistry*, 67(6), 570–574.

Chang, K. C., Lee, C. C., & Brown, G. (1986). Production and nutritional evaluation of high-protein rice flour. *Journal of Food Science*, 51(2), 464–467.

Charles, A. L., Sriroth, K., & Huang, T.-c. (2005). Proximate composition, mineral contents, hydrogen cyanide and phytic acid of 5 cassava genotypes. *Food Chemistry*, 92(4), 615–620.

Chen, X., Liu, Y., Xu, Z., Zhang, C., Liu, X., Sui, Z., & Corke, H. (2021). Microwave irradiation alters the rheological properties and molecular structure of hull-less barley starch. *Food Hydrocolloids*, 120, 106821.

Choi, J. M., Park, C. S., Baik, M. Y., Kim, H. S., Choi, Y. S., Choi, H. W., & Seo, D. H. (2018). Enzymatic extraction of starch from broken rice using freeze-thaw infusion with food-grade protease. *Starch-Stärke*, 70(1–2), 1700007.

Chung, H.-J., Liu, Q., Donner, E., Hoover, R., Warkentin, T. D., & Vandenberg, B. (2008). Composition, molecular structure, properties, and *in vitro* digestibility of starches from newly released Canadian pulse cultivars. *Cereal Chemistry*, 85(4), 471–479.

Contreras-Jiménez, B., Torres-Vargas, O. L., & Rodríguez-García, M. E. (2019). Physicochemical characterization of quinoa (Chenopodium quinoa) flour and isolated starch. *Food Chemistry*, 298, 124982.

Correia, P. R., & Beirão-da-Costa, M. L. (2012). Starch isolation from chestnut and acorn flours through alkaline and enzymatic methods. *Food and Bioproducts Processing*, 90(2), 309–316.

Cox, M. J., MacMasters, M. M., & Hilbert, G. (1944). Effect of the sulfurous acid steep in corn wet milling. *Cereal Chemistry*, 21(6), 447–465.

Dailey Jr, O. D. (2002). Effect of lactic acid on protein solubilization and starch yield in corn wet-mill steeping: A study of hybrid effects. *Cereal Chemistry*, 79(2), 257–260.

Debet, M. R., & Gidley, M. J. (2006). Three classes of starch granule swelling: Influence of surface proteins and lipids. *Carbohydrate Polymers*, 64(3), 452–465.

Dhital, S., Brennan, C., & Gidley, M. J. (2019). Location and interactions of starches in planta: Effects on food and nutritional functionality. *Trends in Food Science & Technology*, 93, 158–166.

Dhital, S., Shrestha, A. K., & Gidley, M. J. (2010). Effect of cryo-milling on starches: Functionality and digestibility. *Food Hydrocolloids*, 24(2–3), 152–163.

Dhital, S., Shrestha, A. K., Hasjim, J., & Gidley, M. J. (2011). Physicochemical and structural properties of maize and potato starches as a function of granule size. *Journal of Agricultural and Food Chemistry*, 59(18), 10151–10161.

Dowd, M. K. (2003). Improvements to laboratory-scale maize wet-milling procedures. *Industrial Crops and Products*, 18(1), 67–76.

Eckhoff, S. R., & Watson, S. A. (2009). Chapter 9 - Corn and Sorghum Starches: Production. In J. BeMiller & R. Whistler (Eds.), *Starch* (Third Edition) (pp. 373–439). San Diego: Academic Press.

El Halal, S. L. M., Kringel, D. H., Zavareze, E. d. R., & Dias, A. R. G. (2019). Methods for extracting cereal starches from different sources: A review. *Starch-Stärke*, 71(11–12), 1900128.

Eliasson, A.-C., & Ljunger, G. (1988). Interactions between amylopectin and lipid additives during retrogradation in a model system. *Journal of the Science of Food and Agriculture*, 44(4), 353–361.

Germani, R., Ciacco, C., & Rodriguez-Amaya, D. (1983). Effect of sugars, lipids and type of starch on the mode and kinetics of retrogradation of concentrated corn starch gels. *Starch-Stärke*, 35(11), 377–381.

Guan, E., Yang, Y., Pang, J., Zhang, T., Li, M., & Bian, K. (2020). Ultrafine grinding of wheat flour: Effect of flour/starch granule profiles and particle size distribution on falling number and pasting properties. *Food Science & Nutrition*, 8(6), 2581–2587.

Guo, K., Lin, L., Fan, X., Zhang, L., & Wei, C. (2018). Comparison of structural and functional properties of starches from five fruit kernels. *Food Chemistry*, 257, 75–82.

Han, X.-Z., & Hamaker, B. R. (2002). Partial leaching of granule-associated proteins from rice starch during alkaline extraction and subsequent gelatinization. *Starch - Stärke*, 54(10), 454–460.

Han, X. Z., Benmoussa, M., Gray, J. A., BeMiller, J. N., & Hamaker, B. R. (2005). Detection of proteins in starch granule channels. *Cereal Chemistry*, 82(4), 351–355.

Haros, M., Tolaba, M. P., & Suárez, C. (2003). Influence of corn drying on its quality for the wet-milling process. *Journal of Food Engineering*, 60(2), 177–184.

Hasjim, J., Li, E., & Dhital, S. (2012). Milling of rice grains: The roles of starch structures in the solubility and swelling properties of rice flour. *Starch - Stärke*, 64(8), 631–645.

Hasjim, J., Li, E., & Dhital, S. (2013). Milling of rice grains: Effects of starch/flour structures on gelatinization and pasting properties. *Carbohydrate Polymers*, 92(1), 682–690.

Hernández-Carmona, F., Morales-Matos, Y., Lambis-Miranda, H., & Pasqualino, J. (2017). Starch extraction potential from plantain peel wastes. *Journal of Environmental Chemical Engineering*, 5(5), 4980–4985.

Hu, A., Li, L., Zheng, J., Lu, J., Meng, X., Liu, Y., & Rehman, R.-u.-. (2014). Different-frequency ultrasonic effects on properties and structure of corn starch. *Journal of the Science of Food and Agriculture*, 94(14), 2929–2934.

Jackowiak, H., Packa, D., Wiwart, M., & Perkowski, J. (2005). Scanning electron microscopy of Fusarium damaged kernels of spring wheat. *International Journal of Food Microbiology*, 98(2), 113–123.

Jaiswal, P., & Kumar, K. J. (2015). Physicochemical properties and release characteristics of starches from seeds of Indian Shahi Litchi. *International Journal of Biological Macromolecules*, 79, 256–261.

Johnston, D. B., & Singh, V. (2004). Enzymatic milling of corn: optimization of soaking, grinding, and enzyme incubation steps. *Cereal Chemistry*, 81(5), 626–632.

Kaur, H., & Gill, B. S. (2019). Effect of high-intensity ultrasound treatment on nutritional, rheological and structural properties of starches obtained from different cereals. *International Journal of Biological Macromolecules*, 126, 367–375.

Kaur, M., & Bhullar, G. K. (2016). Partial characterization of Tamarind (Tamarindus indica L.) kernel starch oxidized at different levels of sodium hypochlorite. *International Journal of Food Properties*, 19(3), 605–617.

Kringel, D. H., Dias, A. R. G., Zavareze, E. d. R., & Gandra, E. A. (2020). Fruit wastes as promising sources of starch: Extraction, properties, and applications. *Starch-Stärke*, 72(3–4), 1900200.

Kushwaha, R., Kaur, S., & Kaur, D. (2021). Potential of Jackfruit (Artocarpus Heterophyllus Lam.) seed starch as an alternative to the commercial starch source: A review. *Food Reviews International*, 1–20.

Langeveld, S. M., van Wijk, R., Stuurman, N., Kijne, J. W., & de Pater, S. (2000). B-type granule containing protrusions and interconnections between amyloplasts in developing wheat endosperm revealed by transmission electron microscopy and GFP expression. *Journal of Experimental Botany*, 51(349), 1357–1361.

Lee, H., Htoon, A., Uthayakumaran, S., & Paterson, J. (2007). Chemical and functional quality of protein isolated from alkaline extraction of Australian lentil cultivars: Matilda and Digger. *Food Chemistry*, 102(4), 1199–1207.

Lee, J. H., Han, J.-A., & Lim, S.-T. (2009). Effect of pH on aqueous structure of maize starches analyzed by HPSEC-MALLS-RI system. *Food Hydrocolloids*, 23(7), 1935–1939.

Lee, Y.-T., Seog, H.-M., Cho, M.-K., & Kim, S.-S. (1996). Physicochemical properties of hull-less barley flours prepared with different grinding mills. *Korean Journal of Food Science and Technology*, 28(6), 1078–1083.

Leewatchararongjaroen, J., & Anuntagool, J. (2016). Effects of dry-milling and wet-milling on chemical, physical and gelatinization properties of rice flour. *Rice Science*, 23(5), 274–281.

León, A. E., Barrera, G. N., Pérez, G. T., Ribotta, P. D., & Rosell, C. M. (2006). Effect of damaged starch levels on flour-thermal behaviour and bread staling. *European Food Research and Technology*, 224(2), 187–192.

Leonel, M., do Carmo, E. L., Fernandes, A. M., Soratto, R. P., Ebúrneo, J. A. M., Garcia, É. L., & dos Santos, T. P. R. (2017). Chemical composition of potato tubers: the effect of cultivars and growth conditions. *Journal of Food Science and Technology*, 54(8), 2372–2378.

Li, E., Dhital, S., & Hasjim, J. (2014). Effects of grain milling on starch structures and flour/starch properties. *Starch-Stärke*, 66(1–2), 15–27.

Li, M., Daygon, V. D., Solah, V., & Dhital, S. (2021). Starch granule size: Does it matter? *Critical Reviews in Food Science and Nutrition*, 1–21.

Li, W., Xiao, X., Guo, S., Ouyang, S., Luo, Q., Zheng, J., & Zhang, G. (2014). Proximate composition of triangular pea, white pea, spotted colored pea, and small white kidney bean and their starch properties. *Food and Bioprocess Technology*, 7(4), 1078–1087.

Lim, S. T., Lee, J. H., Shin, D. H., & Lim, H. S. (1999). Comparison of protein extraction solutions for rice starch isolation and effects of residual protein content on starch pasting properties. *Starch-Stärke*, 51(4), 120–125.

Lim, W. J., Liang, Y. T., Seib, P. A., & Rao, C. S. (1992). Isolation of oat starch from oat flour. *Cereal Chemistry*, 69(3), 233–236.

Liu, J., Wang, Y., Fang, G., Man, Y., & Liu, Y. (2019). Effect of ultrasound-assisted isolation on yield and properties of high-amylose starch from amylomaize. *Starch-Stärke*, 71(9–10), 1800292.

Loubes, M. A., Resio, A. C., Tolaba, M. P., & Suarez, C. (2012). Mechanical and thermal characteristics of amaranth starch isolated by acid wet-milling procedure. *LWT-Food Science and Technology*, 46(2), 519–524.

Lowy, G. D., Sargeant, J. G., & David Schofield, J. (1981). Wheat starch granule protein: The isolation and characterisation of a salt-extractable protein from starch granules. *Journal of the Science of Food and Agriculture*, 32(4), 371–377.

Luo, Z., Fu, X., He, X., Luo, F., Gao, Q., & Yu, S. (2008). Effect of ultrasonic treatment on the physicochemical properties of maize starches differing in amylose content. *Starch - Stärke*, 60(11), 646–653.

Maher, G. (1983). Alkali gelatinization of starches. *Starch-Stärke*, 35(7), 226–234.

Maniñgat, C. C., & Juliano, B. O. (1980). Starch lipids and their effect on rice starch properties. *Starch - Stärke*, 32(3), 76–82.

Marta, H., Cahyana, Y., Djali, M., Arcot, J., & Tensiska, T. (2019). A comparative study on the physicochemical and pasting properties of starch and flour from different banana (Musa spp.) cultivars grown in Indonesia. *International Journal of Food Properties*, 22(1), 1562–1575.

McDonald, A., & Stark, J. (1988). A critical examination of procedures for the isolation of barley starch. *Journal of the Institute of Brewing*, 94(3), 125–132.

Morrison, W. R. (1981). Starch lipids: A reappraisal. *Starch-Stärke*, 33(12), 408–410.

Morrison, W. R. (1988). Lipids in cereal starches: A review. *Journal of Cereal Science*, 8(1), 1–15.

Morrison, W. R., Milligan, T. P., & Azudin, M. N. (1984). A relationship between the amylose and lipid contents of starches from diploid cereals. *Journal of Cereal Science*, 2(4), 257–271.

Myers, D., & Fox, S. (1994). Alkali wet-milling characteristics of pearled and unpearled amaranth seed. *Cereal Chemistry*, 71(1), 96–98.

Nawaz, H., Shad, M. A., Saleem, S., Khan, M. U. A., Nishan, U., Rasheed, T., . . . Iqbal, H. M. N. (2018). Characteristics of starch isolated from microwave heat treated lotus (Nelumbo nucifera) seed flour. *International Journal of Biological Macromolecules*, 113, 219–226.

Palacios-Fonseca, A., Castro-Rosas, J., Gómez-Aldapa, C., Tovar-Benítez, T., Millán-Malo, B., Del Real, A., & Rodríguez-García, M. (2013). Effect of the alkaline and acid treatments on the physicochemical properties of corn starch. *CyTA-Journal of Food*, 11(sup1), 67–74.

Pedersen, M. B., Dalsgaard, S., Arent, S., Lorentsen, R., Knudsen, K. E. B., Yu, S., & Lærke, H. N. (2015). Xylanase and protease increase solubilization of non-starch polysaccharides and nutrient release of corn-and wheat distillers dried grains with solubles. *Biochemical Engineering Journal*, 98, 99–106.

Pérez, G., Ribotta, P., Aguirre, A., Rubiolo, O. J., & León, A. (2004). Changes in proteins and starch granule size distribution during grain filling of triticale. *Agriscientia*, 21(1), 13–21.

Pérez, O., Haros, M., & Suarez, C. (2001). Corn steeping: influence of time and lactic acid on isolation and thermal properties of starch. *Journal of Food Engineering*, 48(3), 251–256.

Puchongkavarin, H., Varavinit, S., & Bergthaller, W. (2005). Comparative study of pilot scale rice starch production by an alkaline and an enzymatic process. *Starch-Stärke*, 57(3-4), 134–144.

Radosavljevic, M., Jane, J., & Johnson, L. (1998). Isolation of amaranth starch by diluted alkaline-protease treatment. *Cereal Chemistry*, 75(2), 212–216.

Raeker, M., Gaines, C., Finney, P., & Donelson, T. (1998). Granule size distribution and chemical composition of starches from 12 soft wheat cultivars. *Cereal Chemistry*, 75(5), 721–728.

Ramírez, E. C., Johnston, D. B., McAloon, A. J., & Singh, V. (2009). Enzymatic corn wet milling: engineering process and cost model. *Biotechnology for Biofuels*, 2(1), 1 9.

Ren, Y., Setia, R., Warkentin, T. D., & Ai, Y. (2021a). Functionality and starch digestibility of wrinkled and round pea flours of two different particle sizes. *Food Chemistry*, 336, 127711.

Ren, Y., Yuan, T. Z., Chigwedere, C. M., & Ai, Y. (2021b). A current review of structure, functional properties, and industrial applications of pulse starches for value-added utilization. *Comprehensive Reviews in Food Science and Food Safety*, 20(3), 3061–3092.

Reyes-Moreno, C., Paredes-López, O., & Gonzalez, E. (1993). Hard-to-cook phenomenon in common beans: A review. *Critical Reviews in Food Science & Nutrition*, 33(3), 227–286.

Roa, D. F., Santagapita, P. R., Buera, M. P., & Tolaba, M. P. (2014). Ball milling of Amaranth starch-enriched fraction. Changes on particle size, starch crystallinity, and functionality as a function of milling energy. *Food and Bioprocess Technology*, 7(9), 2723–2731.

Rooney, L., & Pflugfelder, R. (1986). Factors affecting starch digestibility with special emphasis on sorghum and corn. *Journal of Animal Science*, 63(5), 1607–1623.

Ruales, J., & Nair, B. (1994). Properties of starch and dietary fibre in raw and processed quinoa (Chenopodium quinoa, Willd) seeds. *Plant Foods for Human Nutrition*, 45(3), 223–246.

Schofield, J., & Greenwell, P. (1987). Wheat starch granule proteins and their technological significance. In I. D. Morton (Ed.), *Cereals in a European Context: First European Conference on Food Science urzd Technology* (pp. 407–420). Bournemouth: Australia: Cereal Chemistry Division, Royal Australian Chemical Institute

Shandera, D., & Jackson, D. (1996). Effect of corn wet-milling conditions (sulfur dioxide, lactic acid, and steeping temperature) on starch functionality. *Cereal Chemistry*, 73(5), 632–637.

Shewry, P. (1999). The synthesis, processing, and deposition of gluten proteins in the developing wheat grain. *Cereal Foods World*, 44(8), 587–589.

Shewry, P. R., Wan, Y., Hawkesford, M. J., & Tosi, P. (2020). Spatial distribution of functional components in the starchy endosperm of wheat grains. *Journal of Cereal Science*, 91, 102869.

Shi, J., Zhang, T., Wang, T., & Wu, M. (2021). Effects of glutelin and lipid oxidation on the physicochemical properties of rice starch. *Cereal Chemistry*, 98(3), 683–692.

Simmonds, D. (1971). Morphological and molecular aspects of wheat quality. *Wallerstein Laboratories Communications*, 34(113), 17–34.

Singh, J., Kaur, L., & Singh, H. (2013). Chapter four: Food microstructure and starch digestion. In J. Henry (Ed.), *Advances in Food and Nutrition Research* (pp. 137–179). Academic Press.

Singh, S., Thakur, S., Singh, M., Kumar, A., Kumar, A., Kumar, A., . . . Singh, H. (2018). Influence of different isolation methods on physicochemical and rheological properties of native and heat-moisture-treated chickpea starch. *Journal of Food Processing and Preservation*, 42(2), e13523.

Sit, N., Deka, S. C., & Misra, S. (2014). Combined effect of ultrasound and enzymatic pretreatment on yield and functional properties of taro (Colocasia esculenta) starch. *Starch-Stärke*, 66(11–12), 959–967.

Srichuwong, S., Curti, D., Austin, S., King, R., Lamothe, L., & Gloria-Hernandez, H. (2017). Physicochemical properties and starch digestibility of whole grain sorghums, millet, quinoa and amaranth flours, as affected by starch and non-starch constituents. *Food Chemistry*, 233, 1–10.

Steinke, J., & Johnson, L. (1991). Steeping maize in the presence of multiple enzymes. *Cereal Chemistry*, 68(1), 7–12.

Stevenson, D. G., Johnson, S. R., Jane, J. l., & Inglett, G. E. (2006). Chemical and physical properties of kiwifruit (Actinidia deliciosa) starch. *Starch-Stärke*, 58(7), 323–329.

Tao, H., Lu, F., Zhu, X.-F., Xu, G.-X., Xie, H.-Q., Xu, X.-M., & Wang, H.-L. (2021). Removing surface proteins promote the retrogradation of wheat starch. *Food Hydrocolloids*, 113, 106437.

Tester, R. F., & Morrison, W. R. (1990). Swelling and gelatinization of cereal starches. I. Effects of amylopectin, amylose, and lipids. *Cereal Chemistry*, 67(6), 551–557.

Tosi, P., Parker, M., Gritsch, C. S., Carzaniga, R., Martin, B., & Shewry, P. R. (2009). Trafficking of storage proteins in developing grain of wheat. *Journal of Experimental Botany*, 60(3), 979–991.

Tran, T. T. B., Shelat, K. J., Tang, D., Li, E., Gilbert, R. G., & Hasjim, J. (2011). Milling of rice grains. The degradation on three structural levels of starch in rice flour can be independently controlled during grinding. *Journal of Agricultural and Food Chemistry*, 59(8), 3964–3973.

Vasanthan, T., & Hoover, R. (1992). A comparative study of the composition of lipids associated with starch granules from various botanical sources. *Food Chemistry*, 43(1), 19–27.

Villarreal, M. E., Ribotta, P. D., & Iturriaga, L. B. (2013). Comparing methods for extracting amaranthus starch and the properties of the isolated starches. *LWT-Food Science and Technology*, 51(2), 441–447.

Wang, B., & Wang, J. (2019). Mechanical properties of maize kernel horny endosperm, floury endosperm and germ. *International Journal of Food Properties*, 22(1), 863–877.

Wang, F., Chung, D., Seib, P., & Kim, Y. (2000). Optimum steeping process for wet milling of sorghum. *Cereal Chemistry*, 77(4), 478–483.

Wang, Q., Li, L., & Zheng, X. (2020). A review of milling damaged starch: Generation, measurement, functionality and its effect on starch-based food systems. *Food Chemistry*, 315, 126267.

Wang, S., & Copeland, L. (2012). Effect of alkali treatment on structure and function of pea starch granules. *Food Chemistry*, 135(3), 1635–1642.

Wang, S., Luo, H., Zhang, J., Zhang, Y., He, Z., & Wang, S. (2014). Alkali-induced changes in functional properties and in vitro digestibility of wheat starch: the role of surface proteins and lipids. *Journal of Agricultural and Food Chemistry*, 62(16), 3636–3643.

Wani, I. A., Sogi, D. S., Hamdani, A. M., Gani, A., Bhat, N. A., & Shah, A. (2016). Isolation, composition, and physicochemical properties of starch from legumes: A review. *Starch-Stärke*, 68(9–10), 834–845.

Wei, B., Qi, H., Zou, J., Li, H., Wang, J., Xu, B., & Ma, H. (2021). Degradation mechanism of amylopectin under ultrasonic irradiation. *Food Hydrocolloids*, 111, 106371.

Wittrock, E., Jiang, H., Campbell, M., Campbell, M., Jane, J. l., Anih, E., & Wang, Y. J. (2008). A simplified isolation of high-amylose maize starch using neutral proteases. *Starch-Stärke*, 60(11), 601–608.

Xiong, W., Zhang, B., Huang, Q., Li, C., Pletsch, E. A., & Fu, X. (2018). Variation in the rate and extent of starch digestion is not determined by the starch structural features of cooked whole pulses. *Food Hydrocolloids*, 83, 340–347.

Ye, X., Zhang, Y., Qiu, C., Corke, H., & Sui, Z. (2019). Extraction and characterization of starch granule-associated proteins from rice that affect in vitro starch digestibility. *Food Chemistry*, 276, 754–760.

Zhan, Q., Ye, X., Zhang, Y., Kong, X., Bao, J., Corke, H., & Sui, Z. (2020). Starch granule-associated proteins affect the physicochemical properties of rice starch. *Food Hydrocolloids*, 101, 105504.

Zhang, B., Gilbert, E. P., Qiao, D., Xie, F., Wang, D. K., Zhao, S., & Jiang, F. (2019). A further study on supramolecular structure changes of waxy maize starch subjected to alkaline treatment by extended-q small-angle neutron scattering. *Food Hydrocolloids*, 95, 133–142.

Zhang, X., Shen, Y., Zhang, N., Bao, J., Wu, D., & Shu, X. (2019). The effects of internal endosperm lipids on starch properties: Evidence from rice mutant starches. *Journal of Cereal Science*, 89, 102804.

Zhao, Y., Tan, X., Wu, G., & Gilbert, R. G. (2020). Using molecular fine structure to identify optimal methods of extracting starch. *Starch-Stärke*, 72(5–6), 1900214.

Zheng, G. H., Sosulski, F. W., & Tyler, R. T. (1997). Wet-milling, composition and functional properties of starch and protein isolated from buckwheat groats. *Food Research International*, 30(7), 493–502.

Chapter 3

Starch

Measurement Techniques at Different Length Scales

Kai Wang, Bin Zhang and Sushil Dhital

CONTENTS

DOI: 10.1201/9781003088929-3

3.1 GRANULAR STRUCTURE

There is a wide variation of starch granules synthesized depending on the botanical origin. Variation occurs in granule size (~1–100 µm in diameter), shape (round, lenticular, polygonal), size distribution (unimodal or bimodal) and association (compound or straightforward granules). Some cereal starches like maize, sorghum and barley have pores on their surface, whereas the tuber starches have a smooth surface. Common cereal starches such as wheat and barley contain A- and B-granules: A-granules are disc-like and larger, while B-granules are spherical and smaller in size (Vamadevan & Bertoft, 2015). The internal structure is also heterogeneous, both within the same botanical origin or among different sources. The reason behind the heterogeneity of starch granules is not entirely understood. Different techniques individually or in tandem are used to study the granular structure of starch that largely controls the processing, physicochemical properties and enzyme susceptivity.

3.1.1 Light Microscopy

Light microscopy is an imaging technique for the morphological analysis of starch granules. It is generally used to observe granular structure, growth rings and hilum of starch granules, and to analyze the structural changes subjected to processing or enzymic/bacterial hydrolysis. Microscopic technology can be combined with spectroscopy technology to analyze the structural changes of starch granules and even chemical composition shifts of starch granules during processing (Kowsik & Mazumder, 2018). Light microscopy can also be used to study the distribution of amylose and amylopectin in starch paste through iodine staining, and the changes in polymer–solvent interaction caused by the substitution of polar or nonpolar groups for hydroxyl groups (Blaszczak & Lewandowicz, 2020).

Polarized light microscopy is a tool to investigate local anisotropy of starch, such as absorption and refraction. Optical anisotropy is the result of molecular order in crystals, providing a sensitive tool to analyze the molecular orientation of starch. The refractive index of a given sample depends on the direction of light polarization (termed birefringence), which shows characteristic changes in light intensity. Raw starch granules display a clear polarized cross (Maltese cross) under a polarized light microscope and are widely studied. The birefringence is assumed to be due to the radial orientation of the amylopectin molecule as a semicrystalline lamellar form. Several authors have reviewed as well as studied the granule morphology of cereal, roots, tuber and legume starches (Singh et al., 2003; J. Wang et al., 2018), and highlighted not only the difference in shape and size but also the position of the hilum (nucleus of the starch granules). The hilum on cereal starches is located closer to the center of granules, whereas the tubers and roots are closer to the edge of granules (Pérez & Bertoft, 2010). The polarized microscope can be further used to observe changes in the Maltese cross-reflection of starch during processing. Cai and Wei (2013) used a polarized light microscope to study the destruction of Maltese cross-reflection after heating at different temperatures (as shown in Figure 3.1). Further, the mode of granule hydrolysis could be studied by observing the remnant polarization after enzyme treatment. The birefringence signal of enzyme-treated starch was lighter and finally disappeared in the semicrystalline region of starch granules after hydrolysis (Perez-Rea et al., 2013), suggesting the ability of amylolytic enzymes to hydrolyze the crystalline region of the starch granules along with the amorphous areas.

3.1.2 Scanning Electron Microscopy

Scanning electron microscopy (SEM) has an advantage over the light microscope with higher magnification and resolution, and observation of a three-dimensional view to some extent. A finely focused electron beam scans the sample's surface to excite secondary, backscattered, absorbed and transmitted electrons. The electrons are synchronized as scanning pictures. SEM can also be used to gather information on the physical and chemical properties of the sample, such as the morphology, composition, crystal structure, electronic structure, and internal electric or magnetic field.

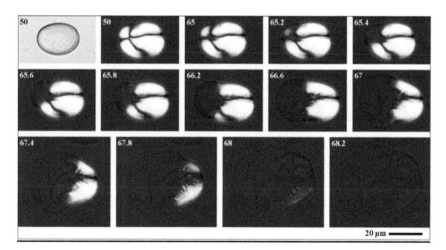

Figure 3.1 Micrographs of potato starch granules viewed under normal and polarizing light in conjunction with a λ plate during heating (adapted from Cai and Wei, 2013). The number in the top left corner of each micrograph represents the corresponding heating temperature (°C).

SEM is also used to observe starch granules at higher magnification to observe morphological changes even at the nanometer scale during processing. Using corn (A-type) and potato (B-type) starch granules as raw materials, Zhang et al. (2014) found via SEM that freeze-drying destroyed the surface "hard shell" of B-type potato starch granules, which was specifically manifested by the production of many microholes on the surface (Figure 3.2). Y. Wang et al. (2019) used SEM to study the surface structural changes of lotus seed and potato starch granules with different gelatinization degrees, and results showed that destruction of starch granules increases and the surface becomes rough as the degree of gelatinization increases. When the degree of gelatinization increases to 90%, only one outline can be observed with starch particles. In addition, the retrogradation behavior of starch can also be measured by scanning electron microscopy. At present, many scholars have performed

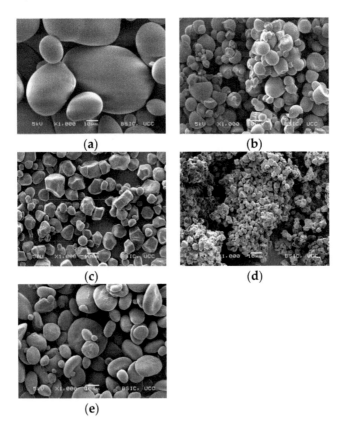

(a)   (b)

(c)   (d)

(e)

Figure 3.2   SEM images of (a) potato starch, (b) tapioca starch, (c) corn starch, (d) rice starch, (e) wheat starch (magnification 1000×) (Horstmann et al., 2016).

structural characterization of the samples of starch gel retrogradation using SEM. After gel storage and dehydration, a fractal network and a cell wall structure are observed with SEM (Charoenrein et al., 2011; Wu et al., 2012). One of the limitations of SEM is the longer sample preparation time and coating with the conductive materials, e.g. gold at 2–6 nm. This coating can mask features that are below 2–6 nm in scale. This has been overcome with Cryo-SEM, where the sample is observed without coating under cryogenic conditions.

3.1.3 Confocal Laser Scanning Microscopy

A confocal laser scanning microscope uses a laser beam as the light source. The laser beam reflects the objective lens through the lighting pinhole and then focuses on the sample and scans every point on the focal plane of the specimen. The fluorescence emitted by the fluorescent substance in the sample tissue after being excited is directly reversed back to the spectroscope through the original incident light path. When the fluorescence signal passes through the detection pinhole, it is focused and collected by the photomultiplier tube (PMT), then the signal is sent to the computer and the image is displayed on the computer monitor after processing.

Compared with ordinary microscopes, confocal laser scanning microscopy (CLSM) has the advantages of high resolution, high sensitivity, optical sectioning, three-dimensional reconstruction and dynamic analysis. As one of the most advanced cell biomedical analysis instruments, the laser scanning confocal microscope has been used to research cell morphology, three-dimensional structure reorganization, dynamic change process, etc., and provides practical research methods such as quantitative fluorescence measurement and quantitative image analysis. For compounds such as starch that do not have autofluorescence, the samples are labeled with fluorochrome. APTS (8-aminopyrene-1,3,6-trisulfonic acid trisodium salt) and FITC (fluorescein isothiocyanate) are used to label starch granules, whereas Nile red and Rhodamine blue are commonly used to label lipids and proteins, respectively (Blonk & Aalst, 1993).

CLSM has been used to visualize starch granules from different plant origins and obtain two-dimensional and three-dimensional images of particles under nondestructive conditions. These images are suitable to characterize the shape, granule size and size distribution of the starches. Van de Velde et al. (2002) studied the characteristics of starch granules from different plant sources and monitored the expansion of starch granules during gelatinization. CLSM has also been used to visualize the morphological change of starch after treatment such as by superheated steam (Ma et al., 2022).

In the research of Mikkel A. Glaring, Christian B. Koch and Andreas Blennow (2006), CLSM was used to analyze the content and detailed distribution of amylose and amylopectin in starch granules. The images of some starch granules indicated that the multiple initiations of new granules are responsible for the compounds or elongated structures observed in these starches. The CLSM optical slices of rice grains showed that the amylose distribution is related to the growth ring structure.

In recent years, many studies have explored the binding sites of enzymes onto starch by labeling enzymes with FITC or tetramethylrhodamine isothiocyanate (TRITC) fluorescent dyes. For example, Dhital et al. (2014) used the above method to label porcine pancreas amylase (AA), and investigated the binding process of AA conjugate to the surface and interior of starch granules under nonhydrolyzed (0°C) and hydrolyzed (37°C) conditions (Figure 3.3). Since the activity of amylase will be retarded after the labelling, some scholars used FITC isomer with molecular size close to the enzyme to simulate the diffusion of the enzyme into starch granules (Guo et al., 2018; M. Wang et al., 2018; Zhang

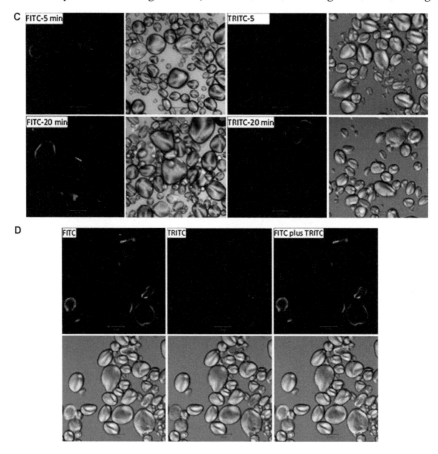

Figure 3.3   Confocal (top panel) and differential interference contrast (bottom panel) images of bound FITC-AA and TRITC-AA conjugate on potato granules incubated for 25 min at 0°C (adapted from Dhital et al.,2014).

et al., 2014). In addition, CLSM was also used to visualize the protein–starch matrix of a noodle after staining the protein and starch with fluorescamine and FITC, respectively (Li et al., 2021).

3.1.4 Transmission Electron Microscopy

A transmission electron microscope (TEM) projects an accelerated and concentrated electron beam onto a very thin sample. Then electrons collide with atoms of the sample to change direction, resulting in solid angle scattering. The size of the scattering angle is related to the density and thickness of the sample, so images with different brightness and darkness can be formed. The images are enlarged and focused, and displayed on imaging devices (e.g., phosphor screens, films and photosensitive coupling components).

TEM is often combined with scanning electron microscopy to determine the structure of starch granules. Sujka and Jamroz (2013) used TEM to observe cracks and depressions on the surface of ultrasonically processed starch granules (Figure 3.4). Jean-Luc Putaux et al. (2000) reported that the starch gel has a

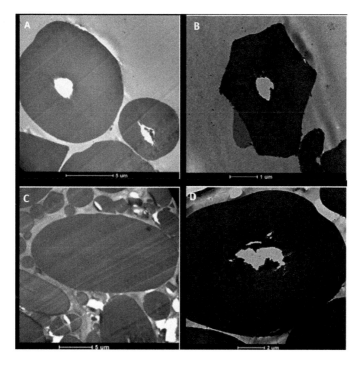

Figure 3.4   TEM micrographs of (A) native corn, (B) rice, (C) wheat and (D) potato starch granules (adapted from Sujka and Jamroz, 2013).

fractal network and a cell wall structure with clear pores through the transmission electron microscopic technique.

3.1.5 Particle Size Analyzer

A particle size analyzer measures the size of powder particles or liquid droplets based on the principle of light scattering. It analyzes the particle size distribution based on the physical phenomenon of laser light scattering. The scattering angle depends on the particle size, smaller particle scattering at a larger angle and larger particle scattering at a smaller angle. The starch granules are mostly dispersed in water, and particle size is measured using instruments such as the Malvern Mastersizer (as shown in Figure 3.5). In cases where the starch is damaged, the damaged starch swells and solubilize in water. The particle size distribution thus can deviate from the true particle size distribution. In these cases, nonpolar solvents such as isopropanol can be used as the dispersant. Modern laser diffraction equipment also has dry cells that use the air as the dispersant, and particle size distribution can be observed without using a

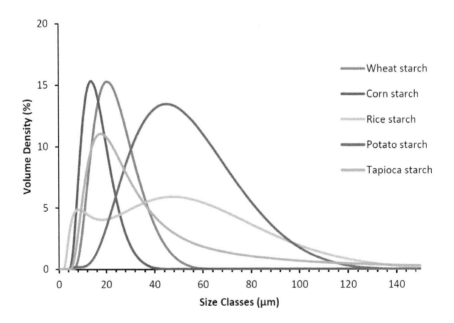

Figure 3.5 Laser light diffraction particle size measurement of potato, tapioca, corn, rice and wheat starch, corresponding to the SEM pictures shown in Figure 3.2 (Horstmann et al., 2016).

liquid dispersant. Laser diffraction techniques further can provide information on volumetric mean diameter, surface mean diameter, median, and 10th and 90th percentile particle size.

3.2 ORDERED STRUCTURE

The ordered structure in starch refers to the molecular order that is obtained from the interaction of amylopectin chains. Although not scientifically proven and highly debated, amylose is considered to be without highly ordered structure in starch. However, amylose can interact with each other in solution (retrogradation) or with a nonstarch component, e.g., lipids, and thus can give the ordered structure. Several techniques are available for characterizing the ordered structure of starch.

3.2.1 X-Ray Diffractometry

During X-ray diffraction, a beam of monochromatic X-ray is incident on a crystal. As the crystal is composed of a unit cell regularly arranged, the distance between these regularly arranged atoms is the same order of magnitude as the wavelength of the incident X-ray, so the difference X-ray scattered by atoms interfere with each other and produce strong X-ray diffraction in particular directions. The orientation and intensity of the diffraction lines in space are closely related to the crystal structure.

X-ray diffractometry (XRD) has been used to characterize the crystalline structure of starch. XRD detects the long-range ordered structure of the double helix regularly and repeatedly, which reflects the starch crystallinity. Natural starch is a semicrystalline material with three different XRD patterns (i.e., A-, B- and C-type) and with various crystallinity. A-type starch has strong diffraction peaks at 15°, 17°, 18° and 23°, and B-type shows classic diffraction peaks at 5.6°, 17°, 22° and 23°. C-type is a combination of A-type and B-type. Cereal starches show an A-type XRD pattern, while tuber and high-amylose starches have a B-type pattern. Starches isolated from pulses, some tubers and fruits show a C-type XRD pattern. The V-type represents the amylose complex with lipid or related compounds (Shi et al., 2021; Zobel, 1988) (Figure 3.6). Waduge et al. (2006) found the relative crystallinity (RC) of regular, high-amylose and waxy barley starches in the range of 20%–36%, 37%–42% and 33%–44%, respectively. The relative crystallinity of different varieties of rice starches was reported to range from 34% to 57% (Dhital et al., 2015).

The definition of crystallinity is based on the theory of a two-phase polymer structure, which assumes that relatively perfect crystalline domains

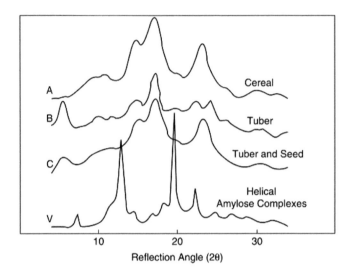

Figure 3.6 X-ray diffraction patterns of different starches. Labelling refers to (A) A-type from cereal starches, (B) B-type from tuber starch, (C) C-type from seed starches, and (V) V-type from helical amylose complexes (Zobel, 1988).

(crystallites) are interspersed with amorphous regions. Thus crystallinity is calculated as the proportion of peak crystalline area corresponding to the total (crystalline plus amorphous) area and expressed in weight percentage or volume percentage. However, the identification (decomposition) of crystalline and amorphous intensity is subjected to error, and diffused scattering from an imperfect crystalline structure cannot be determined. A Gaussian fitting procedure for crystallinity calculations was proposed, where the XRD profile is fitted to an amorphous halo and several discrete crystalline diffraction peaks rather than dividing as the crystalline and amorphous area. The deconvolution of peaks will thus allow the calculation of the contribution from the different crystal polymorphs of starch to the total crystallinity (Lopez-Rubio et al., 2008).

3.2.2 Raman Spectroscopy

Raman spectroscopy is used to probe the internal vibrations of starch molecules ranging from 4000 to 40 cm^{-1}, which is sensitive to the fundamental vibrations of less polar molecular groups and bonds. Therefore, Raman spectroscopy is used to characterize the molecular short-range ordered structure; an example is shown in Figure 3.7 (Flores-Morales et al., 2012). The intensive bands at around 2910 cm^{-1} and 482 cm^{-1} are stretching vibrations of CH- (Gonzalez-Cruz et al. 2018; S. Wang, Li et al., 2015) used to describe the molecular order of starch

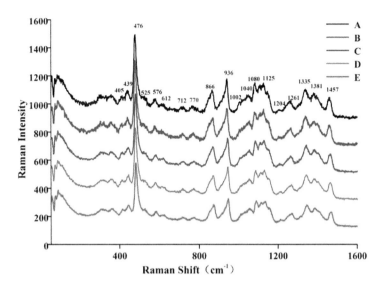

Figure 3.7   Raman spectra of rice starch from different varieties in China, including (A) japonica rice from Heilongjiang, (B) japonica rice from Henan, (C) japonica rice from Shandong, (D) indica rice from Hunan and (E) indica rice from Guangxi. Reprinted with permission from Zhu et al. (2018).

granules or changes in molecular order during gelatinization or retrogradation (S. Wang, Wang et al., 2015). The peak at around 856 and 934 cm^{-1} represents C(1)-H and CH$_2$ deformations and C-O-C skeletal mode vibrations of α-1,4 glycosidic linkages (Bernardino-Nicanor et al., 2017). The band at 1118 cm^{-1} is C-O stretching and C-O-H deformation (S. Wang, Li et al., 2015). The band at 1260 cm^{-1} is CH$_2$OH- related deformation, which is characteristic of V-form amylose (Bernardino-Nicanor et al., 2017). Notably, the full width at half height (FWHH) of the band at 480 cm^{-1} has been used to characterize the structural variation in the short-range molecular order of starch from different botanical sources (S. Wang, Wang et al., 2015). And the FWHH of the band at 480 cm^{-1} was also utilized to characterize revolution in the molecular order of retrograded starch during storage, and the results showed decreasing FWHM values, which indicates the progressively increased degree of short-range molecular order of starch during the retrogradation process (Liu et al., 2021).

3.2.3 Fourier Transform Infrared Spectroscopy

Fourier transform infrared spectroscopy (FT-IR) is another analytical technique used to characterize the chemical bonds and functional groups in molecules. Characteristic carbohydrate peaks are located in the range of 400~4000 cm^{-1}

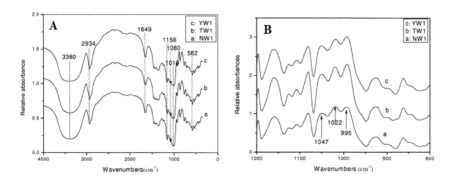

Figure 3.8   FT-IR spectra of three waxy wheat starches: (A) FT-IR spectroscopy; (B) deconvoluted FT-IR spectroscopy from 800 to 1200 cm⁻¹. Reprinted with permission from Wang et al. (2015).

(Figure 3.8). The strong characteristic peak at around 3400 cm⁻¹ represents glucose O-H stretching vibration. The peak at 2930 cm⁻¹ is the antisymmetric stretching vibration of CH_2 (Kačuráková & Mathlouthi, 1996), and that at 1640 cm⁻¹ is the amorphous region vibration of starch. The region from 1500 cm⁻¹ to 800 cm⁻¹ is the fingerprint region of starch. Within this region, the peaks at around 995, 1047 and 1022 cm⁻¹ are the structural features of the crystalline and amorphous of starch, corresponding to the crystalline structure, ordered structure and morphological structure of the starch molecule, respectively (Ding et al., 2020; Ma & Boye, 2018; Zhang et al., 2006). The ratios of integrated areas of 1047/1022 cm⁻¹ and 995/1022 cm⁻¹ reflect the degree of ordered structure and the double helix content of starch, respectively (Chi et al., 2017; Ding et al., 2020; Surendra Babu & Jagan Mohan, 2019). For example, Huang et al. (2021) used the ratio of FT-IR absorbances at 1047/1022 cm⁻¹ to analyze differences in the short-range ordered structure of retrograded starch that were gelatinized under different conditions. FT-IR is also applied to observe the differences in the crystalline and amorphous regions between the complexes and native starch, as well as the bonding forces (Chen et al., 2018). For example, Zheng et al. (2020) suggested that the interaction forces between caffeic acid and maize starch were mainly noncovalent since no new peaks were observed in the FT-IR spectra of the maize starch–caffeic acid complexes. This result is consistent with the literature regarding the longan seed polyphenols and lotus leaf flavonoid on the bonding force changes of starch (He et al., 2021; M. Wang et al., 2018).

3.2.4 Carbon-13 Nuclear Magnetic Resonance

Carbon-13 nuclear magnetic resonance (¹³C NMR) provides information on the short-distance scale molecular organization of starches (Gidley & Bociek, 2002;

Shi et al., 2021). As exhibited in Figure 3.9, the NMR spectrum signals of starch can be classified into several regions: 106–96 ppm for C1; 83–79 ppm for C4; 73–70 ppm for C2, C3 and C5; and 62–59 ppm for C6. Among these carbon sites, C1 and C4 carbons are the most sensitive to polysaccharide conformation, as they were involved in glycosidic linkages (Tan et al., 2007; Yang et al., 2021).

The A-type cereal starches are characterized by a twofold symmetry and result in three inequivalent residues per turn (Mihhalevski et al., 2012), which is confirmed by the distinct three-peak pattern in the C1 region, while the B-type potato starch has a threefold symmetry with respect to adjacent double helices

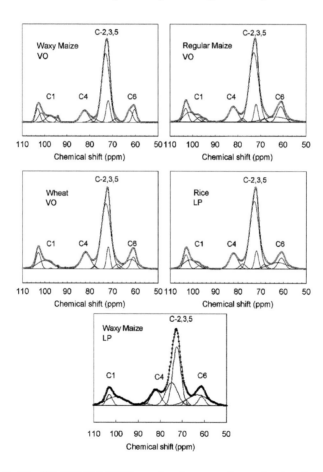

Figure 3.9 The ^{13}C CP/MAS NMR spectra for amorphous starches from different botanical sources and their peak-fitted profiles. LP, lyophilized; VO, vacuum-oven-dried. Experimental data are presented as circles. The peak-fitting correlation coefficient values were at least 0.999. Reprinted with permission from Tan et al. (2007).

resulting in two inequivalent residues per turn and is confirmed by the two-peaks pattern in the C1 region (Gidley & Bociek, 2002). Peaks at around 101.1 and 103 ppm are characteristic of V-type single helices, and high in maize and rice starches but hardly visible in potato (Tan et al., 2007). The absence of these peaks in potato is possibly due to its low lipid content in comparison to cereal starches (Singh et al., 2009) since starch–lipid complexes are known to form V-type single helices (Debet & Gidley, 2006). Peaks at 104.2–105.3 ppm, which were only present for potato, have been associated with the junction zones of the double helices in amylopectin. And peaks at around 106–108 ppm, only visible in the cereal starches, might possibly be a result of cross-linked side chains with better stability shown in the pasting profile (Okyere et al., 2019). Recently, ^{13}C NMR has been applied in combination with Raman spectroscopy and FTIR to characterize changes in the short-range ordered structure of starch during storage, and therefore to unveil the molecular mechanism of starch retrogradation (Flores-Morales et al., 2012; Li et al., 2021; Liu et al., 2021; Yang et al., 2021).

3.2.5 Small-Angle X-Ray Scattering

Small-angle X-ray scattering (SAXS) is an efficient and nondestructive tool to analyze the variations in electron density distributions of amorphous and crystalline lamellae in granular starch (Chen et al., 2017). Compared with other X-ray and neutron scattering methods, small-angle scattering is useful for structural determination at low resolution for systems that do not necessarily possess long-range or crystalline order (Blazek & Gilbert, 2011).

Typical SAXS patterns exhibit a broad scattering peak, representing the repeating crystalline–amorphous lamella structure of starch (Figure 3.10). This broad scattering peak is reciprocally related to alternating crystalline and amorphous regions with a repeat distance of 9–10 nm (Cameron & Donald, 1992; Jenkins & Donald, 1996). In addition to the repeat distance, more information could be obtained from SAXS data using a variety of models, including the paracrystalline model (Cameron & Donald, 1992), the liquid-crystalline model (Daniels & Donald, 2003) and the linear correlation function (Schmidt-Rohr, 2007). These parameters include the thicknesses of crystalline and amorphous lamellae, the electron density difference between the crystalline and amorphous lamellae, and the volume fraction of the crystalline lamellae within the semicrystalline lamellae.

As a powerful technique, SAXS has been applied to analyze the semicrystalline structure of native starches from various botanic sources, and based on these, a number of hypotheses on the location of amylose and amylopectin in crystalline and amorphous lamellae were proposed, which is beneficial for obtaining a better understanding on the starch biosynthetic process in nature (Yuryev et al., 2004). In addition, SAXS has also been used to reveal the effects

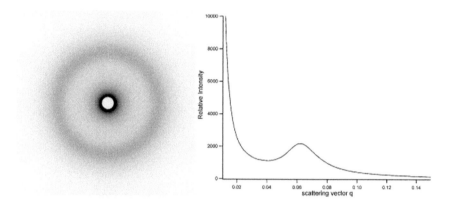

Figure 3.10 Scattering pattern from regular maize starch and corresponding SAXS curve showing relative intensity versus scattering vector. Reprinted from Blazek and Gilbert (2011).

of starch processing or the modification process (heating, extruding, retrogradation, etc.) on starch. Related studies have been extensively reviewed in Blazek and Gilbert (2011).

3.2.6 Atomic Force Microscopy

Atomic force microscopy (AFM) is a powerful tool to characterize biopolymers on the sub-nanometer scale and visualizes the specimen by contouring the surface by controlling forces between a tiny and sharp probe or tip and the specimen surface. It has been successfully used to investigate both the internal and external structures of starch granules close to their native state (Peroni-Okita et al., 2015).

It has been reported that starch from rice and potato showed large aggregate structures. Amylose from potato showed a french fry–like structure with a mean contour length of 504 nm and width ranging 120~135 nm. For amylose from rice, it displayed a well-separated, unbranched structure with a mean contour length of 652 nm and width ranging 4~19 nm (Figure 3.11). The degrees of polymerization was about 1440 and 1860 nm for amylose from potato and rice, respectively, as calculated from the mean contour length (Dang et al., 2006). The polydispersity index (PDI) of amylose was calculated as 1.4, and the height of the chains was measured to be 0.6–0.7 nm. Also, a small number of these chains was found to be branched, and two different patterns could be found in the AFM image, a linear backbone containing a single long branch or several short branches (Gunning et al., 2003). Potato amylopectin had a highly branched and extended structure with a distance of approximately 1 μm between many of the

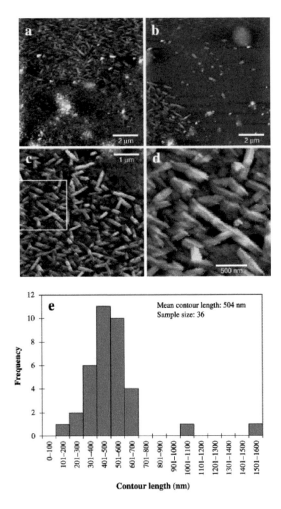

Figure 3.11 Atomic force microscopy images of potato amylose prepared and deposited on mica from a 5 μg mL⁻¹ solution. Image areas are 10 μm × 10 μm in (a) and (b); and 5 μm × 5 μm in (c). The boxed area in (c) is shown at a higher magnification of 2 μm × 2 μm in (d) Contour length distributions of the amylose aggregates shown in (e) were calculated from measurements of 36 structures observed in six separate images (Dang, Braet, & Copeland, 2006).

branching points (Dang et al., 2006). The isoamylolysis of potato starch, however, revealed chains, branches and coiled structures (An et al., 2011). Modifications of starch influenced the structure. It was reported that the chain structure of the starch–iodine complex showed as loose fibrous networks and aggregates with the height of 0.5–1.2 nm, and a superhelix structure was found in reaction between starch and I_2/KI solutions (Du et al., 2014). Under microwave radiation, potato starch formed network structures with the height from 0.3 to 11.0 nm, while corn starch did not exhibit any networks (An et al., 2008).

3.3 MOLECULAR STRUCTURE

The molecular structure of starch consists of two levels: the first is the individual starch chains (branches), while the second level is starch molecules with branches. There are a range of techniques to characterize the molecular structure of starch, and each has distinct advantages and drawbacks.

3.3.1 High-Performance Anion-Exchange Chromatography

High-performance anion-exchange chromatography (HPAEC) has been used to characterize the chain length distribution (CLD) of individual starch chains (branches) after complete debranching of starch molecules using isoamylase, as exhibited in Figure 3.12 (Nilsson, 2013; Wong & Jane, 1997). It separates

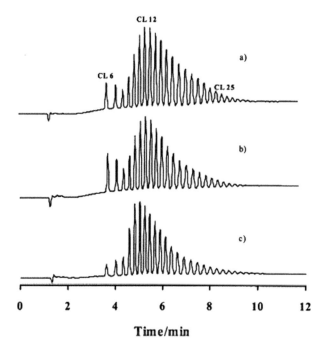

Figure 3.12 Chromatograms obtained from HPAEC-PAD with chain length distribution pattern of starch from (a) PAP injected directly onto the HPAEC-PAD system (0.5 ml/mg), (b) PAP injected via the microdialysis probe (2.0 mg/ml), and (c) waxy maize injected via the microdialysis probe (2.0 mg/ml) (Nilsson, 2013).

molecules based on mass-to-charge ratio and gives baseline resolution. Pulsed amperometric detection (PAD) is usually coupled with HPAEC. As the signal intensity of PAD is mass-dependent, this method has a mass bias that distorts the resulting distribution and is very complicated to correct (Wong & Jane, 1997). To date, HPAEC has been widely used in characterizing CLD of amylopectin chains with DP smaller than 80 (Hanashiro et al., 1996; Wong & Jane, 1997), whereas it cannot give accurate information on amylose and extra-long amylopectin chains due to the low intensity of the detector on large molecules.

3.3.2 Fluorophore-Assisted Carbohydrate Electrophoresis

Fluorophore-assisted carbohydrate electrophoresis (FACE) is another technique to analyze the CLD of starch. Starch molecules are firstly debranched using isoamylase to obtain linear chains, and then the reducing end of each linear chain is labeled with a negatively charged fluorescent dye, for instance, 8-aminopyrene-1,3,6-trisulfonic acid (APTS). Because each linear chain has only one charged group, these chains with various lengths have different mass-to-charge ratios and can be then separated by mass-to-charge ratio using capillary electrophoresis. Signals are obtained by fluorescence detection, and results based on chain length with baseline resolution can be obtained for linear chains with DP up to 135 (O'Shea et al., 1998; Wu et al., 2014).

Compared with HPAEC, which covers amylopectin chains with DP <80, FACE could provide the CLD of almost all amylopectin chains (excluding the extra-long amylopectin chains) and a proportion of amylose chains with the shortest length, as shown in Figure 3.13. Because FACE does not suffer from mass bias, it is more accurate than HPAEC; however the equipment is more costly.

3.3.3 Size Exclusion Chromatography

Size exclusion chromatography (SEC; also known as gel-permeation chromatography, GPC) is also capable of characterizing the CLD of starch chains. In addition to amylopectin chains, SEC can be used for analyzing the CLD of debranched amylose chains, as well as for the molecular size distribution of branched starch molecules, especially amylose molecules, because it does not have length restrictions as HPAED and FACE do.

Three types of detectors are commonly used for SEC. Differential refractive index (DRI) provides the weight distribution of molecules as functions of the hydrodynamic radius (R_h); viscometric detection gives the number distribution; and multi-angle laser light scattering (MALLS) detection provides the distribution of weight average molecular weight and the z-average size radius of gyration (Gray-Weale et al., 2009; K. Wang, Henry, & Gilbert, 2014).

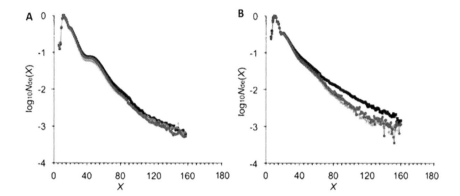

Figure 3.13 CLDs of (A) normal and (B) high-amylose barley starch in the amylopectin range obtained with different dissolution solvents and analyzed with FACE. CLDs obtained by using DMSO/LiBr, 0.05 M NaOH and 0.25 M NaOH are shown in squares, triangles and circles, respectively. Adapted from Wu et al. (2014).

SEC separates molecules by size (R_h), with smaller molecules being retained longer in the pores of the stationary phase than larger ones, and eluted later. For debranched starches that are linear polymers, there is a unique relationship between their molecular size and molecular weight, whereas no relation is present for branched polymers, such as starch molecules. Hence only a DRI detector is sufficient for obtaining the CLDs of debranched amylose and amylopectin.

The elution volume (or elution time) of SEC can be converted to the R_h using narrowly distributed standards, such as dextran or pullulan, through the "universal calibration" principle. However, universal calibration can only be performed within a restricted range of R_h, which is smaller than approximately 50 nm (Cave et al., 2009). This range covers debranched amylose chains, debranched amylopectin chains and whole amylose molecules, whereas it does not cover amylopectin molecules. Although extrapolation of the calibration curve to larger volumes could be used to achieve coverage of larger molecules such as amylopectin molecules, the results for amylopectin molecules may be less accurate (Vilaplana & Gilbert, 2010a).

In addition, amylopectin molecules suffer from shear scission when being separated in SEC columns, leading to degradation of these molecules. It has been reported that this degradation phenomenon is related to the size of eluting molecules. For medium-sized molecules (such as amylose molecules), they are not significantly impacted by shear scission. However, for large-sized molecules (such as amylopectin molecules), degradation induced by shear scission is hardly avoided in current SEC technology (Gray-Weale et al., 2009). Apart from this, the size of amylopectin molecules is normally beyond the optimal separation range of SEC columns, and also because of the unavoidable band broadening issue of SEC, good separation of large amylopectin molecules is hardly achieved. Thus, the resulting molecular size distribution of amylopectin molecules can only qualitative or semiquantitatively provide information of the molecular structure of amylopectin molecules.

Amylose content can also be obtained from SEC results by either using the CLD of debranched chains or using the molecular size distribution of whole starch molecules. Based on the distribution curves, the amylose content is calculated as the proportion of the area under amylose distribution to that of the whole distribution (including amylose and amylopectin). Comparisons among different methods were made in Vilaplana et al. (2012), and it was found that amylose content from the CLD of debranched chains was more reliable. This method works for a range of starches from various sources, as a clear division point between amylose and amylopectin could be seen in the distribution curves (Syahariza et al., 2010; K. Wang, Hasjim et al., 2014; K. Wang et al., 2015). However, there might be difficulties for some types of starch, for instance high-amylose starches, as there is no clear division between amylose and amylopectin distributions.

To overcome this, two-dimensional (2D) SEC distributions were applied, where the first dimension is the CLD individual linear branches and the second is the SEC weight distribution of whole molecules. A clear separation of amylose and amylopectin could be achieved from 2D SEC distributions, allowing a more accurate estimation of amylose content, as shown in Figure 3.14 (Vilaplana et al., 2012). In addition, more detailed structural features of starch molecules that are unresolved using conventional SEC techniques could be unveiled using 2D SEC, for example, the extra-long chains of amylopectin, and the intermediate components that have a molecular size similar to amylose but are highly branched like amylopectin. This allows a better understanding on the structural features of starch molecules and also on the biosynthetic mechanism of starch in nature (Vilaplana & Gilbert, 2010b, 2011; Vilaplana et al., 2012). In recent years, K. Wang et al. (2019) used 2D SEC to study the branching structure of amylose and reported the first measurement of the average number of branches per amylose molecule (from potato tubers) as a function of molecular size. It was found that the molecular weight dispersity, average chain length and average amylose molecular weight increased with increment in the molecular size of amylose. However, the average number of branches of amylose molecules showed weak dependence on the molecular size, with most molecules having fewer than five branches. Variations in the molecular size of amylose mostly resulted from differences in chain length.

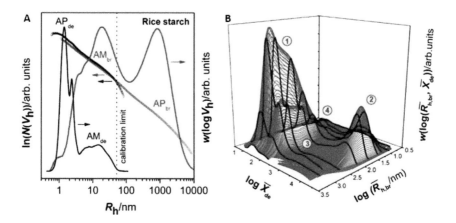

Figure 3.14 (A) SEC weight w(logV$_h$) and number N(V$_h$) distributions for the branched (gray points) and debranched (black points) rice starch. (B) Two-dimensional SEC weight w(logV$_h$) of rice starch based on macromolecular size and branch chain length. Features 1–4 correspond to amylopectin, amylose, amylopectin with extra-long chain branches, and intermediate component, respectively. AM, amylose; AP, amylopectin; br, branched; de, debranched. Reprinted with permission from Vilaplana and Gilbert (2010b).

3.3.4 Field-Flow Fractionation

Field-flow fractionation (FFF), especially asymmetrical flow field-flow fractionation (AF4), has been used for the separate and structure characterization of macromolecules in combination with DRI and MALLS detection in recent years. Similar to SEC, it separates molecules by size (R$_h$). Macromolecules flow through the flat channel and are simultaneously subjected to channel flow and cross flow fields. Compared with SEC, FFF has no stationary phase and the system pressure is relatively lower, hence it has low shear scission. Therefore it has potential to characterize the molecular structure of whole starch molecules, especially large amylopectin molecules without degradation. However, a disadvantage is that the assumption of this technique is made that all particles are of uniform density and spherical (Nilsson, 2013).

Although there has been studies on the application of AF4 coupled with MALLS and RI on the separation and structural characterization of starch, for instance, on amylopectin in waxy maize starch (Juna et al., 2011), normal and waxy barley starch (You et al., 2002), and amylopectin from various botanical sources (Rolland-Sabate et al., 2007), this technique is more commonly applied in analyzing the structural changes of starch after modification or degradation, such as carboxymethylation (Lee et al., 2010) and homogenization

(Rojas et al., 2008). For native starch samples containing amylopectin, it is still a challenge to obtain accurate molecular size characterization using current FFF technologies because it is difficult to obtain fully dissolved starch without degradation and aggregation, and also difficult to obtain adequate signal-to-noise ratio (Gilbert et al., 2013; Vilaplana & Gilbert, 2010a).

3.3.5 Proton Nuclear Magnetic Resonance

Proton nuclear magnetic resonance (^{1}H NMR) is a powerful and reliable technique for determining the degree of branching of starch, which is determined based on variations in anomeric signals α-(1,4) and α-(1,6) linkages. However, the ^{1}H NMR spectra of polysaccharides can sometimes be complicated to interpret in view of the presence of many different hydroxyl groups. Labile protons from OH or any other group containing exchangeable protons generally exhibit broad signals that can hide other peaks of interest. Thus, the sample of starch is performed under conditions in which there is complete and homogeneous dissolution (Tizzotti et al., 2011). And the anomeric signals α-(1,4) and α-(1,6) glycosidic bonds are clearly visible at 5.11 and 4.75 ppm, respectively, as shown in Figure 3.15. The degree of branching (DB) of starch was calculated according to the following equation:

$$DB = \frac{I_{\alpha-(1,6)}}{I_{\alpha-(1,4)} + I_{\alpha-(1,6)}}$$

where $I_{\alpha-(1,4)}$ is the ^{1}H NMR integrals of peak around 5.11 ppm for α-(1,4) glycosidic bonds and $I_{\alpha-(1,6)}$ is the ^{1}H NMR integrals of peak at approximately 4.75 ppm for α-(1,6) glycosidic bonds (Xu et al., 2019).

The relative intensity of peaks indicating α-(1,4) glycosidic bonds of waxy maize starch amylopectin were slightly weakened once it was treated by ultrasonic irradiation, while further increases of the processing time induced no additional changes. The result demonstrated that α-(1,4) glycosidic bonds were degraded during the ultrasonic treatments. Furthermore, the DB of the ultrasonic irradiated amylopectin decreased with the increase of the processing time, which manifested that α-(1,6) glycosidic bonds were also disrupted during the treatments. Whereas, the disruption of α-(1,6) glycosidic bonds was more serious than the disruption of α-(1,4) glycosidic bonds (Wei et al., 2021). The degree of branching increased at intensified reaction conditions and plateaued at approximately 24%. Exhaustively hydrolyzing pyrodextrin by α-amylase and amyloglucosidase significantly decreased the degree of α-(1,4) but not α-(1,6) linkages. The retained α-(1,4) and α-(1,6) linkages were probably protected from enzyme hydrolysis by the nonstarch linkages due to steric hindrance. The resistant starch content was positively correlated with the degree of branching of

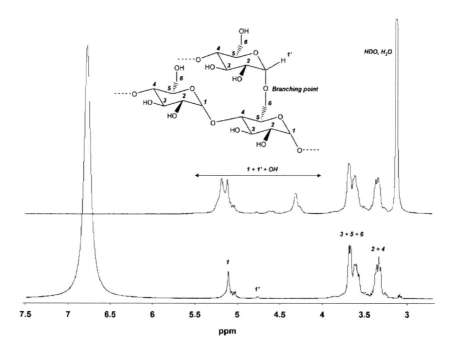

Figure 3.15 ¹H NMR spectrum of maize starch in d_6-DMSO at 70°C prior to (upper panel) and after (lower panel) addition of d_1-TFA. Reprinted with permission from Tizzotti et al. (2011).

pyrodextrin (Chen et al., 2020). Zou et al. (2020) found that Chinese yam starch samples in dormant stage had lower branching degrees than those in expansion stage. And the possible reason was that fewer branches were generated during the dormant stage, which were helpful for starch chain rearrangement.

In addition, ¹H NMR was used to study the interaction of starch with phenolic compounds such as caffeic acid. The resulting spectra showed broader peaks at 5.4 and 5.5 ppm, suggesting changes in the starch hydroxyl groups, which is likely due to hydrogen bonds between starch and caffeic acid (Yu et al., 2021).

3.4 CONCLUSIONS

A range of techniques are currently available for analyzing the multilevel structure of starch, with each having distinct advantages and limitations. With the development of these analytical techniques, the study of the morphology, ordered structure and molecular structure of starch has made great progress in the past a few years. One example is the improved separation of starch achieved using 2D SEC and high-resolution FACE, enabling more precise characterization of the molecular structure of amylose, amylopectin and also the intermediate component of starch. Still, there are gaps that are at least partially related to the limitations in the analytical techniques. For example, the molecular architecture of amylose from different botanical origins and after various physiochemical treatments need to be explored in the future. In addition, our current knowledge on the amorphous regions in granular starch are highly limited. Questions regarding the location of the amorphous region in starch granules, the roles of amylose and long-chain amylopectin in the organization of amorphous portion, and whether there are associations between amylose in the amorphous region with the amylopectin matrix in the crystalline region, etc. need to be answered. The structure of residual short-range ordered structure of gelatinized starch and multilevel structure of retrograded starch have not be fully understood so far. The development of techniques with stable performance, high accuracy and convenient systems, as well as the establishment of standardized measurement and analysis methods, need to be carried out in the future.

REFERENCES

An, H., Liang, H., Liu, Z., Yang, H., Liu, Q., & Wang, H. (2011). Nano-structures of debranched potato starch obtained by isoamylolysis. *Journal of Food Science*, 76(1), N11–14.

An, H., Yang, H., Liu, Z., & Zhang, Z. (2008). Effects of heating modes and sources on nanostructure of gelatinized starch molecules using atomic force microscopy. *LWT: Food Science and Technology*, 41(8), 1466–1471.

Bernardino-Nicanor, A., Acosta-Garcia, G., Guemes-Vera, N., Montanez-Soto, J. L., de Los Angeles Vivar-Vera, M., & Gonzalez-Cruz, L. (2017). Fourier transform infrared and Raman spectroscopic study of the effect of the thermal treatment and extraction methods on the characteristics of ayocote bean starches. *Journal Food Science Technology*, 54(4), 933–943.

Blaszczak, W., & Lewandowicz, G. (2020). Light microscopy as a tool to evaluate the functionality of starch in food. *Foods*, 9(5), 3–15.

Blazek, J., & Gilbert, E. P. (2011). Application of small-angle X-ray and neutron scattering techniques to the characterisation of starch structure: A review. *Carbohydrate Polymers*, 85(2), 281–293.

Blonk;, J. C. G., & Aalst;, H. V . (1993). Confocal scanning light microscopy in food research. *Food Research International*, 26, 297–311.

Cai, C., & Wei, C. (2013). In situ observation of crystallinity disruption patterns during starch gelatinization. *Carbohydr Polymers*, 92(1), 469–478.

Cameron, R. E., & Donald, A. M. (1992). A small-angle X-ray scattering study of the annealing and gelatinization of starch. *Polymer*, 33(12), 2628–2635.

Cave, R. A., Seabrook, S. A., Gidley, M. J., & Gilbert, R. G. (2009). Characterization of starch by size-exclusion chromatography: The limitations imposed by shear scission. *Biomacromolecules*, 10(8), 2245–2253.

Charoenrein, S., Tatirat, O., Rengsutthi, K., & Thongngam, M. (2011). Effect of konjac glucomannan on syneresis, textural properties and the microstructure of frozen rice starch gels. *Carbohydrate Polymers*, 83(1), 291–296.

Chen, J., Xiao, J., Wang, Z., Cheng, H., Zhang, Y., Lin, B., Qin, L., & Bai, Y. (2020). Effects of reaction condition on glycosidic linkage structure, physical-chemical properties and in vitro digestibility of pyrodextrins prepared from native waxy maize starch. *Food Chemistry*, 320, 126491.

Chen, L., Tian, Y., Sun, B., Cai, C., Ma, R., & Jin, Z. (2018). Measurement and characterization of external oil in the fried waxy maize starch granules using ATR-FTIR and XRD. *Food Chemistry*, 242, 131–138.

Chen, P., Xie, F., Zhao, L., Qiao, Q., & Liu, X. (2017). Effect of acid hydrolysis on the multi-scale structure change of starch with different amylose content. *Food Hydrocolloids*, 69, 359–368.

Chi, C., Li, X., Zhang, Y., Chen, L., Li, L., & Wang, Z. (2017). Digestibility and supramolecular structural changes of maize starch by non-covalent interactions with gallic acid. *Food & Function*, 8(2), 720–730.

Dang, J. M., Braet, F., & Copeland, L. (2006). Nanostructural analysis of starch components by atomic force microscopy. *Journal of Microscopy*, 224(Pt 2), 181–186.

Daniels, D. R., & Donald, A. M. (2003). An improved model for analyzing the small angle x-ray scattering of starch granules. *Biopolymers*, 69(2), 165–175.

Debet, M. R., & Gidley, M. J. (2006). Three classes of starch granule swelling: Influence of surface proteins and lipids. *Carbohydrate Polymers*, 64(3), 452–465.

Dhital, S., Dabit, L., Zhang, B., Flanagan, B., & Shrestha, A. K. (2015). In vitro digestibility and physicochemical properties of milled rice. *Food Chemistry*, 172, 757–765.

Dhital, S., Warren, F. J., Zhang, B., & Gidley, M. J. (2014). Amylase binding to starch granules under hydrolysing and non-hydrolysing conditions. *Carbohydr Polymers*, 113, 97–107.

Ding, Y., Liang, Y., Luo, F., Ouyang, Q., & Lin, Q. (2020). Understanding the mechanism of ultrasonication regulated the digestibility properties of retrograded starch following vacuum freeze drying. *Carbohydr Polymers*, 228, 115350.

Du, X., An, H., Liu, Z., Yang, H., & Wei, L. (2014). Probing starch-iodine interaction by atomic force microscopy. *Scanning*, 36(4), 394–400.

Flores-Morales, A., Jiménez-Estrada, M., & Mora-Escobedo, R. (2012). Determination of the structural changes by FT-IR, Raman, and CP/MAS ^{13}C NMR spectroscopy on retrograded starch of maize tortillas. *Carbohydrate Polymers*, 87(1), 61–68.

Gidley, M. J., & Bociek, S. M. (2002). Molecular organization in starches: a carbon 13 CP/MAS NMR study. *Journal of the American Chemical Society*, 107(24), 7040–7044.

Gilbert, R. G., Witt, T., & Hasjim, J. (2013). What is being learned about starch properties from multiple-level characterization. *Cereal Chemistry*, 90(4), 312–325.

Glaring, M. A., Koch, C. B., & Blennow, A. (2006). Genotype-specific spatial distribution of starch molecules in the starch granule: A combined CLSM and SEM approach. *Biomacromolecules*, 7, 2310–2320.

Gonzalez-Cruz, L., Montanez-Soto, J. L., Conde-Barajas, E., Negrete-Rodriguez, M. L. X., Flores-Morales, A., & Bernardino-Nicanor, A. (2018). Spectroscopic, calorimetric and structural analyses of the effects of hydrothermal treatment of rice beans and the extraction solvent on starch characteristics. *International Journal of Biological Macromolecule*, 107(Pt A), 965–972.

Gray-Weale, A. A., Cave, R. A., & Gilbert, R. G. (2009). Extracting physically useful information from multiple-detection size-separation data for starch. *Biomacromolecules*, 10(9), 2708–2713.

Gunning, A. P., Giardina, T. P., Faulds, C. B., Juge, N., Ring, S. G., Williamson, G., & Morris, V. J. (2003). Surfactant-mediated solubilisation of amylose and visualisation by atomic force microscopy. *Carbohydrate Polymers*, 51(2), 177–182.

Guo, P., Yu, J., Wang, S., Wang, S., & Copeland, L. (2018). Effects of particle size and water content during cooking on the physicochemical properties and in vitro starch digestibility of milled durum wheat grains. *Food Hydrocolloids*, 77, 445–453.

Hanashiro, I., Abe, J.-i., & Hizukuri, S. (1996). A periodic distribution of the chain length of amylopectin as revealed by high-performance anion-exchange chromatography. *Carbohydrate Research*, 283, 151–159.

He, T., Wang, K., Zhao, L., Chen, Y., Zhou, W., Liu, F., & Hu, Z. (2021). Interaction with longan seed polyphenols affects the structure and digestion properties of maize starch. *Carbohydrate Polymers*, 256, 117537.

Horstmann, S. W., Belz, M. C., Heitmann, M., Zannini, E., & Arendt, E. K. (2016). Fundamental study on the impact of gluten-free starches on the quality of gluten-free model breads. *Foods*, 5(2), 30.

Huang, S., Chao, C., Yu, J., Copeland, L., & Wang, S. (2021). New insight into starch retrogradation: The effect of short-range molecular order in gelatinized starch. *Food Hydrocolloids*, 120, 106921.

Jenkins, P. J., & Donald, A. M. (1996). Application of small-angle neutron scattering to the study of the structure of starch granules. *Polymer*, 37(25), 5559–5568.

Juna, S., Williams, P. A., & Davies, S. (2011). Determination of molecular mass distribution of amylopectin using asymmetrical flow field-flow fractionation. *Carbohydrate Polymers*, 83(3), 1384–1396.

Kačuráková, M., & Mathlouthi, M. (1996). FTIR and laser-Raman spectra of oligosaccharides in water: characterization of the glycosidic bond. *Carbohydrate Research*, 284(2), 145–157.

Kowsik, P. V., & Mazumder, N. (2018). Structural and chemical characterization of rice and potato starch granules using microscopy and spectroscopy. *Microscopy Research and Technique*, 81(12), 1533–1540.

Lee, S., Kim, S. T., Pant, B. R., Kwen, H. D., Song, H. H., Lee, S. K., & Nehete, S. V. (2010). Carboxymethylation of corn starch and characterization using asymmetrical flow field-flow fractionation coupled with multiangle light scattering. *Journal of Chromatography A*, 1217(27), 4623–4628.

Li, H.-T., Li, Z., Fox, G. P., Gidley, M. J., & Dhital, S. (2021). Protein-starch matrix plays a key role in enzymatic digestion of high-amylose wheat noodle. *Food Chemistry*, 336, 127719.

Li, J., Zou, F., Gui, Y., Guo, L., Wang, N., Liu, P., & Cui, B. (2021). Long-term retrogradation properties of rice starch modified with transglucosidase. *Food Hydrocolloids*, 121, 107053.

Liu, X., Chao, C., Yu, J., Copeland, L., & Wang, S. (2021). Mechanistic studies of starch retrogradation and its effects on starch gel properties. *Food Hydrocolloids*, 120, 106914.

Lopez-Rubio, A., Flanagan, B. M., Gilbert, E. P., & Gidley, M. J. (2008). A novel approach for calculating starch crystallinity and its correlation with double helix content: A combined XRD and NMR study. *Biopolymers: Original Research on Biomolecules*, 89(9), 761–768.

Ma, Y., Zhang, H., Jin, Y., Xu, D., & Xu, X. (2022). Impact of superheated steam on the moisture transfer, structural characteristics and rheological properties of wheat starch. *Food Hydrocolloids*, 122, 107089.

Ma, Z., & Boye, J. I. (2018). Research advances on structural characterization of resistant starch and its structure-physiological function relationship: A review. *Critical Reviews in Food Science and Nutrition*, 58(7), 1059–1083.

Mihhalevski, A., Heinmaa, I., Traksmaa, R., Pehk, T., Mere, A., & Paalme, T. (2012). Structural changes of starch during baking and staling of rye bread. *Journal of Agricultural and Food Chemistry*, 60(34), 8492–8500.

Nilsson, L. (2013). Separation and characterization of food macromolecules using field-flow fractionation: A review. *Food Hydrocolloids*, 30(1), 1–11.

O'Shea, M. G., Samuel, M. S., Konik, C. M., & Morell, M. K. (1998). Fluorophore-assisted carbohydrate electrophoresis (FACE) of oligosaccharides: efficiency of labelling and high-resolution separation. *Carbohydrate Research*, 307(1–2), 1–12.

Okyere, A. Y., Bertoft, E., & Annor, G. A. (2019). Modification of cereal and tuber waxy starches with radio frequency cold plasma and its effects on waxy starch properties. *Carbohydrate Polymers*, 223, 115075.

Pérez, S., & Bertoft, E. (2010). The molecular structures of starch components and their contribution to the architecture of starch granules: A comprehensive review. *Starch-Stärke*, 62(8), 389–420.

Perez-Rea, D., Rojas, C., Carballo, S., Aguilar, W., Bergenståhl, B., & Nilsson, L. (2013). Enzymatic hydrolysis of Canna indica, Manihot esculenta and Xanthosoma sagittifolium native starches below the gelatinization temperature. *Starch - Stärke*, 65(1–2), 151–161.

Peroni-Okita, F. H. G., Gunning, A. P., Kirby, A., Simão, R. A., Soares, C. A., & Cordenunsi, B. R. (2015). Visualization of internal structure of banana starch granule through AFM. *Carbohydrate Polymers*, 128, 32–40.

Putaux, J.-L., Buleón, A., & Chanzy, H. (2000). Network formation in dilute amylose and amylopectin studied by TEM. *Macromolecules*, 33, 6416–6422.

Rojas, C. C., Wahlund, K. G., Bergenstahl, B., & Nilsson, L. (2008). Macromolecular geometries determined with field-flow fractionation and their impact on the overlap concentration. *Biomacromolecules*, 9, 1684–1690.

Rolland-Sabate, A., Colonna, P., Mendez-Montealvo, M. G., & Planchot, V. (2007). Branching features of amylopectins and glycogen determined by asymmetrical flow field flow fractionation coupled with multiangle laser light scattering. *Biomacromolecules*, 8(8), 2520–2532.

Schmidt-Rohr, K. (2007). Simulation of small-angle scattering curves by numerical Fourier transformation. *Journal of Applied Crystallography*, 40(1), 16–25.

Shi, L., Zhou, J., Guo, J., Gladden, I., & Kong, L. (2021). Starch inclusion complex for the encapsulation and controlled release of bioactive guest compounds. *Carbohydrate Polymers*, 274, 118596.

Singh, J., Kaur, L., & Rao, M. A. (2009). Textural and rheological characteristics of raw and cooked potatoes. *Textural and Rheological Characteristics of Raw and Cooked Potatoes*, 9, 249–271.

Singh, N., Singh, J., Kaur, L., Sodhi, N. S., & Gill, B. S. (2003). Morphological, thermal and rheological properties of starches from different botanical sources. *Food Chemistry*, 81(2), 219–231.

Sujka, M., & Jamroz, J. (2013). Ultrasound-treated starch: SEM and TEM imaging, and functional behaviour. *Food Hydrocolloids*, 31(2), 413–419.

Surendra Babu, A., & Jagan Mohan, R. (2019). Influence of prior pre-treatments on molecular structure and digestibility of succinylated foxtail millet starch. *Food Chemistry*, 295, 147–155.

Syahariza, Z. A., Li, E., & Hasjim, J. (2010). Extraction and dissolution of starch from rice and sorghum grains for accurate structural analysis. *Carbohydrate Polymers*, 82(1), 14–20.

Tan, I., Flanagan, B. M., Halley, P. J., Whittaker, A. K., & Gidley, M. J. (2007). A method for estimating the nature and relative proportions of amorphous, single, and double-helical components in starch granules by (13)C CP/MAS NMR. *Biomacromolecules*, 8(3), 885–891.

Tizzotti, M. J., Sweedman, M. C., Tang, D., Schaefer, C., & Gilbert, R. G. (2011). New ^{1}H NMR procedure for the characterization of native and modified food-grade starches. *Journal of Agricultural and Food Chemistry*, 59(13), 6913–6919.

Vamadevan, V., & Bertoft, E. (2015). Structure-function relationships of starch components. *Starch - Stärke*, 67(1–2), 55–68.

van de Velde, F., van Riel, J., & Tromp, R. H. (2002). Visualisation of starch granule morphologies using confocal scanning laser microscopy (CSLM). *Journal of the Science of Food and Agriculture*, 82(13), 1528–1536.

Vilaplana, F., & Gilbert, R. G. (2010a). Characterization of branched polysaccharides using multiple-detection size separation techniques. *Journal of Separation Science*, 33(22), 3537–3554.

Vilaplana, F., & Gilbert, R. G. (2010b). Two-dimensional size/branch length distributions of a branched polymer. *Macromolecules*, 43(17), 7321–7329.

Vilaplana, F., & Gilbert, R. G. (2011). Analytical methodology for multidimensional size/branch-length distributions for branched glucose polymers using off-line 2-dimensional size-exclusion chromatography and enzymatic treatment. *Journal of Chromatography A*, 1218(28), 4434–4444.

Vilaplana, F., Hasjim, J., & Gilbert, R. G. (2012). Amylose content in starches: Toward optimal definition and validating experimental methods. *Carbohydrate Polymers*, 88(1), 103–111.

Waduge, R. N., Hoover, R., Vasanthan, T., Gao, J., & Li, J. (2006). Effect of annealing on the structure and physicochemical properties of barley starches of varying amylose content. *Food Research International*, 39(1), 59–77.

Wang, J., Guo, K., Fan, X., Feng, G., & Wei, C. (2018). Physicochemical properties of C-Type starch from root tuber of *Apios fortunei* in comparison with maize, potato, and pea starches. *Molecules*, 23(9), 3–15.

Wang, K., Hasjim, J., Wu, A. C., Henry, R. J., & Gilbert, R. G. (2014). Variation in amylose fine structure of starches from different botanical sources. *Journal of Agricultural and Food Chemistry*, 62(19), 4443–4453.

Wang, K., Henry, R. J., & Gilbert, R. G. (2014). Causal relations between starch biosynthesis, structure and properties. *Springer Science Reviews*, 2(1–2), 15–33.

Wang, K., Vilaplana, F., Wu, A., Hasjim, J., & Gilbert, R. G. (2019). The size dependence of the average number of branches in amylose. *Carbohydrate Polymers*, 223, 115–134.

Wang, K., Wambugu, P. W., Zhang, B., Wu, A. C., Henry, R. J., & Gilbert, R. G. (2015). The biosynthesis, structure and gelatinization properties of starches from wild and cultivated African rice species (Oryza barthii and Oryza glaberrima). *Carbohydrate Polymers*, 129, 92–100.

Wang, M., Shen, Q., Hu, L., Hu, Y., Ye, X., Liu, D., & Chen, J. (2018). Physicochemical properties, structure and in vitro digestibility on complex of starch with lotus (*Nelumbo nucifera* Gaertn.) leaf flavonoids. *Food Hydrocolloids*, 81, 191–199.

Wang, S., Li, C., Copeland, L., Niu, Q., & Wang, S. (2015). Starch retrogradation: A comprehensive review. *Comprehensive Reviews in Food Science and Food Safety*, 14(5), 568–585.

Wang, S., Wang, J., Zhang, W., Li, C., Yu, J., & Wang, S. (2015). Molecular order and functional properties of starches from three waxy wheat varieties grown in China. *Food Chemistry*, 181, 43–50.

Wang, Y., Chao, C., Huang, H., Wang, S., Wang, S., Wang, S., & Copeland, L. (2019). Revisiting mechanisms underlying digestion of starches. *Journal of Agricultural and Food Chemistry*, 67(29), 8212–8226.

Wei, B., Qi, H., Zou, J., Li, H., Wang, J., Xu, B., & Ma, H. (2021). Degradation mechanism of amylopectin under ultrasonic irradiation. *Food Hydrocolloids*, 111, 106371.

Wong, K. S., & Jane, J. (1997). Quantitative analysis of debranched amylopectin by HPAEC-PAD with a postcolumn enzyme reactor. *Journal of Liquid Chromatography & Related Technologies*, 20(2), 297–310.

Wu, A. C., Li, E., & Gilbert, R. G. (2014). Exploring extraction/dissolution procedures for analysis of starch chain-length distributions. *Carbohydrate Polymers*, 114, 36–42.

Wu, Y., Lin, Q., Chen, Z., Wu, W., & Xiao, H. (2012). Fractal analysis of the retrogradation of rice starch by digital image processing. *Journal of Food Engineering*, 109(1), 182–187.

Xu, X., Chen, Y., Luo, Z., & Lu, X. (2019). Different variations in structures of A-and B-type starches subjected to microwave treatment and their relationships with digestibility. *LWT: Food Science and Technology*, 99, 179–187.

Yang, S., Dhital, S., Shan, C.-S., Zhang, M.-N., & Chen, Z.-G. (2021). Ordered structural changes of retrograded starch gel over long-term storage in wet starch noodles. *Carbohydrate Polymers*, 270, 118367.

You, S., Stevenson, S. G., Izydorczyk, M. S., & Preston, K. R. (2002). Separation and characterization of barley starch polymers by a flow field-flow fractionation technique in combination with multiangle light scattering and differential refractive index detection. *Cereal Chemistry*, 79(5), 624–630.

Yu, M., Liu, B., Zhong, F., Wan, Q., Zhu, S., Huang, D., & Li, Y. (2021). Interactions between caffeic acid and corn starch with varying amylose content and their effects on starch digestion. *Food Hydrocolloids*, 114, 106544.

Yuryev, V. P., Krivandin, A. V., Kiseleva, V. I., Wasserman, L. A., Genkina, N. K., Fornal, J., Blaszczak, W., & Schiraldi, A. (2004). Structural parameters of amylopectin clusters and semi-crystalline growth rings in wheat starches with different amylose content. *Carbohydrate Research*, 339(16), 2683–2691.

Zhang, B., Wang, K., Hasjim, J., Li, E., Flanagan, B. M., Gidley, M. J., & Dhital, S. (2014). Freeze-drying changes the structure and digestibility of B-polymorphic starches. *Journal of Agricultural and Food Chemistry*, 62(7), 1482–1491.

Zhang, G., Venkatachalam, M., & Hamaker, B. R. (2006). Structural basis for the slow digestion property of native cereal starches. *Biomacromolecules*, 7(11), 3259–3266.

Zheng, Y., Tian, J., Kong, X., Yang, W., Yin, X., Xu, E., Chen, S., Liu, D., & Ye, X. (2020). Physicochemical and digestibility characterisation of maize starch–caffeic acid complexes. *LWT: Food Science and Technology*, 121, 108857.

Zhu, L., Sun, J., Wu, G., Wang, Y., Zhang, H., Wang, L., Qian, H., & Qi, X. (2018). Identification of rice varieties and determination of their geographical origin in China using Raman spectroscopy. *Journal of Cereal Science*, 82, 175–182.

Zobel, H. (1988). Starch crystal transformations and their industrial importance. *Starch - Stärke*, 40(1), 1–7.

Zou, J., Xu, M., Wen, L., & Yang, B. (2020). Structure and physicochemical properties of native starch and resistant starch in Chinese yam (Dioscorea opposita Thunb.). *Carbohydr Polymers*, 237, 116188.

Chapter 4

Functional Properties of Starch

Guantian Li and Fan Zhu

CONTENTS

DOI: 10.1201/9781003088929-4

4.1 INTRODUCTION

Starch is the most important energy source for humans and the second most abundant carbohydrate in the biosphere. Starch for human consumption is mainly derived from cereals (such as maize and rice), tubers (potato) and roots (cassava). In most situations, starchy food is cooked before consumption. Heating starch in the presence of excess water induces a series of irreversible changes (cumulatively termed gelatinization) to the starch structure, whereas the cooling of the cooked starch leads to the formation of ordered structures, a process typically referred to as retrogradation. During the heating and cooling, the functional properties (such as thermal, swelling and gel properties) of starch alter significantly. These eventually affect the quality of starch-based foods (such as texture and mouthfeel). Although starch is essential for human beings, starch that is digested rapidly could significantly raise the postprandial blood glucose level associated with the risks of cardiovascular diseases and diabetes. Thus, the functional properties of starch gelatinization and retrogradation are important factors affecting the quality and health of starch-based foods.

4.1.1 Structure of Starch

Native starch exists as granules that are insoluble in water. The size of the granules is generally in the range between 1 and 100 μm. There are also other minor components interacting with the granules, such as proteins and lipids. The proteins are mostly enzymes related to starch biosynthesis, while the lipids can be found in the form of phospholipids in cereal starches or phosphate monoesters in tuber and root starches (Vamadevan & Bertoft, 2015). Inside the granules, there are repeated layers of the micrometer scale, which are termed growth rings. This structure is further classified into amorphous growth rings and semicrystalline growth rings based on the localized density. The amorphous materials in amorphous growth rings are less ordered and easier to hydrate compared to those in the semicrystalline region (Tetlow & Bertoft, 2020). The semicrystalline growth rings are formed by lamellar structures with a length of around 9 nm. Each lamellar structure is made by crystalline lamellae with higher electron density and amorphous lamellae with lower electron density (Blazek & Gilbert, 2011).

On the molecular level, starch is composed of two biopolymers: amylose and amylopectin. Amylose is mostly a linear polymer made of D-glucose via an α (1→4) glycosidic bond. Due to the linear nature of amylose, it has a high tendency to form an aggregation or network structure via an H-bond. The molecular weight of amylopectin (degree of polymerization, DP, in the magnitude of 10^4–10^5) is significantly larger than that of amylose (DP in the magnitude of 10^2–10^3). In contrast to amylose, amylopectin is highly branched. The linear part is joined by an α (1→4) glycosidic bond, while the branch point is formed via an α (1→6) glycosidic bond. Due to the large molecular weight (in the magnitude of 10^7–10^8 g/mol), the starch molecules are hard to dissolve in water and they tend to form hydrated colloidal suspensions upon heating.

The relative ratio and structure of amylose and amylopectin vary significantly among different species or genotypes. The variance in the ratio or structure of amylose and amylopectin can profoundly affect the functional properties of starch, such as the thermal, swelling, pasting and gel properties, which also affect the final quality of starch-based products.

4.1.2 Gelatinization

When the starch is heated in water, a series of irreversible events occur, which is termed gelatinization. Starch gelatinization is a phase transition process of starch granules from ordered structures to disordered ones (Hermansson & Svegmark, 1996). This process involves granule hydration, swelling and leaking of amylose, which results in the loss of crystalline order and granular architecture. The gelatinization causes significant changes in appearance, thermal properties and rheological properties.

The gelatinization process occurs in several stages. When the starch is suspended in water, the water is reversibly bound in the amorphous lamellae and amorphous growth rings at a slower rate (Kuang et al., 2017). Due to a low degree of order of materials in amorphous growth rings, the water molecules might be firstly absorbed by the materials in amorphous growth rings followed by the amorphous lamellae (Cameron & Donald, 1992; Waigh et al., 2000; Vermeylen et al., 2006). As a result, the electron density between amorphous and crystalline lamellae increases with time or temperature as water molecules are absorbed by amorphous lamellae. The increased water content significantly enhances the mobility of the materials in amorphous parts, which induces a transition process among double helices in the crystalline region from the nematic to the smectic phase (Vermeylen et al., 2006; Waigh et al., 2000).

As the temperature keeps rising, hydrogen bonds of the amorphous materials break. The amorphous growth rings swell, and amylose leaches out of the granules. A considerable amount of water that enters the amorphous parts provides enough pressure to break the semicrystalline structure (Blazek & Gilbert, 2011). The rising temperature also causes the melting of the crystalline region due to the disruption of the double-helical structure formed by external chains of amylopectin. These two processes result in the loss of crystallinity. It has been hypothesized that during gelatinization, the swelling of the amorphous region and the unwinding of the double-helical structure might happen simultaneously or successively (Blazek & Gilbert, 2011; Waigh et al., 2000). Also, the viscosity of the medium increases due to the swelled starch granules and the leaching of the polymer molecules (particularly the amylose) (Wani et al., 2012).

4.1.3 Retrogradation

The behavior of starch upon cooling and storage after gelatinization is usually termed retrogradation. During retrogradation, the gelatinized starch shows significant changes in physicochemical properties, such as the increase in viscosity and turbidity, the enhancement of crystallinity, the reduction in opacity, and exudation of water (Wang et al., 2015; Karim et al., 2000).

There are short-term and long-term retrogradation processes. The short-term retrogradation is mainly due to the reassociation of amylose molecules, while the long-term retrogradation is the result of recrystallization of the external chains (DP ~15) of amylopectin. During short-term retrogradation, the amylose forms folded lamellar microcrystalline "junction zones," which results in a three-dimensional network and provides elasticity to starch gels (Singh et al., 2012). Some researchers proposed that the outer chains in amylopectin could interact with amylose molecules in the gel network via the formation of junction zones (Fu et al., 2015). In long-term retrogradation, the double-helical structures are reassembled between the external chains of amylopectin. The amount of recrystallized short chains could be estimated by the melting energy required (enthalpy change during melting of retrograded starch) (Wang et al., 2015).

For the starch that contains both amylose and amylopectin, the gel structure formed after short-term retrogradation is composed of a three-dimensional network of amylose filled with swollen amylopectin-enriched granules. As the storage time increases, the recrystallization of amylopectin increases the rigidity of swollen granules and enhances the gel network (Karim et al., 2000). Indeed, the waxy starch gel is soft due to the lack of network with only aggregates (Wang et al., 2015).

On the molecular level, the retrogradation process could be separated by three stages, namely, nucleation, propagation and maturation (Silverio et al., 2000). By analyzing the endotherm caused by retrogradation using the Avrami equation, Miles et al. (1985) proposed the process of recrystallization as instantaneous nucleation followed by a rod like growth of the crystalline units.

4.2 THERMAL PROPERTIES

The thermal properties of starch can strongly influence the applications of starch. The thermal properties of starches are mainly measured by differential scanning calorimetry (DSC), which observes the heat changes during starch gelatinization (Zhang et al., 2019a).

4.2.1 DSC Analysis of Starch Gelatinization

The DSC parameters of starches with various botanical backgrounds, polymorphous types and crystallinity studied in recent years are summarized in Table 4.1. The most common parameters include the onset temperature (T_o, the intersection point of tangents to the DSC profile before maximum heat flow), the peak temperature or melting temperature (T_p, the temperature of maximum heat flow), the conclusion temperature (T_c, the intersection point of tangents to the trace at the upslope after T_p and the estimate of baseline), the gelatinization temperature range (ΔT, the difference between T_c and T_o) and the enthalpy change (ΔH, the area under the line drawn from the start temperature to the end temperature) (Ren et al., 2020).

Gelatinization temperatures (T_o, T_p, T_c) reflect the perfectness of starch crystallites (Tester & Morrison, 1990). Higher transition temperatures are linked to more organized crystalline structures, which inhibits starch gelatinization (Zou et al., 2020). The ΔT is linked to the heterogeneity of the microcrystallites in starch granules, which contributes to form a semicrystalline structure inside the granules. A broad melting range suggests crystals with a great variation in size and stability, while a narrow range might imply crystals of a more homogeneous quality and similar stability (Zhang et al., 2019b). The DSC endothermic peak is related to the energy required for gelatinization, which can be a reflection of the loss of double-helical and crystalline order of starch granules (Cooke & Gidley, 1992).

4.2.2 Factors Affecting the Thermal Properties of Gelatinization

4.2.2.1 Structure of Amylopectin

The thermal properties of starch gelatinization are highly affected by the chain-length distribution and organization of amylopectin (Vamadevan & Bertoft,

TABLE 4.1   CRYSTALLINITY AND THERMAL PROPERTIES OF STARCH FROM VARIOUS SOURCES

Starch source	No.	Polymorphous	Crystallinity (%)	Conc. (w/w)	Temperature range (°C)	Heating rate (°C/min)	T_o (°C)	T_p (°C)	T_c (°C)	ΔT (°C)	ΔH (J/g)	Reference
Acorn fruit kernel (*Quercus fabri*)	4	C	48.16–66.85	25%	20–120	10	53.85–60.49	58.70–63.77	66.51–70.39	9.71–12.66	10.2–11.4	(Zhang et al., 2020)
Adzuki bean (*Vigna angularis*)	1	C_A	26.27	25%	30–120	5	62.99	69.84	76.63	13.65	2.39	(Liu et al., 2020)
Adzuki bean (*Vigna angularis*)	1	C	18.79	20%	30–120	10	61.22	68.35	78.99	17.66	6.76	(Zhang et al., 2019b)
Adzuki bean (*Vigna angularis*)	13	—	—	30%	20–100	10	59.58–63.40	64.43–68.93	70.03–75.53	—	4.69–8.47	(Yadav et al., 2019)
Amaranth (*Amaranthus cruentus*)	2	A	13.97–14.20	30%	30–130	10	66.77–67.06	71.42–72.12	80.67–82.51	—	3.57–3.59	(Sindhu & Khatkar, 2018)
Batata-de-teiú root (*Jatropha elliptica* (Pohl) Muell Arg.)	1	A	60.23	—	40–120	10	70.1	73.84	85.52	—	12.12	(Bento et al., 2020)
Bamboo culm (*Dendrocalamus asper*)	3	A	23.87 (SB) 27.73 (SM) 24.90 (ST)	25%	25–125	5	78.58 (SB) 80.16 (SM) 80.88 (ST)	82.76 (SB) 84.07 (SM) 84.67 (ST)	86.16 (SB) 87.07 (SM) 87.84 (ST)	—	4.25 (SB) 6.35 (SM) 5.19 (ST)	(Felisberto et al., 2019)
Banana*	7	B/C	22.4–40.4	30%	30–130	10	73.79–82.38	78.15–86.73	86.06–93.71	9.32–13.72	13.98–19.43	(Wang et al., 2019a)
Barley (*Hordeum vulgare*)	2	A	44.16–52.46	33%	25–100	10	65.30–65.45	67.77–69.20	73.08–73.23	7.63–7.93	9.66–9.80	(Chen et al., 2020)
Barley (*Hordeum vulgare*)	6	—	—	25%	20–100	5	53–58 (0)[A] 57–61 (2)[A] 57–60 (4)[A]	59–63 (0)[A] 63–65 (2)[A] 63–65 (4)[A]	67–71 (0)[A] 69–72 (2)[A] 69–71 (4)[A]	—	5–6 (0)[A] 4–5 (2)[A] 4–5 (4)[A]	(Quek et al., 2019)
Barley (waxy) (*Hordeum vulgare*)	2	A	56.83–60.95	33%	25–100	10	66.42–66.67	69.80–70.25	74.82–75.68	8.40–9.02	10.52–10.71	(Chen et al., 2020)

(Continued)

TABLE 4.1 (CONTINUED) CRYSTALLINITY AND THERMAL PROPERTIES OF STARCH FROM VARIOUS SOURCES

Starch source	No.	Polymorphous	Crystallinity (%)	Conc. (w/w)	Temperature range (°C)	Heating rate (°C/min)	T_o (°C)	T_p (°C)	T_c (°C)	ΔT (°C)	ΔH (J/g)	Reference
Buckwheat (*Fagopyrum esculentum*)	5	A	24.74–27.19	33%	30–100	10	62.59–66.52	65.68–71.50	70.19–77.08	—	6.83–8.30	(Gao et al., 2020)
Buckwheat (*Fagopyrum esculentum*)	2	A	14.96–16.16	30%	30–130	10	64.92–66.94	69.65–72.18	77.90–85.15	—	2.70–2.97	(Sindhu & Khatkar, 2018)
Chinese chestnut (*Castanea mollissima*)	9	C	18.5 – 21.2	25%	25–130	10	60.4–63.9	64.8–68.3	70.5–74.5	10.0–12.0	12.8–13.6	(Zhang et al., 2018)
Chinese chestnut (*Castanea mollissima*)	8	C	19.2–20.3(L) 16.2–18.2(S)	25%	RT–110	10	58.0–60.4(L) 54.0–55.5(S)	62.6–65.7(L) 60.3–61.7(S)	68.3–72.2(L) 68.0–72.5(S)	10.2–11.8 (L) 12.7–18.5(S)	12.5–13.7 (L) 10.1–11.7 (S)	(Liu et al., 2019)
Chinese yam (*Dioscorea opposita*)	2	A	20.44–29.06	25%	20–120	5	70.06–72.77	80.86–81.20	88.59–89.64	18.53–16.87	8.18–9.30	(Zou et al., 2020)
Cowpea (*Vigna unguiculata*)	3	C	32.1–32.7	33%	5–150	5	63.8–69.0	69.6–75.3	82.3–84.6	—	14.7–15.4	(Kim et al., 2018)
Djulis (*Chenopodium formosanum*)	3	A	35.25(HP) 36.17(WP) 37.42(NP)	25%	40–100	5	59.45(HP) 60.27(WP) 60.73(NP)	64.35(HP) 65.21(WP) 66.02(NP)	72.19(HP) 72.95(WP) 70.74(NP)	12.74(HP) 12.68(WP) 10.01(NP)	9.24(HP) 8.51(WP) 6.95(NP)	(Lu et al., 2019)
Dolichos bean (*Lablab purpureus*)	1	C_A	24.08	25%	30–120	5	63.03	75.44	91.11	28.08	8.44	(Liu et al., 2020)
Durian seed (*Durio zibethinus*)	2	—	—	25%	30–100	10	71.6–74.3	76.0–78.7	79.9–83.1	—	14–14.6	(Baraheng & Karrila, 2019)
Foxtail millet (*Setaria italica*)	8	A	33.2–36.4	25%	20–100	10	65.4–68.0	70.9–73.2	75.6–80.3	8.9–12.3	11.8–15.2	(Qi et al., 2020)
Ginger (*Rhizoma curcumae longae*)	1	A	39.62	32%	40–100	5	75.9	84.7	92.1	—	25.32	(Li, Chen et al., 2020)
Ginkgo (*Ginkgo biloba*)	7	—	—	25%	20–140	10	70.19–71.97	75.72–77.38	85.13–89.56	—	13.58–15.82	(Lu et al., 2020)

(Continued)

TABLE 4.1 (CONTINUED) CRYSTALLINITY AND THERMAL PROPERTIES OF STARCH FROM VARIOUS SOURCES

Starch source	No.	Polymorphous	Crystallinity (%)	Conc. (w/w)	Temperature range (°C)	Heating rate (°C/min)	T_o (°C)	T_p (°C)	T_c (°C)	ΔT (°C)	ΔH (J/g)	Reference
Jicama (*Pachyrhizus erosus*)	1	—	—	14%	30–120	7.5	53.8	59.6	72.3	9.7	15.8	(Contreras-Jiménez et al., 2019)
Kiwifruit (*Actinidia chinensis*)	6	B	26.78–27.47(OP) 25.47–26.55 (CO)	25%	25–90	10	59.31–60.45 (OP) 57.22–58.13 (CO)	62.61–63.60 (OP) 60.80–61.46 (CO)	68.10–69.10 (OP) 65.49–66.57 (CO)	—	20.95–21.49 (OP) 15.86–16.83 (CO)	(Li & Zhu, 2017a)
Lily rhizomes (*Hedychium coronarium*)	1	B	19.3	25%	40–120	10	72.3	76.78	91.21	18.91	16.82	(Bento et al., 2019)
Longan seeds (*Dimocarpus longan*)	3	A	28.6–29.2	25%	30–110	10	69.6–72.8	76.7–80.5	84.4–88.3	—	13.5–16.4	(Hu et al., 2018)
Lycoris (*Lycoris* spp.)	13	C/C$_A$	25.6–32.7	25%	30–110	10	55.4–60.1	58.8–69.7	62.6–78.0	—	7.5–24.7	(Li et al., 2020a)
Maca (*Lepidium meyenii*)	3	B	22.2–24.3	25%	25–90	10	47.1–47.5	50.8–52.1	57.4–58.0	—	14.2–14.6	(Zhang et al., 2017)
Maize (waxy) (*Zea mays*)	8	—	—	33%	20–100	10	67.1–69.4	72.6–74.7	80.3–81.9	—	10.8–12.7	(Wang et al., 2019b)
Maize (waxy) (*Zea mays*)	1	A	33.9	33/40%	10–140	10	66.9 (33)[B] 65.1 (40)[B]	75.1 (33)[B] 73.4 (40)[B]	90.9 (33)[B] 89.7 (40)[B]	24.0 (33)[B] 24.0 (40)[B]	14.5 (33)[B] 15.1 (40)[B]	(Hsieh et al., 2019)
Mango kernel*	1	A	—	24%	30–120	10	69.89	75.89	86.3	16.4	9.04	(Patiño-Rodríguez et al., 2020)
Wild mango seeds (*Cordyla africana*)	1	C	—	25%	35–100	10	—	77.3	—	—	—	(Ngobese et al., 2018)
Mashua (*Tropaeolum tuberosum*)	1	—	—	30%	20–95	10	48.7	54.9	60.7	—	4.3	(Pacheco et al., 2019)
Melloco (*Ullucus tuberosus*)	1	—	—	30%	20–95	10	46.4	52.3	57.5	—	12.3	(Pacheco et al., 2019)
Mung bean (*Vigna radiata*)	2	C	29.5–30.4	33%	5–150	5	56.3–59.2	67.0–67.8	79.9–80.8	—	13.7–15.3	(Kim et al., 2018)

(*Continued*)

TABLE 4.1 (CONTINUED) CRYSTALLINITY AND THERMAL PROPERTIES OF STARCH FROM VARIOUS SOURCES

Starch source	No.	Polymorphous	Crystallinity (%)	Conc. (w/w)	Temperature range (°C)	Heating rate (°C/min)	T_o (°C)	T_p (°C)	T_c (°C)	ΔT (°C)	ΔH (J/g)	Reference
Mung bean (*Vigna radiata*)	4	C_A	17.23–19.41	25%	20–120	10	62.15–64.45	67.40–68.40	72.40–73.10		19.38–21.30	(Yao et al., 2019)
Oca (*Oxalis tuberosa*)	2	—	—	25%	15–90	10	54.7–55.2	59.4–59.8	65.1–65.6	—	54.4–55.8	(Zhu & Cui, 2020)
Pitomba endocarp (*Talisia esculenta*)	1	—	—	22%	20–120	10	66.92	72.84	78.1	—	4.7	(de Castro et al., 2019)
Potato (*Solanum tuberosum*)	14	—	—	25%	30–110	10	64.0–64.8 (E1) / 64.3–66.3(E2)	68.0–69.2 (E1) / 68.0–70.2(E2)	75.3–76.8 (E1) / 75.7–76.9 (E2)	—	15.7– 20.1 (E1) / 14.9– 18.6 (E2)	(Ahmed et al., 2019)
Potato (*Solanum tuberosum*)	7	—	—	2%	20–150	10	55.34–58.27	60.29–63.80	69.09–73.67	—	7.95–8.88	(Choi et al., 2020)
Potato (waxy) (*Solanum tuberosum*)	1	B	32.6	33/40%	10–140	10	63.5 (33)B / 62.1 (40)B	71.5 (33)B / 69.8 (40)B	83.3 (33)B / 82.0 (40)B	19.9 (33)B / 19.9 (40)B	18.3 (33)B / 18.8 (40)B	(Hsieh et al., 2019)
Potato (*Solanum tuberosum*)	3	C_B	—	25%	40–120	10	60.32–64.81	63.90–68.91	74.71–79.01	12.43–13.72		(Wang, et al., 2020a)
Potatoes of Andean region (*Solanum tuberosum, Solanum × juzepczukii*)	3	B	36.3	24%	30–120	10	56.8–59.5	60.2–63.1	66.3–69.5	9.5–11.4	15.6–15.8	(Martínez et al., 2019)
Proso millet (*Panicum miliaceum*)	95	—	—	25%	30–110	10	67.4–75.5	71.5–79.0	76.5–84.0	—	11.9–17.6	(Li et al., 2020b)
Quinoa (*Chenopodium quinoa*)	4	A	21.00–29.67	25%	10–110	10	57.89–61.76	63.77–67.44	71.86–77.24	13.55–16.54	7.79–11.76	(Jiang et al., 2020)
Rice (*Oryza sativa*)	1	—	—	8%	50–90	15	—	68.91	—	63.87–78.65		(Desam et al., 2020)

(Continued)

TABLE 4.1 (CONTINUED) CRYSTALLINITY AND THERMAL PROPERTIES OF STARCH FROM VARIOUS SOURCES

Starch source	No.	Polymor-phous	Crystallinity (%)	Conc. (w/w)	Temperature range (°C)	Heating rate (°C/min)	T_o (°C)	T_p (°C)	T_c (°C)	ΔT (°C)	ΔH (J/g)	Reference
Rice (*Oryza sativa*)	2	A	15.7 (RES$_f$) 10.7 (RBS$_f$)	26%	20–120	10	65.07 (RBS$_f$) 52.31 (RES$_f$)	65.07 (RBS$_f$) 52.31 (RES$_f$)	97.23 (RBS$_f$) 90.7 (RES$_f$)	—	12.52 (RBS$_f$) 8.68 (RES$_f$)	(Singh & Sogi, 2018)
Rice (*Oryza sativa*)	12	A	32.08–36.06 (3)^c 32.45–36.38(6)^c 33.12–37.00(9)^c 33.31–36.70(12)^c	25%	10–120	10	61.87–64.45(3)^c 61.15–64.24(6)^c 60.95–64.35(9)^c 59.36–64.71(12)^c	66.76–70.06(3)^c 66.24–69.77(6)^c 66.60–70.12(9)^c 65.97–69.8(12)^c	72.29–76.23(3)^c 71.92–75.98(6)^c 72.27–75.58(9)^c 71.65–76.27(12)^c	10.32–12.18(3)^c 10.77–11.91(6)^c 10.59–12.36(9)^c 11.43–12.30(12)^c	10.64–12.25(3)^c 10.57–10.76(6)^c 10.55–10.72(9)^c 9.24–11.03(12)^c	(Gu et al., 2019)
Rice (*Oryza sativa*)	4	—	—	30%	25–100	10	58.0–63.8	65.8–72.2	74.8–79.5	—	13.2–15.4	(Tangsrianugul et al., 2019)
Rice (waxy) (*Oryza sativa*)	1	—	—	8%	50–90	15		63.17		58.86–73.74	—	(Desam et al., 2020)
Rice (waxy) (*Oryza sativa*)	1	A	38.2	33/40%	10–140	10	58.5 (33)^B 58.8 (40)^B	65.3 (33)^B 65.3 (40)^B	80.9 (33)^B 84.5 (40)^B	22.4 (33)^B 25.7 (40)^B	15.8 (33)^B 14.2 (40)^B	(Hsieh et al., 2019)
Rice mutants (*Oryza sativa*)	9	A	23.3–31.9	25%	20–130	10	55.7–64.4	63.9–75.7	74.2–81.8	13.8–19.5	7.8–10.9	(Zhang et al., 2019)
Sorghum (*Sorghum bicolor*)	1	A	32.9	33%	RT–110	10	67	71.8	78	—	10.3	(Yang et al., 2019)

(Continued)

TABLE 4.1 (CONTINUED)CRYSTALLINITY AND THERMAL PROPERTIES OF STARCH FROM VARIOUS SOURCES

Starch source	No.	Polymorphous	Crystallinity (%)	Conc. (w/w)	Temperature range (°C)	Heating rate (°C/min)	T_o (°C)	T_p (°C)	T_c (°C)	ΔT (°C)	ΔH (J/g)	Reference
Sweetpotato (*Ipomoea batatas*)	7	–	–	25%	30–120	10	54.7–73.2	66.8–81.7	79.1–88.9	–	8.9–15.3	(Tong et al., 2020)
Talipot palm (*Corypha umbraculifera*)	1	A	16.35	25%	20–100	10	79.71	82.22	87.62	–	11.19	(Navaf et al., 2020)
Tapioca (waxy) (*Manihot esculenta*)	1	A	35.4	33/40%	10–140	10	63.9 (33)[B] 64.2 (40)[B]	69.7 (33)[B] 70.0 (40)[B]	83.6 (33)[B] 86.5 (40)[B]	19.7 (33)[B] 22.3 (40)[B]	15.8 (33)[B] 14.6 (40)[B]	(Hsieh et al., 2019)
Tea seed (*Camellia sinensis*)	3	A	9.53–19.19	25%	30–110	10	60.84–68.56	64.99–77.04	70.47–82.74	9.63–14.58	12.80–12.94	(Huang et al., 2019)
Wheat (*Triticum aestivum*)	2	A	22.40–26.40	33%	20–120	10	61.6–61.7	65.7–65.8	72.8–73.2	–	9.7–10.2	(Li et al., 2020a)
Wheat (waxy) (*Triticum aestivum*)	1	A	33.3	33/40%	10–140	10	59.5 (33)[B] 57.2 (40)[B]	69.7 (33)[B] 67.7 (40)[B]	87.9 (33)[B] 88.2 (40)[B]	28.4 (33)[B] 31.0 (40)[B]	14.0 (33)[B] 14.7 (40)[B]	(Hsieh et al., 2019)

No.: number of samples studied; Conc.: concentration of starch (w/w); RT: room temperature; SB: starch extracted from bottom part; SM: starch extracted from middle part; ST: starch extracted from top part; L: large starch granules; S: small starch granules; HP: starch isolated by hydrochloric acid procedure; WP: starch isolated by deionized water procedure; NP: starch isolated by sodium hydroxide procedure; OP: starch isolated from outer pericarp of kiwifruit; CO: starch isolated from core of kiwifruit; E1: starch isolated from potato grown at Yiwu, China; E2: starch isolated from potato grown at Jinhua, China; RBS₁: rice bran starch fraction; RES₂: rice endosperm starch fraction; ^A: the number represents the days of germination; ^B: the number represents the percentage of starch; ^C: the number represents the months of storage; LA: cassava grown in low altitude; IA: cassava grown in intermediate altitude; *: scientific name not provided.

2015; Li & Zhu, 2017b). Kong et al. (2015) suggested that the T_o and T_p are negatively affected by the number of short chains within the DP between 6 and 12. The correlations between the weight-based percentage of amylopectin unit chains and internal chains and thermal properties were calculated and plotted against the degree of polymerization (Figure 4.1a, b). The result suggested that the gelatinization temperature was negatively controlled by short chains (Li & Zhu, 2017b). The short chains including A_{fp}-chains are too short to form double-helical structures, and thus induce defects and result in low gelatinization temperatures (Bertoft, 2004; Zhu, 2018). The gelatinization temperatures are also affected by the internal structure of amylopectin (Zhu, 2018). The gelatinization temperature was found positively correlated with the amount of long B-chains but negatively correlated with those of A-chains and short B-chains (Kong et al., 2008; Qi et al., 2020). The short internal chains might affect the packing of double-helical structure and reduce the perfection of the crystalline structure, which results in low gelatinization temperatures (Zhu, 2018).

The double-helical structures are mainly formed by the external chains of amylopectin (Bertoft, 2004). A positive correlation between clustered A-chains with ΔH further confirmed this (Zhu & Liu, 2020). Thus, a longer average chain length of external chains is responsible for a higher ΔH of gelatinization. Amylopectin with longer external chains could form more stable double-helical structures and require more energy for disruption. The average chain length of long internal chains in amylopectin was also found positively correlated with the ΔH. The long internal chains, which act as the backbone in the amorphous region, could induce flexibility and help the formation of double-helical structures in crystalline lamellae, which increases the relative crystallinity and the ΔH (Zhu, 2018; Qi et al., 2020; Li & Zhu, 2017b).

4.2.2.2 Other Factors

In addition to the effect of amylopectin structure, the thermal properties of gelatinization are also affected by other factors, such as granule morphology, amylose content and minor components (Hu et al., 2018). For example, amylose could entangle with other starch molecules and bind the granules, thereby increasing the thermal stability (Srichuwong & Jane, 2007). Nonstarch components such as phosphates, proteins and lipids could also play roles in the thermal properties of starch (Srichuwong & Jane, 2007).

The botanical sources of starch as well as location within the same source also affect the thermal properties of starch. The rice starch from tiller panicles showed significantly higher ΔH than the starch from main stem panicles (Wang et al., 2020c). The starch isolated from three different parts of young bamboo culm also showed significant differences in gelatinization temperatures, enthalpy change and retrogradation properties (Felisberto et al., 2020). When the starch granules are heated in an excess amount of water at a

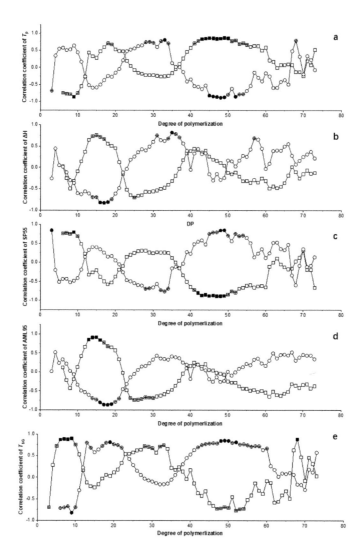

Figure 4.1   The correlation coefficients (Y-axis) between weight-based percentages of individual units and internal chains of amylopectin and (a) T_p, (b) ΔH, (c) SP55 (swelling power at 55°C), (d) AML95 (amylose leaching at 95°C), and (e) T_SG (sol/gel transition temperature). The X-axis represents the degree of polymerization of individual amylopectin units and internal chains; square points are of the unit chain and round points are of the internal chains of amylopectin. Significant correlations at $p < 0.05$ and $p < 0.01$ are represented by crossed and filled black points, respectively. Reprinted with permission from Elsevier (Li and Zhu, 2017b, 2018).

temperature between the glass transition temperature and the onset temperature of gelatinization, a phenomenon called annealing happens, which induces changes in physicochemical properties of starches (Blazek & Gilbert, 2011). The annealing of starch can significantly increase the gelatinization temperatures and reduce the ΔT, indicating the formation of more ordered structures (Laity & Cameron, 2008). The T_p and T_c were negatively related to the granule size in potato starch, while the ΔH was positively related to the granule size (Dhital et al., 2011). The gelatinization range decreased with the increasing starch granule size in chestnut starch (Lu et al., 2020). The experimental conditions such as the starch/water ratio and the heating rate could also affect the thermal parameters (Hu et al., 2018). When the starch is gelatinized in a limited amount of water (for example, below 40% w/w), less water is available for the plasticization of the backbone and spacers in the amorphous region. As a result, the T_o increased significantly with less water (Waigh et al., 2000).

4.2.3 DSC Analysis of Starch Retrogradation

Starch retrogradation can be defined as a process where amylose and amylopectin chains, which are disaggregated during gelatinization, reassociate to develop ordered structures. Both amylose and amylopectin are involved in the retrogradation process. Amylose retrogradation is a fast process, while amylopectin retrogradation needs several days to form more organized structures (Punia et al., 2020). The difference mainly came from the length of linear structures. Upon cooling, the amylose molecules shifted from random-coil confirmation to ordered crystalline structure simultaneously or before any phase separation. The branched structure of amylopectin might hinder the process of reordering, thus needing longer time to occur (Fu et al., 2015). The ordered structure formed by amylose was more stable than that of amylopectin, thus it required a high temperature of up to 160°C to be melted (Koehler & Wieser, 2013; Vamadevan & Bertoft, 2015). The thermal properties of retrograded starch studied by DSC mostly relate to the recrystallization of amylopectin.

The thermal parameters of retrograded starches from various sources with different storage conditions and concentrations studied in recent years are summarized in Table 4.2. The melting temperatures of retrograded amylopectin are significantly lower than the gelatinization temperatures of native starch (Table 4.1). The retrogradation enthalpy change (ΔH_r) also showed a significantly lower value than ΔH. During retrogradation, small or imperfect crystals are formed, which are observed with lower gelatinization temperatures and enthalpy change. Increased ΔT indicated higher heterogeneity of the crystals reformed (Patiño-Rodríguez et al., 2020).

TABLE 4.2   RETROGRADATION PROPERTIES OF STARCH FROM VARIOUS SOURCES

Sample	No.	Conc. (w/w)	Storage conditions	T_{or} (°C)	T_{pr} (°C)	T_{er} (°C)	ΔT_r (°C)	ΔH_r (J/g)	R (%)	Reference
Adzuki bean (*Vigna angularis*)	13	30%	4°C, 7 days	46.86–48.41	56.32–57.98	64.72–66.50	—	3.86–6.36	—	(Yadav et al., 2019)
Bamboo culm (*Dendrocalamus asper*)	3	25%	4°C, 7 days	43.70(SB) 41.61(SM) 43.07(ST)	56.52(SB) 55.66(SM) 55.38(ST)	66.19(SB) 65.24(SM) 66.46(ST)	—	2.32(SB) 3.89(SM) 3.89(ST)	54.57(SB) 61.30(SM) 74.90(ST)	(Felisberto et al., 2019)
Lycoris (*Lycoris* spp.)	13	25%	4°C, 7 days	—	—	—	—	—	3.5–14.8	(Li et al., 2020b)
Maize (waxy) (*Zea mays*)	8	33%	4°C, 7 days	—	—	—	—	5.0–7.0	44.3–58.8	(Wang et al., 2019b)
Maize (waxy) (*Zea mays*)	1	33%/40%	4°C, 7 days	47.8(33)[A] 46.3(40)[A]	57.9(33)[A] 56.7(40)[A]	69.0(33)[A] 68.0(40)[A]	25.2(33)[A] 21.7(40)[A]	6.6(33)[A] 6.7(40)[A]	—	(Hsieh et al., 2019)
Mango kernel*	1	24%	4°C, 7/14 days	53.85(7)[B] 47.13(14)[B]	61.31(7)[B] 59.25(14)[B]	68.7(7)[B] 70.65(14)[B]	9.89(7)[B] 23.51(14)[B]	1.01(7)[B] 3.33(14)[B]	10.92(7)[B] 35.93(14)[B]	(Patiño-Rodríguez et al., 2020)
Oca (*Oxalis tuberosa*)	2	25%	4°C, 4 weeks	—	54.4–55.8	—	—	—	33–47	(Zhu & Cui, 2020)
Potato (*Solanum tuberosum*)	14	25%	4°C, 7 days	—	—	—	—	1.9–5.2 (E1) 2.3–4.3 (E2)	12.1–29.7 (E1) 15.0–22.9 (E2)	(Ahmed et al., 2019)
Potato (waxy) (*Solanum tuberosum*)	1	33%/40%	4°C, 7 days	46.5(33)[A] 44.8(40)[A]	61.6(33)[A] 57.6(40)[A]	76.5(33)[A] 75.0(40)[A]	30.1(33)[A] 30.1(40)[A]	8.5(33)[A] 9.6(40)[A]	—	(Hsieh et al., 2019)

(Continued)

TABLE 4.2   (CONTINUED) RETROGRADATION PROPERTIES OF STARCH FROM VARIOUS SOURCES

Sample	No.	Conc. (w/w)	Storage conditions	T_{or} (°C)	T_{pr} (°C)	T_{cr} (°C)	ΔT_r (°C)	ΔH_r (J/g)	R (%)	Reference
Proso millet (*Panicum miliaceum*)	95	25%	4°C, 7 days	–	–	–	–	1.5–3.7	0.1–28.3	(Li et al., 2020c)
Rice (*Oryza sativa*)	16	40%	4°C, 5 days	36.4–38.8	48.8–52.7	61.2–67.4	23.2–29.4	4.2–8.1	–	(Gong et al., 2019)
Rice (waxy) (*Oryza sativa*)	1	33%/40%	4°C, 7 days	46.9(33)^A 44.9(40)^A	56.4(33)^A 53.3(40)^A	62.6(33)^A 61.9(40)^A	15.8(33)^A 17.0(40)^A	0.4(33)^A 4.7(40)^A	–	(Hsieh et al., 2019)
Tapioca (waxy) (*Manihot esculenta*)	1	33%/40%	4°C, 7 days	45.6(33)^A 44.5(40)^A	57.0(33)^A 56.9(40)^A	69.3(33)^A 67.2(40)^A	23.7(33)^A 22.7(40)^A	2.0(33)^A 5.0(40)^A	–	(Hsieh et al., 2019)
Tea seed (*Camellia sinensis*)	3	25%	4°C, 7 days	46.60–48.22	57.40–58.34	66.62–66.93	18.40–20.33	4.29–6.09	–	(Huang et al., 2019)
Wheat (*Triticum aestivum*)	2	33%	4°C, 7 days	48.2–48.5	56.9–57.1	64.0–64.8		0.4–0.8	4.3–8.2	(Li et al., 2020a)
Wheat (waxy) (*Triticum aestivum*)	1	33%/40%	4°C, 7 days	47.6(33)^A 44.2(40)^A	57.1(33)^A 54.0(40)^A	67.6(33)^A 64.5(40)^A	20.0(33)^A 21.2(40)^A	3.3(33)^A 5.8(40)^A	–	(Hsieh et al., 2019)

No.: number of samples studied; Conc.: concentration of starch (w/w); SB: starch extracted from bottom part; SM: starch extracted from middle part; ST: starch extracted from top part; ^{A}: the number represents the percentage of starch; ^{B}: the number represents the days of storage; *: scientific name not provided.

4.2.4 Factors Affecting the Thermal Properties of Retrogradation

4.2.4.1 Structure of Amylopectin

Similar to the gelatinization, the thermal properties of retrograded starch are significantly affected by the fine structure of amylopectin. The retrogradation temperatures (T_{pr}, T_{cr}) and retrogradation enthalpy (ΔH_r) were found negatively correlated with the number of short B-chains, while the number of long B-chains showed positive relationships with T_{pr} and T_{cr}. The retrogradation degree (calculated as the percentage of $\Delta H_r/\Delta H$) was also positively correlated to the amount of long internal chains (Zhu & Liu, 2020). Vamadevan and Bertoft (2018) studied the relationship between cluster structure and long-term retrogradation of different starches. The melting temperatures and ΔH_r are found positively related to the interblock chain length. The long chains in amylopectin acting as backbones could increase the flexibility of starch chains and facilitate the realignment and recrystallization during retrogradation (Zhu & Liu, 2020). The average external chain length of amylopectin was positively related to the retrogradation temperatures and ΔH_r. The longer external chain length could facilitate the formation of a double-helical structure during retrogradation (Vamadevan & Bertoft, 2018). In general, the average chain lengths of both internal chains and external chains were positively correlated with the retrogradation temperatures and ΔH_r. The structure of amylose might also have effects on the thermal properties of starch retrogradation via the process of nucleation.

4.2.4.2 Other Factors

Besides the amylopectin structure, other factors such as amylose content, water/starch ratio, minor components and storage conditions can also affect the thermal properties of retrograded starches (Wang et al., 2015). Minor components such as proteins, fibers, lipids and sugars are found in most food products. Proteins might interact with amylose via hydrogen bonds, reducing the extent of amylose retrogradation. The addition of sodium caseinate reduced the crystallinity of starch-based films after storage (Fu et al., 2015). The hydrophilic nature of protein might indirectly affect the reordering process by reducing the amount of available water. During heating, the amylose is able to form V-type inclusion complexes with polar lipids such as lysophospholipids. As the complexes cannot participate in the reordering process, the presence of lipids significantly retards the retrogradation of starch (Koehler & Wieser, 2013). In addition, the complexes could prevent the retrogradation of amylopectin. However, the detailed mechanism is still to be investigated (Fu et al., 2015). Sugars such as glucose, ribose, fructose, maltose, sucrose and soluble maltodextrins could retard the retrogradation of starch. The effect of oligosaccharides, however, do not have a consistent effect on starch retrogradation (Wang et al., 2015). The possible mechanism might be related to the changes in available free water and mobility of starch molecular chains (Fu et al., 2015). Some nonstarch polysaccharides, such as pectin, corn fiber gum, guar gum and carboxymethyl cellulose, could also affect the starch retrogradation in various manners. The nonstarch hydrocolloids may inhibit long-term retrogradation by affecting the amylopectin–amylopectin interactions, and facilitate the short-term retrogradation by promoting amylose–amylose interactions (Wang et al., 2015).

4.3 SWELLING AND SOLUBILITY OF STARCH

4.3.1 The Definitions of WSI, SP and AML

During hydration and thermal treatment, the swelling properties of starch and the solubility of starch granules are mostly represented by swelling power (SP), water solubility index (WSI) and amylose leaching (AML). The WSI indicates the amount of soluble fraction that is released from the starch in excess of water (Tsai, 1997). The WSI is an indicator of the solubility of small molecules. The AML represents the amount of amylose dissolved in water due to the phase separation from immiscible amylopectin during heating (Koehler & Wieser, 2013). The WSI is mostly contributed by leached amylose. As a result, there are significant positive correlations between the two (Lan et al., 2015; Li & Yeh, 2001).

The degree of granular swelling relates to the swelling factor (measures only intra-granular water) or swelling power (measures both inter- and intragranular water) (Leach et al., 1959; Tester & Morrison, 1990). In this chapter, only SP is discussed, as it is more widely used. The SP is an indicator of swelling property and the ability of starch to keep its integrity upon heating. It is calculated from the sediment paste weight and WSI (Li et al., 2020d). The SP of the starch represents the inter associative force that keeps the starch granular structure (Gourilekshmi et al., 2020). The greater swelling ability of the starch granules, the weaker the binding forces (Hoover & Manuel, 1996). It is believed that SP is strongly linked with the free space within and outside starch molecules as well as lamellar structures (Lan et al., 2015).

Both SP and WSI give an idea about the extent of the interactions among starch chains within the crystalline and amorphous regions. The WSI and SP of starch with various botanical backgrounds, amylose content and experimental conditions studied in recent years are summarized in Table 4.3.

TABLE 4.3 AMYLOSE CONTENT, SWELLING POWER AND WATER SOLUBILITY OF STARCH FROM VARIOUS SOURCES

Starch source	No.	AM (%)	Conc. (w/w)	Temperature (°C)	SP (g/g)	WSI (%)	Reference
Acorn fruit kernel (*Quercus fabri*)	4	—	1%	90	13.5–25.04	4.87–15.22	(Zhang et al., 2020)
Adzuki bean (*Vigna angularis*)	1	30.61 [B]	1%	90	10.4	14.81	(Zhang et al., 2019b)
Adzuki bean (*Vigna angularis*)	13	9.95–23.99 [A]	2%	90	11.12–15.26	7.85–19.08	(Zhang et al., 2019b)
Adzuki bean (*Vigna angularis*)	1	26.64 [B]	1%	90	16.69	17.43	(Liu et al., 2020)
Avocado seed (*Persea americana*)	1	30.41 [A]	7%	90	7.68	2	(Macena et al., 2020)
Batata-de-teiú root (*Jatropha elliptica* (Pohl) Muell Arg.)	1	32.82 [A]	5%	90	~10	~5	(Bento et al., 2020)
Baiyue bean (*Phaseolus vulgaris*)	1	32.64 [B]	1%	90	11.26	12.62	(Zhang et al., 2019b)
Barley (*Hordeum vulgare*)	2	17.74–20.08 [A]	2%	95	29.59–32.98	13.2–14.67	(Chen et al., 2020)
Barley (waxy) (*Hordeum vulgare*)	2	8.81–9.65 [A]	2%	95	38.05–42.44	19.72–19.75	(Chen et al., 2020)

(Continued)

TABLE 4.3   (CONTINUED) AMYLOSE CONTENT, SWELLING POWER AND WATER SOLUBILITY OF STARCH FROM VARIOUS SOURCES

Starch source	No.	AM (%)	Conc. (w/w)	Temperature (°C)	SP (g/g)	WSI (%)	Reference
Buckwheat (*Fagopyrum esculentum*)	5	–	3%	90	24.45–28.38	13.95–15.15	(Gao et al., 2020)
Cassava (*Manihot esculenta*)	4	18.25–25.85[A]	1%	90	~20–30	–	(Oyeyinka et al., 2020)
Cassava (waxy) (*Manihot esculenta*)	18	–	1%	75	52.2–73.8(LA) 47.6–75.4(IA)	5.2–17.6(LA) 4.1–11.1(IA)	(Karlström et al., 2019)
Cassava (*Manihot esculenta*)	20	–	1%	75	24.7–36.1(LA) 23.5–35.1(IA)	11.8–16.6(LA) 12.1–17.1(IA)	(Karlström et al., 2019)
Chickpea (*Cicer arietinum*)	1	32.61[B]	1%	90	18.25	10.21	(Zhang et al., 2019b)
Cowpea (*Vigna unguiculata*)	3	35.7–36.8[A]	1%	80	14.4–19.9(SF)	–	(Kim et al., 2018)
Curcuma karnatakensis	2	84.6–86.6[A]	1%	60	5.03–5.06	0.6–0.9	(Tejavathi et al., 2020)
Dolichos bean (*Lablab purpureus*)	1	31.85[B]	1%	90	13.46	14.50	(Liu et al., 2020)
Durian seed (*Durio zibethinus*)	2	22.9–23.3[A]	9%	95	~9–11	~22–27	(Baraheng & Karrila, 2019)

(Continued)

TABLE 4.3 (CONTINUED) AMYLOSE CONTENT, SWELLING POWER AND WATER SOLUBILITY OF STARCH FROM VARIOUS SOURCES

Starch source	No.	AM (%)	Conc. (w/w)	Temperature (°C)	SP (g/g)	WSI (%)	Reference
Faba bean (*Vicia faba*)	1	33.55[B]	1%	90	12.67	9.92	(Zhang et al., 2019b)
Foxtail millet (*Setaria italica*)	8	16.8–26.8[A]	4%	92.5	15.9–22.6	—	(Qi et al., 2020)
Ginger (*Rhizoma curcumae longae*)	1	—	9%	90	~10	~8	(Li et al., 2020d)
Inhambu tuber (*Dioscorea trifida*)	1	36.8[A]	1%	90	20.52	33	(Silva et al., 2020)
Kiwifruit (*Actinidia chinensis*)	6	24.5–27.5(OP)[A] 32.4–36.7(CO)[A] 15.5–17.8(OP)[B] 20.7–23.3(CO)[B]	1%	95	~20–25	~75–80	(Li & Zhu, 2017a)
Lily rhizome (*Hedychium coronarium*)	1	59.16[A]	5%	90	15.89	10.02	(Bento et al., 2019)
Maca (*Lepidium meyenii*)	3	21.0–21.3[A]	1%	85	~30–45	~80–95	(Zhang et al., 2017)
Mung bean (*Vigna radiata*)	2	36.9–38.5[A]	1%	80	16.3–16.5(SF)		(Kim et al., 2018)
Oca (*Oxalis tuberosa*)	2	—	1%	85	37.1–41.6	60.6–61.0	(Zhu & Cui, 2020)

(Continued)

TABLE 4.3   (CONTINUED) AMYLOSE CONTENT, SWELLING POWER AND WATER SOLUBILITY OF STARCH FROM VARIOUS SOURCES

Starch source	No.	AM (%)	Conc. (w/w)	Temperature (°C)	SP (g/g)	WSI (%)	Reference
Pitomba endocarp (*Talisia esculenta*)	1	19.25[A]	0.25%	90	~3	2	(de Castro et al., 2019)
Potato (*Solanum tuberosum*)	1	25	8%	80	~60	—	(Desam et al., 2020)
Queen sago (*Cycas circinalis*)	3	31.10(WS)[A] 24.26(AS)[A] 34.19(ES)[A]	5%[C]	90	~9–13	~6–9	(Noora et al., 2019)
Quinoa (*Chenopodium quinoa*)	4	9.43–10.90[A]	1%	95	21.83–28.15	6.45–17.57	(Jiang et al., 2020)
Rice (*Oryza sativa*)	1	18.6	8%	80	~10	—	(Desam et al., 2020)
Rice (*Oryza sativa*)	3	1.6–18.9[B]	1%	100	9.6–32.6	4.9–6.8	(Wang et al., 2020b)
Rice (*Oryza sativa*)	2	0.49(RBS$_f$)[A] 10.25(RES$_f$)[A]	2%	90	~10(RES$_f$) ~6(RBS$_f$)	~15%(RES$_f$) ~12%(RBS$_f$)	(Singh & Sogi, 2018)
Rice (waxy) (*Oryza sativa*)	1	2	8%	80	~20	—	(Desam et al., 2020)
Sweetpotato (*Ipomoea batatas*)	1	25.84–29.52[A]	9%	95	8.42	8	(Vithu et al., 2020)

(Continued)

TABLE 4.3   (CONTINUED) AMYLOSE CONTENT, SWELLING POWER AND WATER SOLUBILITY OF STARCH FROM VARIOUS SOURCES

Starch source	No.	AM (%)	Conc. (w/w)	Temperature (°C)	SP (g/g)	WSI (%)	Reference
Talipot palm (*Corypha umbraculifera*)	1	28.05[A]	5%	90	~8	~7	(Navaf et al., 2020)
Tea seed (*Camellia sinensis*)	3	27.06–33.17[A]	3%	95	36.25–40.80	2.27–9.70	(Huang et al., 2019)
Thai pigmented rice (*Oryza sativa*)	4	2.77–12.09[B]	1%[C]	90	31.0–55.6	4.7–7.8	(Tangsrianugul et al., 2019)
Wheat (waxy) (*Triticum aestivum*)	2	<2	2%	95	30.0–31.4	18.8–22.1	(Li et al., 2020c)

No.: number of samples studied; AM, amylose content; Conc.: concentration of starch (w/w); WSI: water solubility index; SP: swelling power; Temp.: temperature which SP and WSI estimated; LA: cassava grown in low altitude; IA: cassava grown in intermediate altitude; SF: swelling factor instead of swelling power is recorded; WS: starch isolated by water steeping method; AS: starch isolated by alkali steeping method; ES: starch isolated by enzymatic extraction method; RBS: rice bran starch fraction; RES: rice endosperm starch fraction; OP: starch isolated from outer pericarp of kiwifruit; CO: starch isolated from core of kiwifruit; [A]: apparent amylose content estimated by iodine-based method; [B]: amylose content estimated by concanavalin A precipitation based method. [C]: concentration is expressed as (w/v).

4.3.2 Factors Affecting WSI, SP and AML

The swelling and solubility of starch granules are influenced by various factors, including temperature, starch composition, the structures of amylopectin and amylose, and the presence of minor components such as phosphate monoesters and phospholipids.

4.3.2.1 Temperature

The increases in WSI and SP upon heating are due to the rupture of the breakage of intermolecular bonds, letting the freed binding sites form hydrogen bonds with the water molecules. Therefore, increasing temperature causes the weakening of intramolecular bonds, bringing about higher amylose leaching, and as a result, increases the solubility (Bento et al., 2020). The SP and solubility of rice starches with different amylose content were studied between 65°C and 125°C (Vandeputte et al., 2003). The result suggested that the incubation temperature significantly changed the extent of swelling and amylose leaching (Figure 4.2). The SP and WSI of starch usually increase with the temperature because the internal associative forces that maintain the starch granule itself weaken, while the increase has been reported to be nonlinear (Kong et al., 2015). In some cases, the SP decreased at high temperatures due to the rupture of the starch granules in the process of gelatinization following amylose leaching. With the breakage of these granules and the increased WSI, the SP decreased (Silva et al., 2020).

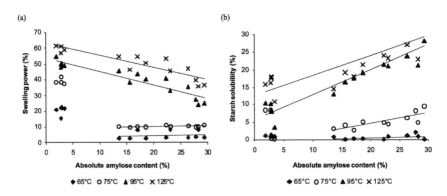

Figure 4.2   The swelling power (a) and solubility (b) of rice starch at different temperatures as a function of amylose content. Reprinted with permission from Elsevier (Vandeputte et al. 2003).

4.3.2.2 Amylose Content

The swelling capability of the starch granules is significantly linked with the proportion of amylose and amylopectin (Hoover, 2001; Desam et al., 2020). The high extent of swelling in waxy starch showed that low amylose content is linked with greater swelling (Punia, 2020). The SP of rice starches showed negative correlations with the amylose content (Figure 4.2) (Vandeputte et al., 2003). Bean varieties with low amylose content were found to have higher SP than those with higher amylose content (Yadav et al., 2019).

The amylose performs both as a diluent and an inhibitor of swelling, resulting in insoluble complexes during swelling and gelatinization (Tester & Morrison, 1990). The amylose molecules could form entanglements with amylopectin, which limits the swelling of starch (Liu et al., 2020). The increased SP in low-amylose starches might be due to the reduced interactions between amylose and amylopectin as well as weakened intramolecular bonds (Li et al., 2020d).

4.3.2.3 Structure of Amylopectin

Previous studies have shown that the different swelling properties can be attributed to the differences in amylopectin fine structure (Figure 4.1c, d) (Li & Zhu, 2017b). Short chains in amylopectin could enhance swelling and long chains may hinder swelling (Vamadevan & Bertoft, 2015). The long AP chains (DP >18) in potato and cassava starches were seen to prevent swelling, while short chains (DP <14) encouraged swelling (Gomand et al., 2010). The finger-print A-chains (A_{fp}-chains) and other short chains could cause defects in the crystalline domain, allowing water molecules to easily penetrate into crystalline lamellae (Vamadevan & Bertoft, 2015). Similarly, a high amount of short internal chains in amylopectin could increase the SP and WSI. The possible reason is the short internal chains could hinder the packing of double helices in the crystalline lamellae, resulting in easier hydration and swelling (Zhu, 2018).

4.3.2.4 Proteins and Lipids

Swelling properties for high-amylose starches can be significantly controlled by the amylose content, while the effect of protein and lipid content is more obvious in other starches (Debet & Gidley, 2006). The formation of amylose–lipid complexes within starch granules could limit the amylose leaching. It has been found that the polysaccharides leached out of starch granules were positively correlated with the swelling, while the content of lipid-bound amylose was negatively correlated with the swelling of starch (Tester & Morrison, 1990). The addition of lipids in starch–water systems can inhibit water penetration into granules and reduce swelling during the heating process (Wang et al., 2015). On the other hand, the removal of lipids during the production of starch enhanced amylose leaching (Tangsrianugul et al., 2019). Besides the

lipid content, the protein content could significantly inhibit swelling and amylose leaching by pronounced protein–starch interactions (Noora et al., 2019). By
removing both the proteins and lipids, the swelling rates and extents of wheat
and maize starches increased significantly (Debet & Gidley, 2006).

4.3.2.5 Other Factors

The SP was positively related to the amount of phosphate groups, due to the
electrostatic repulsions between neighboring amylopectin molecules, which is
the main reason of elevated swelling power in potato starch for its high content
of negatively charged phosphate monoester groups (Dhital et al., 2015a;
Hoover, 2001). The SP was also observed to increase with decreasing granule
size, which might be related to the expansion rate of starches (Hu et al., 2020).
The decreased tendency in SP and WSI with increased granule size was also
observed (Choi et al., 2020). The starches isolated from different genotypes of
the same plant might also have significant variances in SP and WSI. The growth
conditions, such as nitrogen fertilization rates, were found to affect the SP and
WSI (Wang et al., 2019b). The starch isolated from different parts and different
stages in the growth cycle were also found to have significant variances in SP
and WSI (Gourilekshmi et al., 2020; Singh & Sogi, 2018; Li & Zhu, 2017a).

4.4 RHEOLOGICAL PROPERTIES

Rheology relates to the flow of matter in various states. The rheological
properties of food matters deal with the relationships between strain and
stress. Stress is the force applied on the matter per unit area, while the strain
is the extent of the deformation (Kasapis & Bannikova, 2017). Depending on
whether the strain is within the linear viscoelastic region or not, the rheological
experiments could be classified into large and small deformation tests (Mezger,
2006). In a large deformation test, the strain applied to the samples is beyond the
linear viscoelastic region, which eventually changes its structure irreversibly.
By contrast, small deformation tests only apply small strain to the sample,
thereby allowing the viscoelastic properties to be measured nondestructively
(Kasapis & Bannikova, 2017).

4.4.1 Large Deformation Analysis

The large deformation analysis is useful in studying the viscous and elastic
properties of materials outside the linear viscoelastic region, which is close to
the application and processing of food materials. The results of large deformation analysis could be used to estimate the sensory evaluation outcomes, study
the structure and mechanical properties of the food system, and evaluate the

quality of food products (Ahmed, 2016). The large deformation tests are usually conducted by rheometer, texture analyzer, Brabender Amylograph or Rapid Visco Analyser (RVA). The most common methods include pasting analysis, texture profile analysis (TPA) and steady-flow analysis (Li & Zhu, 2018; Wang et al., 2015).

4.4.1.1 Pasting

Pasting properties are studied by observing viscosity changes during the process of heating a starch suspension in a programmed heating and cooling procedure. As discussed earlier, when starch is heated in the presence of water and above gelatinization temperature, the starch granules start to swell and the amylose begins to leach out into the water, increasing its viscosity. The swollen granules become susceptible to shear stress, which could lead to granule rupture and dispersion of amylose and amylopectin. During the heating, cooling and shearing, the starch paste is characterized by its rheological properties (Punia et al., 2020).

The pasting properties are usually estimated using RVA, Amylograph, or a rheometer with starch cell. These instruments continuously record the viscosity change of starch pastes under controlled temperature and shearing as a function of time. The most common instrument applied is RVA (Balet et al., 2019).

Under the most common working conditions, the samples are first heated to 90°C–95°C. During this process, the temperature at which the viscosity of starch suspension starts to rise is termed pasting temperature (PT). As the temperature keeps rising, a significant number of starch granules undergo rapid swelling and partial amylose leaching, after which a significant rise in viscosity can be observed. The viscosity and temperature of the peak are termed peak viscosity (PV) and peak temperature (PKT), respectively. With both heating and shearing, the swollen granules start to be disrupted and the leached amylose starts to align under shearing. These normally lead to a reduction in viscosity in most cases. In order to achieve complete gelatinization, a holding time is applied when the temperature reaches maximum. The lowest value during the holding period is usually termed trough viscosity (TV) or hot paste viscosity. A cooling process is applied in order to study the increase of viscosity due to the decrease of temperature and retrogradation of amylose. The viscosity at the end of the cooling stage is termed final viscosity (FV) or cold paste viscosity (Ronda et al., 2017; Vithu et al., 2020; Wang et al., 2015).

The experimental settings and results of pasting analysis of starch with various background and experimental conditions are summarized in Table 4.4. When the viscosity reaches a peak, the rates of granular expansion and disintegration are the same. The PV is related to the SP of starch (Wang et al., 2020b; Zhang et al., 2020). The breakdown (BD) is the gap between PV and TV.

TABLE 4.4   PASTING PROPERTIES OF STARCH FROM VARIOUS SOURCES

Source of starch	No.	Instrument	Conc. (w/w)	Unit	PV	TV	BD	FV	SB	PT	Reference
Acorn fruit kernel (Quercus fabri)	4	RVA	7%	cP	485–1037	361–978	41–124	732–1291	164–371.5	—	(Zhang et al., 2020)
Adzuki bean (Vigna angularis)	13	RVA	8%	cP	2485–3777	2219–2980	105–798	3998–5217	1780–2485	72.28–74.7	(Yadav et al., 2019)
Adzuki bean (Vigna angularis)	1	RVA	7%	cP	4930	3605	1325	4764	1159	75.6	(Zhang et al., 2019b)
Avocado seed (Persea americana)	1	BV	7%	BU	616	—	—	831	—	83.5	(Macena et al., 2020)
Batata-de-teiú root (Jatropha elliptica (Pohl) Muell Arg.)	1	RVA	9%	cP	1960	641	1318.5	1193	551.5	69.07	(Bento et al., 2020)
Baiyue bean (Phaseolus vulgaris)	1	RVA	7%	cP	3286	2367	919.5	5378	3011	73.1	(Zhang et al., 2019)
Banana*	7	BV	6%	BU	89–203	87–195	1–8	136–264	48–69	74.9–80.9	(Wang et al., 2019a)
Barley (Hordeum vulgare)	2	RVA	10%	cP	4186–4503	880–947	3306–3556	2576–2705	1696–1758	—	(Chen et al., 2020)
Barley (Hordeum vulgare)	10	RVA	12%	RVU	430–430	29–1946	401–1335	57–3718	28–1777	68–89	(Balet et al., 2020)
Barley (waxy) (Hordeum vulgare)	2	RVA	11%	cP	4790–4893	1056–1134	3734–3759	1488–1845	1696–1758	—	(Chen et al., 2020)
Buckwheat (Fagopyrum esculentum)	5	RVA	11%	cP	2289–2423	1903–2268	122–969	2465–3429	651–2267	75.53–79.60	(Gao et al., 2020)

(Continued)

TABLE 4.4   (CONTINUED) PASTING PROPERTIES OF STARCH FROM VARIOUS SOURCES

Source of starch	No.	Instrument	Conc. (w/w)	Unit	PV	TV	BD	FV	SB	PT	Reference
Buckwheat (Tartary) (*Fagopyrum tataricum*)	1	RVA	11%	cP	4166	2397	1769	4396	1999	73.6	(Yang et al., 2019)
Buckwheat (*Fagopyrum esculentum*)	1	RVA	11%	cP	3875	2267	1608	4212	1945	74.3	(Yang et al., 2019)
Cassava (*Manihot esculenta*)	20	RVA	5%	cP	500.1–745.4(LA)– 680.0–938.1(IA)	302.1–531.4(LA)– 375.9–589.1(IA)	177.5–256.0(LA)– 297.5–456.0(IA)	363.6–710.4(LA)– 436.4–875.6(IA)	(–)1365–(–)1000(LA) (–)281.0–(–)1360(IA)	70.4–73.7(LA)– 65.7–71.8(IA)	(Karlström et al., 2019)
Chinese chestnut (*Castanea mollissima*)	9	RVA	11%	mPa·s	5524–6505	3042–3616	2205–2954	4378–494	1326–1788	—	(Zhang et al., 2018)
Chickpea (*Cicer arietinum*)	1	RVA	7%	cP	3286	2367	919.5	5378	3011	73.1	(Zhang et al., 2019b)
Cowpea (*Vigna unguiculata*)	3	RVA	7%	cP	1750–2236	—	379–685	2624–3355	1373–1855	74.7–79.9	(Kim et al., 2018)
Djulis (*Chenopodium formosanum*)	3	RVA	8%	cP	4674.67(HP)– 5200.01(WP)– 4397.67(NP)	3974.33(HP)– 3683.67(WP)– 3248.00(NP)	772.67(HP)– 1516.33(WP)– 1149.67(NP)	5113.17(HP)– 5462.33(WP)– 5104.73(NP)	1139.00(HP)– 1778.67(WP)– 1856.33(NP)	71.70(HP) 72.80(WP) 69.53(NP)	(Lu et al., 2019)
Durian seed (*Durio zibethinus*)	2	RVA	11%	mPa·s	1300–1330	1040–1060	260–270	2020–2120	970–1060	~81	(Baraheng & Karrila, 2019)
Faba bean (*Vicia faba*)	1	RVA	7%	cP	3524	2277	1247	4814	2536	70.2	(Zhang et al., 2019b)
Foxtail millet (*Setaria italica*)	8	RVA	11%	cP	2780–3685	1380–1694	1400–2121	2838–3679	1265–2191	77.0–81.0	(Qi et al., 2020)
Ginger (*Rhizoma curcumae longae*)	1	BV	11%	RVU	136	114	22	248	134	—	(Li et al., 2020d)

(*Continued*)

TABLE 4.4 (CONTINUED) PASTING PROPERTIES OF STARCH FROM VARIOUS SOURCES

Source of starch	No.	Instrument	Conc. (w/w)	Unit	PV	TV	BD	FV	SB	PT	Reference
Ginkgo (Ginkgo biloba)	7	RVA	8%	cP	1356.0–1624.5	270.5–428.5	1064.5–1211.5	297.0–445.5	(−)1167.5–(−)1032.5	79.7–81.2	(Lu et al., 2020)
Jicama (Pachyrhizus erosus)	1	RH	14%	mPa·s	12447	–	2835	4497	–	64.74	(Contreras-Jiménez et al., 2019)
Lily rhizomes (Hedychium coronarium)	1	RVA	9%	cP	1917	1907	–	3987	2079	73.92	(Bento et al., 2019)
Lycoris (Lycoris spp.)	13	RVA	7%	cP	1000–1528	769–1077	197–587	1180–1575	(−)161–350	74.9–79.5	(Li et al., 2020)
Maca (Lepidium meyenii)	3	RH	9%	Pa·s	4.12–5.25	1.54–2.20	2.45–3.72	3.32–3.94	1.70–1.79	–	(Zhang et al., 2017)
Maize (waxy) (Zea mays)	8	RVA	7%	cP	1751–2119	1033–1143	697–1022	1092–1325	50–183	74.6–76.8	(Wang et al., 2019b)
Maize (waxy) (Zea mays)	1	BV	6%	BU	879	–	508	473	129	70.2	(Hsieh et al., 2019)
Wild mango seeds (Cordyla africana)	1	RVA	6%	cP	2811	1609	1202	3336	1727	80	(Ngobese et al., 2018)
Mung bean (Vigna radiata)	2	RVA	7%	cP	2059–2122	–	501–630	3033–3046	1580–1739	74	(Kim et al., 2018)
Mung bean (Vigna radiata)	1	RVA	11%	cP	7176	3391	3786	4432	1059	73.6	(Yang et al., 2019)
Oca (Oxalis tuberosa)	2	RH	9%	Pa·s	13.0–13.7	2.15–2.16	10.8–11.6	4.02–3.98	1.86–1.83	64.3	(Zhu & Cui, 2020)
Pea (Pisum sativum)	1	RVA	11%	cP	3221	2286	935	4990	2704	78.1	(Yang et al., 2019)

(Continued)

TABLE 4.4   (CONTINUED) PASTING PROPERTIES OF STARCH FROM VARIOUS SOURCES

Source of starch	No.	Instrument	Conc. (w/w)	Unit	PV	TV	BD	FV	SB	PT	Reference
Pitomba endocarp (*Talisia esculenta*)	1	RVA	9%	cP	2531	1056	1475	1607	551	77.5	(de Castro et al., 2019)
Potato (*Solanum tuberosum*)	3	RVA	11%	cP	7589–17439	4481–5579	2799–12957	4928–6375	487–911	67.9–72.4	(Wang et al., 2020a)
Potato (waxy) (*Solanum tuberosum*)	1	BV	6%	BU	1482	—	866	778	209	66.4	(Hsieh et al., 2019)
Proso millet (*Panicum miliaceum*)	95	RVA	6%	cP	2215–3787	1444–2177	511–1437	2577–3373	(–)1435–752	—	(Li et al., 2020c)
Quinoa (*Chenopodium quinoa*)	4	RVA	8%	cP	2983–3551	2250–3205	313–733	2692–3705	442–730	72.60–72.63	(Jiang et al., 2020)
Rice (*Oryza sativa*)	4	RVA	8%	RVU	190.2–226.8	—	8.4–121.3	114.3–224.8	16.8–42.8	67.2–75.7	(Tangsrianugul et al., 2019)
Rice (*Oryza sativa*)	2	RVA	11%	cP	2734–3657	763–972	1971–2685	1443–1447	471–684	73.45–78.10	(Chen et al., 2019)
Rice (waxy) (*Oryza sativa*)	6	RVA	9%	cP	1331–2393 (Fresh) 1805–2774 (Aged)	818–1575 (Fresh) 1036–1868 (Aged)	513–989 (Fresh) 769–1144 (Aged)	1118–1923 (Fresh) 1384–2354 (Aged)	300–349 (Fresh) 343–486 (Aged)	69.1–79.8 (Fresh) 68.3–78.8 (Aged)	(Wu et al., 2019)
Rice (waxy) (*Oryza sativa*)	1	BV	6%	BU	704	—	330	483	123	65.5	(Hsieh et al., 2019)
Sorghum (*Sorghum bicolor*)	1	RVA	11%	cP	4503	1433	3070	2195	762	75.5	(Yang et al., 2019)
Sweetpotato (*Ipomoea batatas*)	3	RVA	11%	cP	6099–6337	2263–2918	3193–4074	3096–4106	833–1490	78.8–79.8	(Wang et al., 2020a)

(Continued)

TABLE 4.4   (CONTINUED) PASTING PROPERTIES OF STARCH FROM VARIOUS SOURCES

Source of starch	No.	Instrument	Conc. (w/w)	Unit	PV	TV	BD	FV	SB	PT	Reference
Sweetpotato (*Ipomoea batatas*)	7	RVA	7%	RVU	115.5–193.7	71.9–101.5	33.0–104.0	102.1–146.7	(–)62.2–8.3	70.4–82.0	(Tong et al., 2020)
Sweetpotato (*Ipomoea batatas*)	1	RVA	7%	cP	3035.33	1786.66	1248.66	2960	1173.33	—	(Vithu et al., 2020)
Talipot palm (*Corypha umbraculifera*)	1	RVA	12%	cP	3646	1765	1881	1647	2562	86.6	(Navaf et al., 2020)
Tapioca (waxy) (*Manihot esculenta*)	1	BV	6%	BU	922	—	491	548	151	68.6	(Hsieh et al., 2019)
Tea seed (*Camellia sinensis*)	3	RVA	8%	cP	5203–5404	2256–3329	1981–3003	4095–6764	1839–3435	78.57–80.73	(Huang et al., 2019)
Wheat (waxy) (*Triticum aestivum*)	1	BV	6%	BU	783	—	452	436	121	66.8	(Hsieh et al., 2019)
Chinese yam (*Dioscorea opposita*)	2	RVA	11%	mPa·s	4271–4860	3686–4471	389–584	4886–5725	1254–1200	—	(Zou et al., 2020)

No.: number of samples studied; Conc.: concentration of starch (w/w); RH: rheometer equipped with a starch cell; RVA: Rapid Visco Analyser with RVU as the viscosity unit; BV: Brabender viscograph with BU as the viscosity unit; cP: centipoise; HP: starch isolated by hydrochloric acid procedure; WP: starch isolated by deionized water procedure; NP: starch isolated by sodium hydroxide procedure; Fresh: starch isolated from rice at the beginning of storage; Aged: starch isolated from rice after 12 months of storage; PT: pasting temperature; PV: peak viscosity; TV (HPV): trough viscosity (hot paste viscosity); FV (CPV): final viscosity (cold paste viscosity); BD: breakdown; SB: setback; *: scientific name not provided.

It represents the stability of starch paste under heating and shearing (Wang et al., 2020). The ability of starch paste to maintain its viscosity upon heating and shear stress is an important factor for starch processing. The FV represents the rearrangement of amylose molecules. It indicates the gel-forming capacity of starch after cooking. It has significant correlations with the gel texture properties and the sensory properties, which makes it an important parameter for food processing (Li et al., 2020a). Setback (SB) is the difference between TV and FV. It indicates the stability of the cold paste and the tendency of a gelatinized starch to retrograde to a gel-type structure (Wang et al., 2020b; Wang et al., 2015). High SB value usually means a high retrogradation tendency of amylose (Li et al., 2020a).

The pasting of starch is multifactor-controlled, which involves the composition (amylose content, lipid content, protein, and fiber content), molecular structure (amylopectin chain length distribution, amylopectin internal chain length distribution) and granular structure (e.g., particle size distribution and morphology) (Chen et al., 2019).

4.4.1.1.1 Effect of Amylose Content on Pasting Properties
Higher amylose contents are usually related to higher PT (Wang et al., 2020b). The high PV and SB were suggested to be due to the low amylose content (Li et al., 2020a). Indeed, the waxy starches have high peak viscosities followed by a more significant BD and a low SB during pasting. This is probably due to the lack of amylose molecules, which inhibits granule swelling and maintains the swollen starch granules (Hsieh et al., 2019; Li et al., 2020a). Similarly, a significant negative correlation was found between amylose content and BD in rice starch (Wang et al., 2020b). The high SB of normal starch compared with waxy starch is because the amylose molecules with linear structure can reassociate and form gel-type structures (Hsieh et al., 2019). The appearance of various starch pastes produced by pasting analysis was compared (Figure 4.3). Waxy maize, wheat, rice and potato starches showed smooth and cohesive textures, while the normal starches exhibited thick and discontinuous textures (Hsieh et al., 2019). However, other research suggested that the PV and the TV of starch from a high-amylose rice cultivar were the highest compared with other cultivars. Lower amylose content was related with higher BD (Park et al., 2020). The role of amylose in pasting behavior is complex and might depend on other factors, such as the source of starch or the structure of amylopectin.

4.4.1.1.2 Effect of Amylopectin Structure on Pasting Properties
Amylopectin fine structure has significant impact on the pasting properties of starch. Amylopectin chain fractions such as fa, fb1 and fb2 were found significantly correlated with pasting parameters such as PV, CPV and BD (Ahmed et al., 2019). The PT was positively correlated with longer B-chains, suggesting the

Figure 4.3 Texture and appearance of starch gels made from waxy and normal starches (6%, w/w). Reprinted with permission from Elsevier (Hsieh et al., 2019).

longer internal chains may provide interactions to inhibit the swelling (Qi et al., 2020). Similarly, in starches from rice cultivars with similar amylose content, low PV and TV were observed in those with a high amount of long amylopectin chains (Park et al., 2020). The PV was also negatively correlated with the amylopectin chains of DP between 13 and 24 (Lu et al., 2020). The BD was positively correlated with long chains, while the SB had negative correlations with them. These results suggested that the long-chain fractions play important roles in the pasting properties of starches (Lu et al., 2020; Zhu & Xie, 2018).

The internal structure of amylopectin also has significant impact on the pasting properties. The internal chain categories were significantly correlated with the pasting parameters such as PV, SB, TV and FV (Kong et al., 2008; Zhu et al., 2011). More short internal chains tend to cause more unparalleled packing of double helices, which results in lower structural integrity of swelled granules (Zhu, 2018).

4.4.1.1.3 Other Factors Affecting the Pasting Properties

Besides the amylose content and amylopectin structure, other factors such as size distribution, minor components, isolation method and starch source can affect the pasting properties. The PT was positively correlated to the average granule size, indicating that large starch granules absorbed water slower than small starch granules (Qi et al., 2020). The growth condition affected the pasting properties of cassava starch (Karlström et al., 2019). The proteins and lipids could significantly restrict the swelling of starches, while removing them led to higher extents and rates of swelling (Debet & Gidley, 2006). The removal of granule surface proteins had a significant impact on pasting properties such as the PV and FV of rice starch (Zhan et al., 2020). The removal of lipids during isolation could markedly alter the pasting and gelling properties of starch (Solaesa et al., 2019). The pasting properties, such as the SB and FV of starch, obtained using an alkaline-based isolation method are significantly higher compared to those isolated by water and enzyme methods (Noora et al., 2019). In other studies, compared with the acid isolation and water isolation methods, the alkali isolation method gave the lowest PV of the resulting starch. The possible explanation was that the alkali might irreversibly damage the starch structure and reduce the size of starch molecules (Lu et al., 2019).

4.4.1.2 Texture Analysis

The gel texture is an important property affecting the quality of starch-based products. Generally, the texture is based on judgment from sensory evaluation. However, methods such as TPA provide a fast and cost-effective way of evaluating the textural characteristics (Kasapis & Bannikova, 2017). TPA analysis involves a two-step compression carried out successively. Five main characteristics of a gel that could be estimated are hardness, cohesiveness, adhesiveness, elasticity (also called springiness) and fracturability. In addition, several texture parameters, such as gumminess (Hardness × Cohesiveness) and chewiness (Hardness × Cohesiveness × Springiness), can be calculated (Muthukumarappan & Swamy, 2017; Wang et al., 2015). The experimental settings and results of texture properties of starch are summarized in Table 4.5. Hardness is the peak force that occurs during the first compression. In most cases, hardness is expressed in units of force (pound, gram or newton). The area of the negative peak following the first compression is termed adhesiveness. The length and area of the second peak divided by those of the first peak are termed springiness and cohesiveness, respectively (Kasapis & Bannikova, 2017). Adhesiveness and chewiness indicate the energy required for disintegration and mastication of semisolid foods, respectively (Qian et al., 2019). Springiness is an indication of elasticity in food products and represents the recovery rate of the samples from the two compressions. Cohesiveness indicates the internal binding capacity of the gel. Both springiness and cohesiveness are related to the network structure formed by the starch gel (Qian et al., 2019; Noora et al., 2019). The texture properties of the gels depend on many factors, including the starch constituents, amylose content, volume and the deformation of the granules, the interactions between continuous and dispersed phases, water content, and storage conditions (Macena et al., 2020).

TABLE 4.5 GEL TEXTURE OF STARCH FROM VARIOUS SOURCES

Starch source	No.	Conc. (w/w)	Storage condition	HD (g)	AD (g·s)	COH	Gum	SPR	Reference
Amaranth (*Amaranthus cruentus*)	2	11%	Refrigeration temperature, 24 h	0.90–1.10	1.70–1.90	0.57	—	0.94	(Sindhu & Khatkar, 2018)
Avocado seed (*Persea americana*)	1	4%		195	—	0.32	63.11	—	(Macena et al., 2020)
Buckwheat (*Fagopyrum esculentum*)	2	11%	Refrigeration temperature, 24 h	9.26–9.63	1.70–2.01	0.41–0.49	—	1.35–1.59	(Sindhu & Khatkar, 2018)
Cowpea (*Vigna unguiculata*)	3	7%	4°C, 24 h	768.5–1008.2	4.35–16.35	—	0.76–0.87	0.96–0.98	(Kim et al., 2018)
Durian seed (*Durio zibethinus*)	2	12%	RT, 3 h	3.4–3.54	—	—	—	—	(Baraheng & Karrila, 2019)
Foxtail millet (*Setaria italica*)	8	11%	4°C, 7 day	156.5–570.2	—	—	—	—	(Qi et al., 2020)
Lycoris (*Lycoris* spp.)	13	6%	4°C, 24 h	8.1–23.8	32.7–17.3	0.460–0.540	—	—	(Li et al., 2020b)
Maca (*Lepidium meyenii*)	3	9%	4°C, 24 h	21.1–34.3	130–155	0.57–0.62	—	—	(Zhang et al., 2017)
Wild mango seeds (*Cordyla africana*)	1	6%	25°C, 2 h	114.7^	—	—	—	—	(Ngobese et al., 2018)
Mung bean (*Vigna radiata*)	2	7%	4°C, 24 h	590.9–784.4	4.00–4.84	—	0.86–0.87	0.98	(Kim et al., 2018)
Oca (*Oxalis tuberosa*)	2	9%	4°C, 24 h	53–48	286–297	0.52–0.51	24–27	—	(Zhu & Cui, 2020)
Potato (*Solanum tuberosum*)	3	8%	4°C, 12 h	247.85–380.29	141.5–323.52	0.19–0.25	62.93–89.67	0.96–1.00	(Wang et al., 2020a)
Sweetpotato (*Ipomoea batatas*)	7	7%	4°C, 24 h	21.1	21.1	0.5	10.6	—	(Tong et al., 2020)
Sweetpotato (*Ipomoea batatas*)	3	8%	4°C, 12 h	123.73–329.16	135.92–535.75	0.30–0.41	37.99–136.43	0.95–1.00	(Wang et al., 2020a)
Potato (*Solanum tuberosum*)	14	6%	4°C, 24 h	21.3–30.6(E1) 20.8–30.0(E2)	—	—	—	—	(Ahmed et al., 2019)
Proso millet (*Panicum miliaceum*)	95	6%	4°C, 24 h	1.7–61.8	94.4–0.8	0.122–0.856	—	—	(Li et al., 2020)
Queen sago (*Cycas circinalis*)	3	11%	20°C, 24 h	4.02(WS)^ 11.46(AS)^ 13.05(ES)^	—	0.34(WS) 0.45(AS) 0.49(ES)	—	0.92(WS) 0.84(AS) 0.96(ES)	(Noora et al., 2019)

No.: number of samples studied; RT: room temperature; Conc.: concentration of starch (w/w); HD: hardness; AD: adhesiveness (the values are expressed as positives); COH: cohesiveness; GUM: gumminess; SPR: springiness; E1: starch isolated from potato grown at Yiwu, China; E2: starch isolated from potato grown at Jinhua, China; WS: starch isolated by water steeping method; AS: starch isolated by alkali steeping method; ES: starch isolated by enzymatic extraction method; ^: the unit for hardness is Newton.

4.4.1.2.1 Effect of Amylose Content on Texture Properties

The amylose content is an important factor in gel texture properties. It is positively correlated with hardness, which is the result of promoted short-term retrogradation by forming a rigid gel network (Qi et al., 2020; Li et al., 2020a). It also acts as a binding material of aggregates like fragmented swollen granules (Wang et al., 2020a). The network formed by amylose provides the gels with high strength and elasticity against deformation, while the gels formed by amylopectin aggregates without amylose tend to have higher adhesiveness (Figure 4.3). The low amylose content also decreased long-range intermolecular interactions, which decreased the cohesiveness (Wang et al., 2015). Apart from the content, the amylose with longer chain length could enhance the hardness of sweet potato starch gels (Tong et al., 2020).

4.4.1.2.2 Effect of Amylopectin Structure on Texture Properties

The high amount of long unit chains was linked to the high value of hardness in different studies (Zhu & Cui, 2020; Wang et al., 2020c; Li and Zhu, 2017b). These studies suggested that the long chains of amylopectin also provide inter-/intramolecular interactions in a similar way to amylose, thus enhancing the rigidity of the starchy foods (Lan et al., 2017). However, other reports also suggested that the long chains in amylopectin had a negative relationship with hardness (Wu et al., 2019). The variance could be due to the different starch samples studied. Besides the effect on hardness, Li et al. (2017) suggested that the stickiness of rice was negatively correlated with the amount of short chains in amylopectin while being positively correlated to the molecular size of amylopectin.

4.4.1.2.3 Other Factors Affecting the Texture Properties

Apart from the amylose and amylopectin, other components have significant effects on the gel texture properties. The nonstarch components like proteins and lipids could have significant effects on the formation and properties of starch gels (Noora et al., 2019). The phosphate groups could reduce the interactions and realignment of starch molecules, thus retarding the gel formation. The lipids could form complexes with amylose and enhance the starch retrogradation (Zhu & Cui, 2020).

The starch contents and storage conditions play an essential role in gel formation and gel properties. A schematic diagram of the gel formation under diluted and concentrated starch solution is presented (Figure 4.4). Changing starch content could result in significant variations of gel properties (Fonseca-Florido et al., 2017). The gel hardness increased with increasing storage time. The increase was more obvious in the first few days than in the following days (Lan et al., 2017). The high ion strength could affect the availability of water molecules, and the extreme pH could reduce the molecular size of starch, inhibiting starch retrogradation and reducing gel hardness (Macena et al., 2020). The

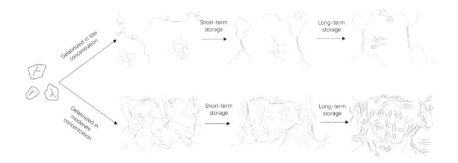

Figure 4.4   Schematic diagram showing the effect of starch concentration and storage period on gelatinized starch solution.

source and isolation method of starch also matter. The starches isolated from main rice stem and tiller panicles differed significantly in gel properties (Wang et al., 2020c). The starch isolated with the enzyme method showed significantly higher hardness than those isolated by the alkali method and water method (Noora et al., 2019).

4.4.1.3 Steady-Flow Analysis

Steady-flow analysis is a widely applied large deformation test in which the viscosity of the sample under different shear rates is obtained. Starch gels from different sources showed shear-thinning behaviors in which the apparent viscosity of all samples decreased with the increasing shear rate (Li and Zhu, 2018; Wang et al., 2020b; Punia et al., 2020). These properties could be explained by the conformational changes of starch chains during shearing. As the shear rate increased, the entangled starch molecules started to unravel and align with each other. The viscosity of the paste thus decreased (Wang et al., 2020b, Punia et al., 2020). However, it should be pointed out that several starch pastes could exhibit temporary shear-thickening behaviors under certain conditions (Fang et al., 2019).

This kind of test is usually done on a rotary rheometer. The flow behaviors of starch could be fitted using mathematic models. It has been suggested that four mathematic models – the Herschel–Bulkley, Heinz–Casson, Mizrahi–Berk and Robertson–Stiff models – could effectively describe the flow properties of quinoa starch gels (Li and Zhu, 2018). The Herschel–Bulkley model (Equation 4.1) is the most widely applied model:

$$\tau = \sigma_0 + K \cdot \gamma^n \qquad (4.1)$$

where τ is the shear stress, K refers to the consistency coefficient, γ represents shear rate, n stands for flow behavior index and σ_0 is the yield stress.

The *n* indicates the extent of shear-thinning behavior from Newtonian fluids. The consistency index (*K*) reflects the values of the viscosity at the shear rate of 1 s^{-1}. The yield stress represents the minimum force needed to drive certain pastes to flow (Li & Zhu, 2018; Zhu & Cui, 2020). The higher yield stress and consistency index of oca (*Oxalis tuberosa*) starch than maize starch was attributed to the longer average chain length of oca amylopectin. The long chains could induce more entanglements and thus enhance the gel network (Zhu & Cui, 2020). The granule size was also positively related to the yield stress (Li & Zhu, 2018).

The degradation of gel structure under shear could also be estimated by the area of the hysteresis loop (measured by the area contained between the upward and downward curves of shear stress versus shear rate data). A higher area of the hysteresis loop suggested a larger extent of structural degradation of a starch gel. The hysteresis loop areas of gelatinized waxy starches followed the order of waxy potato > waxy maize > waxy rice (Fang et al., 2019).

4.4.2 Small Deformation Analysis

The rheological analysis within the linear viscoelastic range is termed small deformation analysis. This kind of test is useful in the characterization of the viscoelastic properties of materials as a function of time, temperature, strain or frequency (Kasapis & Bannikova, 2017). It allows us to estimate the structure without causing irreversible changes. The parameters obtained from small amplitude oscillatory analysis include storage modulus (*G′*), loss modulus (*G″*) and loss tangent (tan δ, *G″/G′*). *G′* refers to the energy stored in a substance and represents the elastic nature of the sample. *G″*, by contrast, represents the energy lost during the deformation of each cycle and represents the viscous nature of the sample. Tan δ is the ratio of *G″* and *G′*, and represents the relative contribution of elasticity and viscosity to the viscoelastic properties of the samples (Wang et al., 2020b).

4.4.2.1 Amplitude Sweep
Before carrying out the small deformation analysis, the linear viscoelastic range of a particular sample has to be determined. An amplitude sweep (strain or stress) is performed with constant frequency and temperature; the moduli are measured as a function of increasing strain (Sanz et al., 2017; Muthukumarappan & Swamy, 2017). The objective of an amplitude sweep test is to determine the critical point of strain beyond which the moduli become affected by the strain. For four types of maize starch samples under the concentrations of 6%–10%, the strains of 5%–10% were the boundaries beyond which the *G′* and *G″* started to be affected by the amplitude (Rosalina & Bhattacharya, 2002).

4.4.2.2 Temperature Sweep

Temperature sweep studies the values of G' and G'' as a function of temperature at fixed frequency and strain (or stress) (Muthukumarappan & Swamy, 2017). This test is suitable to estimate the gelatinization properties during heating and the gel formation during cooling. The temperature sweep results of kiwifruit starch were significantly different from those of maize starch, while sharing several similarities with those of potato starch. The viscoelastic properties of kiwifruit starch upon heating were more similar to that of potato starch (Li & Zhu, 2017a). The temperature sweep is also useful to investigate the phase transition during gelatinization and retrogradation (Muthukumarappan & Swamy, 2017). One of the methods to determine the transition point (or gel point) from liquid (sol) to solid (gel) is based on the crossover of G' and G'' where tan δ equals 1 (Dickinson, 2011). When tan δ is below 1 ($G' > G''$), the system could be recognized as a solid or gel state, and the system could be characterized as a liquid state if this value is above 1 ($G' < G''$). For example, the temperature sweep was applied to study the elastic behaviors of retrograded waxy potato starch in comparison with those of waxy maize and waxy rice starch upon heating. The tan δ of retrograded waxy potato starch remained stable but increased significantly when the temperature was above 55°C. The results suggested that retrograded waxy potato starch transits to a more liquid-like behaviors above this temperature, probably due to the melting of junction zones (Fang et al., 2019). It has been reported that the quinoa starch encountered a four-stage process including two phase transitions (sol to gel, and gel to sol). The sol/gel transition temperature was similar to T_o estimated by DSC, while the gel/sol transition temperature was lower than the T_p obtained by DSC (Li & Zhu, 2018). The amylopectin unit chain distribution had significant relationships with the sol/gel transition temperatures (Figure 4.1e).

4.4.2.3 Frequency Sweep

In the frequency sweep test, strain (or stress) with fixed amplitude is applied to the material, and the dynamic moduli are determined over a range of frequencies. Frequency sweep estimates the viscous and elastic properties of the material at the rate of application of strain, while the amplitude remains constant (Muthukumarappan & Swamy, 2017; Kasapis & Bannikova, 2017). According to the relative value and frequency dependency of G' and G'', a specific sample could be classified into the following four categories: strong gel (G' is significantly higher than G'', and both moduli are independent of frequency), weak gel (G' is slightly higher than G'', and both moduli are slightly dependent on frequency), entangled polymer solution (G' is lower than G'' at low frequencies, but both moduli increase with increasing frequency; there is a crossover point such that, at higher frequencies, G' is higher than G'') and nonentangled polymer solution (G' is much lower than G'' at all frequencies, and both moduli are strongly dependent on frequency) (Kaneda, 2016). The starch-based gel could be mostly classified into a strong or weak gel, depending on the gel strength. To better estimate the relationship between two dynamic moduli and frequency, power-law type functions are usually used to fit the data obtained (Equations 4.2 and 4.3) (Li & Zhu, 2018; Ahmed, 2016; Zhu et al., 2016):

$$G' = K_{G'} \times \omega^{n_{G'}} \tag{4.2}$$

$$G'' = K_{G''} \times \omega^{n_{G''}} \tag{4.3}$$

The ω refers to angular frequency. The n is the frequency exponent, which is in the range between 0 and 1. A lower n represents less dependency of the two moduli on the frequency. The quinoa starch gel was studied by frequency sweep at 25°C from 0.1 to 40 Hz. The values of n were between 0.0123 to 0.0519, which suggested a gel structure having low frequency dependency (Li & Zhu, 2018). The significantly higher G' value than G'' was the result of high degrees of cross-linking and entanglements. The G' was higher than G'' of potato, quinoa and oca gels during the frequency range between 0.1 and 10 Hz, suggesting their solid-like behaviors (Zhu & Cui, 2020; Li & Zhu, 2018; Hu et al., 2020). The results of the frequency sweep are multifactor controlled. The amylose content, amylopectin structure, origin of starch, concentration of gel and particle size distribution can impact the results (Hu et al., 2020; Li and Zhu, 2018; Ahmed et al., 2018; Zhu & Cui, 2020). For example, the amounts and extents of entanglements in quinoa starch gel were positively correlated to the linear fraction in quinoa starches including amylose and super-long chains of amylopectin (Li & Zhu, 2018). The loss tangent was significantly affected by the granule size of potato starches (Hu et al., 2020).

4.5 EFFECT OF STARCH GELATINIZATION AND RETROGRADATION ON STARCH DIGESTION

By applying enzymatic hydrolysis, the starch in food can be classified into three fractions, namely, rapidly digestible starch (RDS), slowly digestible starch (SDS) and resistant starch (RS) (Englyst et al., 1992). Practically, the RDS and SDS are the fractions that can be completely digested within 20 and 120 min, respectively. The remaining fraction after the digestion is termed RS (Englyst et al., 1992).

The RS has diverse health beneficial effects, and it is further classified into four categories: RSI, physically inaccessible starch; RSII, intact granular starches that have slow digestion rates; RSIII, starch that has been processed and partially recrystallized; and RSIV, chemically modified starch (Bird et al., 2009).

4.5.1 RS in Native and Cooked Starches

The starches isolated from plants with various genetic backgrounds have significant differences in their RS content. The predicted glycemic index (GI) of Chinese yam starch isolated during the dormant stage was significantly lower than that isolated from the expansion stage, indicating a higher amount of RS in the former (Zou et al., 2020). The starch isolated from the kernel of unripe mango showed high RS content in native (73.73%) and cooked (28.75%) samples (Patiño-Rodríguez et al., 2020). It has been suggested that starch with high amylose content tends to have several beneficial health effects. Indeed, the high amylose content of starch isolated from *Curcuma karnatakensis* (84.6%–86.6%) related to the low digestion rate of the starch (Tejavathi et al.,2020). Thus, it is possible to use starch from different sources to achieve a slow digestion rate and a high RS content. The enzymatic hydrolysis rate of native starch with the same genetic background was related to the granule size distribution, surface pores and channels, and polymorph type. Rice starch with smaller particle size (Figure 4.5A) showed a significantly higher enzymatic digestion rate compared to the large ones (Figure 4.5B) (Dhital et al., 2015a).

Most starch consumed by humans has been processed or cooked. The susceptibility of cooked starch to human digestion is mainly affected by the extent of disruption of starch structure during gelatinization (Wang et al., 2015). To reduce the amount of RDS and increase the RS content, it is necessary to

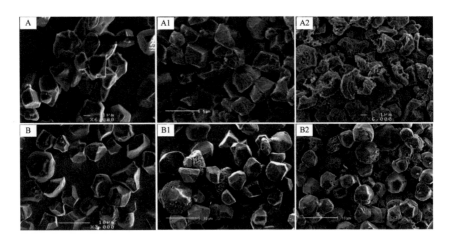

Figure 4.5 SEM images of rice starch granules with different sizes (A and B) before and (A1, B1, A2, B2) after enzyme hydrolysis. (A) Small granule size fraction; (B) large granule size fraction. The number after A and B represents the hydrolysis period of (A1 and B1) 30 min and (A2 and B2) 3 h. Reprinted with permission from Elsevier (Dhital et al., 2015a).

maintain the extent of starch gelatinization at a relatively low level. The partial gelatinization of starch could reduce the accessible fraction of glucans and result in a low digestion rate (Li et al., 2019).

4.5.2 RS in Retrograded Starches

Aggregations and gel networks could prevent enzymes from attacking retrograded starch (Wang et al., 2015). Thus, factors that enhance the starch retrogradation could also reduce the digestion rate of starch. It has been suggested that a high amount of short to medium chains in amylose as well as medium chains in amylopectin could decrease the digestibility of cooked rice starch after short-term retrogradation (Li & Hu, 2020). The possible mechanism is the short and medium chains in amylose might form densely packed cells in the gel network. These cells might inhibit the enzymatic digestion (Gong et al., 2019). The schematic diagram of this process is shown in Figure 4.6. As the storage time increased, the digestibility gradually decreased with retrogradation of amylopectin (Wang et al., 2015). An increase in the length/amount of long internal chains of amylopectin could lead to a decrease in the hydrolysis rate of retrograded starches due to the formation of more ordered structures (Zhu & Liu, 2020). The amylose content or the amylopectin structure are not the only factors affecting the digestibility of cooked starch-based food. Other factors

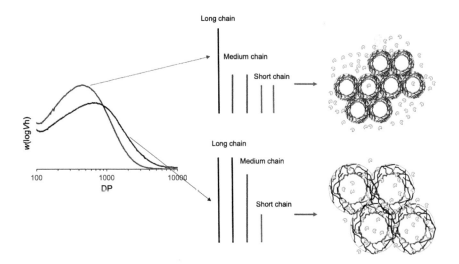

Figure 4.6   Schematic diagram of the relationship between amylose structure and the network formed by retrograded starch. The half moon symbol represents the enzyme. Reprinted with permission from Elsevier (Gong et al., 2019).

such as fibers and proteins can also affect the rate of digestion (Dhital et al., 2015b; Dhital et al., 2017).

4.5.3 SDS of Starches

SDS represents the fraction that is digested in the small intestine and gives a slow and prolonged rise in blood glucose level. In most cases, the SDS coexists with RS. The increase in the proportions of RS and SDS reduces the amount of RDS. RDS is the primary source of high postprandial glycemic response (Punia, 2020; Zhang & Hamaker, 2009). The foods that contain a high amount of SDS are likely to be beneficial in reducing metabolic syndrome-related diseases such as diabetes and heart disease (Zhang & Hamaker, 2009). Starch processing could alter the content of SDS. This SDS value increased significantly after autoclave processing and low-temperature storage in ginger starch, rising from 15.35% to 50.45% (Li et al., 2020d). Similar trends have been reported in mung bean starch in which the SDS content raised beyond 50% after cooking (Yao et al., 2019). The results suggested great potential for healthy food formulations by using these starches as ingredients. The molecular structure of starch could also affect the amount of SDS. The amounts of SDS ranged from 35.29% to 65.84% in four rice cultivars with different amylose contents. The cultivars with high amylose content showed a reduced amount of SDS and an increased ratio of SDS to RS (Park et al., 2020). Previous investigation using maize mutants indicated that the amylopectin with a high amount of short chains or long chains tended to have high values of SDS. The high amount of short chains with more frequently branched structures could retard the enzymatic digestion, while the long chains of amylopectin were associated with the imperfect crystallites formed by fast retrogradation after cooking (Zhang & Hamaker, 2009). The defatted quinoa sample showed a lower SDS value compared to native quinoa starch, which suggested that the minor components also affected the composition of SDS (Solaesa et al., 2019).

4.6 CONCLUSIONS

Starch is an important nutritional source for humans. The application of starch-based food is closely related to the functional properties of starch, such as thermal, swelling and rheological properties. The gelatinization and retrogradation, as the most important physicochemical aspects of starch, have been studied using different approaches. The functional properties, such as thermal properties, are closely related to the structural factors, such as unit chain length distribution. However, it is still hard to present a complete picture of the structure–function relationships. Most of the functional properties are multifactor controlled, and studies with various experimental conditions might lead to different results. Nevertheless, understanding the functional properties and the relationship to other factors could provide a theoretical basis for the design, processing and application of starch-based foods.

ABBREVIATIONS

A_{fp}-chains:	fingerprint A-chains (DP 6–8)
AML:	amylose leaching
BD:	breakdown viscosity
DP:	degree of polymerization
FV(CPV):	final viscosity (cold paste viscosity)
G':	storage modulus
G'':	loss modulus
GI:	glycemic index
PKT:	peak temperature of pasting
PT:	pasting temperature
PV:	peak viscosity
RDS:	rapidly digestible starch
RS:	resistant starch
RVA:	Rapid Visco Analyser
SB:	setback viscosity
SDS:	slowly digestible starch
SP:	swelling power
tan δ:	loss tangent
$T_{c(r)}$:	conclusion temperature of gelatinization (retrogradation)
$T_{o(r)}$:	onset temperature of gelatinization (retrogradation)
$T_{p(r)}$:	peak temperature of gelatinization (retrogradation)
TPA:	texture profile analysis
TV(HPV):	trough viscosity (hot paste viscosity)
WSI:	water solubility index
$\Delta H_{(r)}$:	enthalpy change of gelatinization (retrogradation)
$\Delta T_{(r)}$:	temperature range of gelatinization (retrogradation)

REFERENCES

Ahmed, J. (2016). *Advances in Food Rheology and Its Applications* (J. Ahmed, Ed.). Woodhead Publishing.

Ahmed, J., Thomas, L., Arfat, Y. A., & Joseph, A. (2018). Rheological, structural and functional properties of high-pressure treated quinoa starch in dispersions. *Carbohydrate Polymers*, 197, 649–657.

Ahmed, S., Ru, W., Han, H., et al. (2019). Fine molecular structure and its effects on physicochemical properties of starches in potatoes grown in two locations. *Food Hydrocolloids*, 97, 105172.

Balet, S., Guelpa, A., Fox, G., & Manley, M. (2019). Rapid Visco Analyser (RVA) as a tool for measuring starch-related physiochemical properties in cereals: A review. *Food Analytical Methods*, 12, 2344–2360.

Balet, S., Gous, P., Fox, G., Lloyd, J., & Manley, M. (2020). Characterisation of starch quality from barley varieties grown in South Africa. *International Journal of Food Science and Technology*, 55, 443–452.

Baraheng, S., & Karrila, T. (2019). Chemical and functional properties of durian (*Durio zibethinus* Murr.) seed flour and starch. *Food Bioscience*, 30, 100412.

Bento, J. A. C., Ferreira, K. C., de Oliveira, A. L. M., et al. (2019). Extraction, characterization and technological properties of white garland-lily starch. *International Journal of Biological Macromolecules*, 135, 422–428.

Bento, J. A. C., Fidelis, M. C., de Souza Neto, M. A., et al. (2020). Physicochemical, structural, and thermal properties of 'batata-de-teiu' starch. *International Journal of Biological Macromolecules*, 145, 332–340.

Bertoft, E. (2004). On the nature of categories of chains in amylopectin and their connection to the super helix model. *Carbohydrate Polymers*, 57, 211–224.

Bird, A. R., Lopez-Rubio, A., Shrestha, A. K., & Gidley, M. J. (2009). CHAPTER 14 - Resistant starch in vitro and in vivo: Factors determining yield, structure, and physiological relevance. In S. Kasapis, I. T. Norton, & J. B. Ubbink (Eds.), *Modern Biopolymer Science* (pp. 449–510). Academic Press.

Blazek, J., & Gilbert, E. P. (2011). Application of small-angle X-ray and neutron scattering techniques to the characterisation of starch structure: A review. *Carbohydrate Polymers*, 85, 281–293.

Cameron, R. E., & Donald, A. M. (1992). A small-angle X-ray scattering study of the annealing and gelatinization of starch. *Polymer*, 33, 2628–2635.

Chen, X., Chen, M., Lin, G., et al. (2019). Structural development and physicochemical properties of starch in caryopsis of super rice with different types of panicle. *BMC Plant Biology*, 19, 482.

Chen, X., Shao, S., Chen, M., et al. (2020). Morphology and physicochemical properties of starch from waxy and non-waxy barley. *Starch/Stärke*, 72, 1900206.

Choi, I., Chun, J., Choi, H.-S., et al. (2020). Starch characteristics, sugars and thermal properties of processing potato (*Solanum tuberosum* L.) cultivars developed in Korea. *American Journal of Potato Research*, 97, 308–317.

Contreras-Jiménez, B., Vázquez-Contreras, G., de los Ángeles Cornejo-Villegas, M., del Real-López, A., & Rodríguez-García, M. E. (2019). Structural, morphological, chemical, vibrational, pasting, rheological, and thermal characterization of isolated jicama (*Pachyrhizus* spp.) starch and jicama starch added with Ca(OH)$_2$. *Food Chemistry*, 283, 83–91.

Cooke, D., & Gidley, M. J. (1992). Loss of crystalline and molecular order during starch gelatinization: Origin of the enthalpic transition. *Carbohydrate Research*, 227, 103–112.

Debet, Martine R., and Michael J. Gidley. (2006). Three classes of starch granule swelling: Influence of surface proteins and lipids. *Carbohydrate Polymers*, 64, 452–465.

de Castro, D. S., Moreira, I. D., Silva, L. M. D., et al. (2019). Isolation and characterization of starch from pitomba endocarp. *Food Research International*, 124, 181–187.

Desam, G. P., Li, J., Chen, G., Campanella, O., & Narsimhan, G. (2020). Swelling kinetics of rice and potato starch suspensions. *Journal of Food Process Engineering*, 43, e13353.

Dhital, S., Shrestha, A. K., Hasjim, J., & Gidley, M. J. (2011). Physicochemical and structural properties of maize and potato starches as a function of granule size. *Journal of Agricultural and Food Chemistry*, 59, 10151–10161.

Dhital, S., Butardo, V. M., Jr., Jobling, S. A., & Gidley, M. J. (2015a). Rice starch granule amylolysis Differentiating effects of particle size, morphology, thermal properties and crystalline polymorph. *Carbohydrate Polymers*, 115, 305–316.

Dhital, S., Dabit, L., Zhang, B., Flanagan, B., & Shrestha, A. K. (2015b). *In vitro* digestibility and physicochemical properties of milled rice. *Food Chemistry*, 172, 757–765.

Dhital, S., Warren, F. J., Butterworth, P. J., Ellis, P. R., & Gidley, M. J. (2017). Mechanisms of starch digestion by α-amylase: Structural basis for kinetic properties. *Critical Reviews in Food Science and Nutrition*, 57, 875–892.

Dickinson, E. (2011). Food colloids research: Historical perspective and outlook. *Advances in Colloid and Interface Science*, 165, 7–13.

Englyst, H. N., Kingman, S., & Cummings, J. (1992). Classification and measurement of nutritionally important starch fractions. *European Journal of Clinical Nutrition*, 46(Supplement 2), S33–50.

Fang, F., Tuncil, Y. E., Luo, X., et al. (2019). Shear-thickening behavior of gelatinized waxy starch dispersions promoted by the starch molecular characteristics. *International Journal of Biological Macromolecules*, 121, 120–126.

Felisberto, M. H. F., Beraldo, A. L., Costa, M. S., et al. (2019). Characterization of young bamboo culm starch from *Dendrocalamus asper*. *Food Research International*, 124, 222–229.

Felisberto, M. H. F., Beraldo, A. L., Costa, M. S., et al. (2020). Bambusa vulgaris starch: Characterization and technological properties. *Food Research International*, 132, 109102.

Fonseca-Florido, H. A., Gómez-Aldapa, C. A., Velazquez, G., et al. (2017) Gelling of amaranth and achira starch blends in excess and limited water. *LWT-Food Science and Technology*, 81, 265–273.

Fu, Z., Chen, J., Luo, S. J., Liu, C., & Liu, W. (2015). Effect of food additives on starch retrogradation: A review. *Starch/Stärke*, 67, 69–78.

Gao, L., Wang, H., Wan, C., et al. (2020). Structural, pasting and thermal properties of common buckwheat (*Fagopyrum esculentum* Moench) starches affected by molecular structure. *International Journal of Biological Macromolecules*, 156, 120–126.

Gomand, S. V., Lamberts, L., Visser, R. G. F., & Delcour, J. A. (2010). Physicochemical properties of potato and cassava starches and their mutants in relation to their structural properties. *Food Hydrocolloids*, 24, 424–433.

Gong, B., Cheng, L., Gilbert, R. G., & Li, C. (2019). Distribution of short to medium amylose chains are major controllers of *in vitro* digestion of retrograded rice starch. *Food Hydrocolloids*, 96, 634–643.

Gourilekshmi, S. S., Jyothi, A. N., & Sreekumar, J. (2020). Physicochemical and structural properties of starch from cassava roots differing in growing duration and ploidy level. *Starch/Stärke*, 72, 1900237.

Gu, F., Gong, B., Gilbert, R. G., et al. (2019). Relations between changes in starch molecular fine structure and in thermal properties during rice grain storage. *Food Chemistry*, 295, 484–492.

Hermansson, A.-M., & Svegmark, K. (1996). Developments in the understanding of starch functionality. *Trends in Food Science & Technology*, 7, 345–353.

Hoover, R. (2001). Composition, molecular structure, and physicochemical properties of tuber and root starches: A review. *Carbohydrate Polymers*, 45, 253–267.

Hoover, R., & Manuel, H. (1996). The effect of heat-moisture treatment on the structure and physicochemical properties of normal maize, waxy maize, dull waxy maize and amylomaize V starches. *Journal of Cereal Science*, 23, 153–162.

Hsieh, C., Liu, W., Whaley, J., & Shi, Y. (2019). Structure and functional properties of waxy starches. *Food Hydrocolloids*, 94, 238–254.

Hu, C., Xiong, Z., Xiong, H., et al. (2020). Effects of granule size on physicochemical and digestive properties of potato powder. *Journal of the Science of Food and Agriculture*, 100, 4005–4011.

Hu, Z., Zhao, L., Hu, Z., & Wang, K. (2018). Hierarchical structure, gelatinization, and digestion characteristics of starch from longan (*Dimocarpus longan* Lour.) seeds. *Molecules*, 23, 3262.

Huang, J., Yuan, M., Kong, X., et al. (2019). A novel starch: Characterizations of starches separated from tea (*Camellia sinensis* (L.) O. Ktze) seed. *International Journal of Biological Macromolecules*, 139, 1085–1091.

Jiang, F., Du, C., Guo, Y., et al. (2020). Physicochemical and structural properties of starches isolated from quinoa varieties. *Food Hydrocolloids*, 101, 105515.

Kaneda, I. (2016). *Rheology of Biological Soft Matter: Fundamentals and Applications* (I. Kaneda, Eds.). Springer.

Karim, A. A., Norziah, M. H., & Seow, C. C. (2000). Methods for the study of starch retrogradation. *Food Chemistry*, 71, 9–36.

Karlström, A., Belalcazar, J., Sánchez, T., et al. (2019). Impact of environment and Genotype-by-Environment interaction on functional properties of amylose-free and wildtype cassava starches. *Starch/Stärke*, 71, 1700278.

Kasapis, S., & Bannikova, A. (2017). Chapter 2: Rheology and food microstructure. In P. Ptaszek & S. Basu (Eds.), *Advances in Food Rheology and Its Applications* (pp. 7–46). Woodhead Publishing.

Kim, Y. Y., Woo, K. S., & Chung, H. J. (2018). Starch characteristics of cowpea and mungbean cultivars grown in Korea. *Food Chemistry*, 263, 104–111.

Koehler, P., & Wieser, H. (2013). Chemistry of cereal grains. In M. Gobbetti & M. Gänzle (Eds.), *Handbook on Sourdough Biotechnology* (pp. 11–45). Springer.

Kong, X., Bertoft, E., Bao, J., & Corke, H. (2008). Molecular structure of amylopectin from amaranth starch and its effect on physicochemical properties. *International Journal of Biological Macromolecules*, 43, 377–382.

Kong, X., Zhu, P., Sui, Z., & Bao, J. (2015). Physicochemical properties of starches from diverse rice cultivars varying in apparent amylose content and gelatinisation temperature combinations. *Food Chemistry*, 172, 433–440.

Kuang, Q., Xu, J., Liang, Y., et al. (2017). Lamellar structure change of waxy corn starch during gelatinization by time-resolved synchrotron SAXS. *Food Hydrocolloids*, 62, 43–48.

Laity, P. R., & Cameron, R. E. (2008). A small-angle X-ray scattering study of powder compaction. *Powder Technology*, 188, 119–127.

Lan, X., Li, Y., Xie, S., & Wang, Z. (2015). Ultrastructure of underutilized tuber starches and its relation to physicochemical properties. *Food Chemistry*, 188, 632–640.

Lan, X., Liu, X., Yang, Y., Wu, J., & Wang, Z. (2017). The effect of lamellar structure ordering on the retrogradation properties of canna starch subjected to thermal and enzymatic degradation. *Food Hydrocolloids*, 69, 185–192.

Leach, H. W., McCowen, L. D., & Schoch, T. J. (1959). Structure of the starch granule. 1. Swelling and solubility patterns of various starches. *Cereal Chemistry*, 36, 534–544.

Li, C., & Hu, Y. (2020). Combination of parallel and sequential digestion kinetics reveals the nature of digestive characteristics of short-term retrograded rice starches. *Food Hydrocolloids*, 108, 106071.

Li, C., Zhou, D., Fan, T., et al. (2020a). Structure and physicochemical properties of two waxy wheat starches. *Food Chemistry*, 318, 126492.

Li, D., & Zhu, F. (2017a). Physicochemical properties of kiwifruit starch. *Food Chemistry*, 220, 129–136.

Li, G., & Zhu, F. (2017b). Amylopectin molecular structure in relation to physicochemical properties of quinoa starch. *Carbohydrate Polymers*, 164, 396–402.

Li, G., & Zhu, F. (2018). Rheological properties in relation to molecular structure of quinoa starch. *International Journal of Biological Macromolecules*, 114, 767–775.

Li, H., Fitzgerald, M. A., Prakash, S., Nicholson, T. M., & Gilbert, R. G. (2017). The molecular structural features controlling stickiness in cooked rice, a major palatability determinant. *Scientific Reports*, 7, 1–12.

Li, H., Dhital, S., Gidley, M. J., & Gilbert, R. G. (2019). A more general approach to fitting digestion kinetics of starch in food. *Carbohydrate Polymers*, 225, 115244.

Li, J., & Yeh, A. (2001). Relationships between thermal, rheological characteristics and swelling power for various starches. *Journal of Food Engineering*, 50, 141–148.

Li, K., Li, Q., Jin, C., et al. (2020b). Characterization of morphology and physicochemical properties of native starches isolated from 12 lycoris species. *Food Chemistry*, 316, 126263.

Li, K., Zhang, T., Narayanamoorthy, S., et al. (2020c). Diversity analysis of starch physicochemical properties in 95 proso millet (*Panicum miliaceum* L.) accessions. *Food Chemistry*, 324, 126863.

Li, X., Chen, W., Chang, Q., et al. (2020d). Structural and physicochemical properties of ginger (*Rhizoma curcumae longae*) starch and resistant starch: A comparative study. *International Journal of Biological Macromolecules*, 144, 67–75.

Liu, D., Tang, W., Xin, Y., et al. (2020). Comparison on structure and physicochemical properties of starches from adzuki bean and dolichos bean. *Food Hydrocolloids*, 105, 105784.

Liu, T., Ma, M., Guo, K., et al. (2019). Structural, thermal, and hydrolysis properties of large and small granules from C-type starches of four Chinese chestnut varieties. *International Journal of Biological Macromolecules*, 137, 712–720.

Lu, W., Chan, Y., Tseng, F., Chiang, P., & Li, P. (2019). Production and physicochemical properties of starch isolated from djulis (*Chenopodium formosanum*). *Foods*, 8, 551.

Lu, Y., Zhang, X., Yang, Y., et al. (2020). Relationship between structure and physicochemical properties of ginkgo starches from seven cultivars. *Food Chemistry*, 314, 125082.

Macena, J. F. F., de Souza, J. C. A., Camilloto, G. P., & Cruz, R. S. (2020). Physico-chemical, morphological and technological properties of the avocado (*Persea americana* Mill. cv. Hass) seed starch. *Ciencia e Agrotecnologia*, 44, e001420.

Martínez, P., Peña, F., Bello-Pérez, L. A., et al. (2019). Physicochemical, functional and morphological characterization of starches isolated from three native potatoes of the Andean region. *Food Chemistry: X*, 2, 100030.

Mezger, T. G. (2006). *The Rheology Handbook: For Users of Rotational and Oscillatory Rheometers* (2nd revised edition ed). Vincentz Network GmbH & Co. KG.

Miles, M. J., Morris, V. J., Orford, P. D., & Ring, S. G. (1985). The roles of amylose and amylopectin in the gelation and retrogradation of starch. *Carbohydrate Research*, 135, 271–281.

Muthukumarappan, K., & Swamy, G. J. (2017). Chapter 10: Rheology, microstructure, and functionality of cheese. In J. Ahmed, P. Ptaszek, & S. Basu (Eds.), *Advances in Food Rheology and Its Applications* (pp. 245–276). Woodhead Publishing.

Navaf, M., Sunooj, K. V., Aaliya, B., Sudheesh, C., & George, J. (2020). Physico-chemical, functional, morphological, thermal properties and digestibility of talipot palm (*Corypha umbraculifera* L.) flour and starch grown in Malabar region of South India. *Journal of Food Measurement and Characterization*, 14, 1601–1613.

Ngobese, N. Z., Wokadala, O. C., Du Plessis, B., et al. (2018). Physicochemical and morphological properties of a small granule legume starch with atypical properties from wild mango (*Cordyla africana* L.) seeds: A comparison to maize, pea, and kidney bean starch. *Starch/Stärke*, 70, 1700345.

Noora, B., Sudheesh, C., Sangeetha, N., & Sunooj, K. V. (2019). Effect of isolation methods on the crystalline, pasting, thermal properties and antioxidant activity of starch from queen sago (*Cycas circinalis*) seed. *Journal of Food Measurement and Characterization*, 13, 2147–2156.

Oyeyinka, S. A., Salako, M. O., Akintayo, O. A., et al. (2020). Structural, functional, and pasting properties of starch from refrigerated cassava root. *Journal of Food Processing and Preservation*, 44, e14476.

Pacheco, M. T., Moreno, J., Moreno, R., Villamiel, M., & Hernandez-Hernandez, O. (2019). Morphological, technological and nutritional properties of flours and starches from mashua (*Tropaeolum tuberosum*) and melloco (*Ullucus tuberosus*) cultivated in Ecuador. *Food Chemistry*, 301, 125268.

Park, J., Oh, S. K., Chung, H. J., & Park, H. J. (2020). Structural and physicochemical properties of native starches and non-digestible starch residues from Korean rice cultivars with different amylose contents. *Food Hydrocolloids*, 102, 105544.

Patiño-Rodríguez, O., Agama-Acevedo, E., Ramos-Lopez, G., & Bello-Pérez, L. A. (2020). Unripe mango kernel starch: Partial characterization. *Food Hydrocolloids*, 101, 105512.

Punia, S. (2020). Barley starch: Structure, properties and *in vitro* digestibility: A review. *International Journal of Biological Macromolecules*, 155, 868–875.

Punia, S., Sandhu, K. S., Dhull, S. B., et al. (2020). Oat starch: Physico-chemical, morphological, rheological characteristics and its applications: A review. *International Journal of Biological Macromolecules*, 154, 493–498.

Qi, Y., Wang, N., Yu, J., et al. (2020). Insights into structure-function relationships of starch from foxtail millet cultivars grown in China. *International Journal of Biological Macromolecules*, 155, 1176–1183.

Qian, S., Tang, M., Gao, Q., et al. (2019). Effects of different modification methods on the physicochemical and rheological properties of Chinese yam (*Dioscorea opposita* Thunb.) starch. *LWT-Food Science and Technology*, 116, 108513.

Quek, W. P., Yu, W., Tao, K., Fox, G. P., & Gilbert, R. G. (2019). Starch structure-property relations as a function of barley germination times. *International Journal of Biological Macromolecules*, 136, 1125–1132.

Ren, F., Wang, J., Xie, F., et al. (2020). Applications of ionic liquids in starch chemistry: A review. *Green Chemistry*, 22, 2162–2183.

Ronda, F., Pérez-Quirce, S., & Villanueva, M. (2017). Chapter 12 - Rheological properties of gluten-free bread doughs: Relationship with bread quality. In J. Ahmed, P. Ptaszek, & S. Basu (Eds.), *Advances in Food Rheology and Its Applications* (pp. 297–334). Woodhead Publishing.

Rosalina, I., & Bhattacharya, M. (2002). Dynamic rheological measurements and analysis of starch gels. *Carbohydrate Polymers*, 48, 191–202.

Sanz, T., Salvador, A., & Hernández, M. J. (2017). Chapter 11: Creep–recovery and oscillatory rheology of flour-based systems. In J. Ahmed, P. Ptaszek and S. Basu (Eds.), *Advances in Food Rheology and Its Applications* (pp. 277–295). Woodhead Publishing.

Silva, G. M. D., Veloso, C. M., Santos, L. S., et al. (2020). Extraction and characterization of native starch obtained from the inhambu tuber. *Journal of Food Science and Technology-Mysore*, 57, 1830–1839.

Silverio, J., Fredriksson, H., Andersson, R., Eliasson, A. C., & Åman, P. (2000). The effect of temperature cycling on the amylopectin retrogradation of starches with different amylopectin unit-chain length distribution. *Carbohydrate Polymers*, 42, 175–184.

Sindhu, R., & Khatkar, B. S. (2018). Thermal, structural and textural properties of amaranth and buckwheat starches. *Journal of Food Science and Technology-Mysore*, 55, 5153–5160.

Singh, H., Lin, J., Huang, W., & Chang, Y. (2012). Influence of amylopectin structure on rheological and retrogradation properties of waxy rice starches. *Journal of Cereal Science*, 56, 367–373.

Singh, T. P., & Sogi, D. S. (2018). Comparison of physico-chemical properties of starch isolated from bran and endosperm of rice (*Oryza sativa* L.). *Starch/Stärke*, 70, 1700242.

Solaesa, A. G., Villanueva, M., Beltrán, S., & Ronda, F. (2019). Characterization of quinoa defatted by supercritical carbon dioxide starch enzymatic susceptibility and structural, pasting and thermal properties. *Food and Bioprocess Technology*, 12, 1593–1602.

Srichuwong, S., & Jane, J. (2007). Physicochemical properties of starch affected by molecular composition and structures: A review. *Food Science and Biotechnology*, 16, 663–674.

Tangsrianugul, N., Wongsagonsup, R., & Suphantharika, M. (2019). Physicochemical and rheological properties of flour and starch from Thai pigmented rice cultivars. *International Journal of Biological Macromolecules*, 137, 666–675.

Tejavathi, D. H., Sujatha, B. S., & Karigar, C. S. (2020). Physicochemical properties of starch obtained from *Curcuma karnatakensis*: A new botanical source for high amylose content. *Heliyon*, 6, e03169.

Tester, R. F., & Morrison, W. R. (1990). Swelling and gelatinization of cereal starches. 1. effects of amylopectin, amylose, and lipids. *Cereal Chemistry*, 67, 551–557.

Tetlow, I. J., & Bertoft, E. (2020). A review of starch biosynthesis in relation to the building block-backbone model. *International Journal of Molecular Sciences*, 21, 7011.

Tong, C., Ru, W., Wu, L., Wu, W., & Bao, J. (2020). Fine structure and relationships with functional properties of pigmented sweet potato starches. *Food Chemistry*, 311, 126011.

Tsai, M., Li, C., & Lii, C.. (1997). Effects of granular structures on the pasting behaviors of starches. *Cereal Chemistry*, 74, 750–757.

Vamadevan, V., & Bertoft, E. (2015). Structure-function relationships of starch components. *Starch/Stärke*, 67, 55–68.

Vamadevan, V., & Bertoft, E. (2018). Impact of different structural types of amylopectin on retrogradation. *Food Hydrocolloids*, 80, 88–96.

Vandeputte, G. E., V. Derycke, J. Geeroms, & J. A. Delcour. (2003). Rice starches. II. Structural aspects provide insight into swelling and pasting properties. *Journal of Cereal Science*, 38, 53–59.

Vermeylen, R., Derycke, V., Delcour, J. A., et al. (2006). Gelatinization of starch in excess water: Beyond the melting of lamellar crystallites. A combined wide- and small-angle X-ray scattering study. *Biomacromolecules*, 7, 2624–2630.

Vithu, P., Dash, S. K., Rayaguru, K., Panda, M. K., & Nedunchezhiyan, M. (2020). Optimization of starch isolation process for sweet potato and characterization of the prepared starch. *Journal of Food Measurement and Characterization*, 14, 1520–1532.

Waigh, T. A., Gidley, M. J., Komanshek, B. U., & Donald, A. M. (2000). The phase transformations in starch during gelatinisation: A liquid crystalline approach. *Carbohydrate Research*, 328, 165–176.

Wang, H., Yang, Q., Gao, L., et al. (2020a). Functional and physicochemical properties of flours and starches from different tuber crops. *International Journal of Biological Macromolecules*, 148, 324–332.

Wang, J., Wang, A., Ma, W., et al. (2019a). Comparison of physicochemical properties and *in vitro* digestibility of starches from seven banana cultivars in China. *International Journal of Biological Macromolecules*, 121, 279–284.

Wang, J., Wen, Z., Fu, P., Lu, W., & Lu, D. (2019b). Effects of nitrogen rates on the physicochemical properties of waxy maize starch. *Starch/Stärke*, 71, 1900146.

Wang, L., Gong, Y., Li, Y., & Tian, Y. (2020b). Structure and properties of soft rice starch. *International Journal of Biological Macromolecules*, 157, 10–16.

Wang, S., Li, C., Copeland, L., Niu, Q., & Wang, S. (2015). Starch retrogradation: A comprehensive review. *Comprehensive Reviews in Food Science and Food Safety*, 14, 568–585.

Wang, W., Ge, J., Xu, K., et al. (2020c). Differences in starch structure, thermal properties, and texture characteristics of rice from main stem and tiller panicles. *Food Hydrocolloids*, 99, 105341.

Wani, A. A., Singh, P., Shah, M. A., et al. (2012). Rice starch diversity: Effects on structural, morphological, thermal, and physicochemical properties: A review. *Comprehensive Reviews in Food Science and Food Safety*, 11, 417–436.

Wu, P., Li, C., Bai, Y., Yu, S., & Zhang, X. (2019). A starch molecular basis for aging-induced changes in pasting and textural properties of waxy rice. *Food Chemistry*, 284, 270–278.

Yadav, U., Singh, N., Arora, S., & Arora, B. (2019). Physicochemical, pasting, and thermal properties of starches isolated from different adzuki bean (*Vigna angularis*) cultivars. *Journal of Food Processing and Preservation*, 43, e14163.

Yang, Q., Zhang, W., Luo, Y., et al. (2019). Comparison of structural and physicochemical properties of starches from five coarse grains. *Food Chemistry*, 288, 283–290.

Yao, M., Tian, Y., Yang, W., et al. (2019). The multi-scale structure, thermal and digestion properties of mung bean starch. *International Journal of Biological Macromolecules*, 131, 871–878.

Zhan, Q., Ye, X., Zhang, Y., et al. (2020). Starch granule-associated proteins affect the physicochemical properties of rice starch. *Food Hydrocolloids*, 101, 105504.

Zhang, G., & Hamaker, B. R. (2009). Slowly digestible starch: Concept, mechanism, and proposed extended glycemic index. *Critical Reviews in Food Science and Nutrition*, 49, 852–867.

Zhang, L., Li, G., Wang, S., Yao, W., & Zhu, F. (2017). Physicochemical properties of maca starch. *Food Chemistry*, 218, 56–63.

Zhang, L., Liu, T., Hu, G., Guo, K., & Wei, C. (2018). Comparison of physicochemical properties of starches from nine Chinese chestnut varieties. *Molecules*, 23, 3248.

Zhang, S., Li, Z., Lin, L., Zhang, L., & Wei, C. (2019a). Starch components, starch properties and appearance quality of opaque kernels from rice mutants. *Molecules*, 24, 4580.

Zhang, Z., Saleh, A. S. M., Wu, H., et al. (2020). Effect of starch isolation method on structural and physicochemical properties of acorn kernel starch. *Starch/Stärke*, 72, 1900122.

Zhang, Z., Tian, X., Wang, P., Jiang, H., & Li, W. (2019b). Compositional, morphological, and physicochemical properties of starches from red adzuki bean, chickpea, faba bean, and baiyue bean grown in China. *Food Science & Nutrition*, 7, 2485–2494.

Zhu, F. (2018). Relationships between amylopectin internal molecular structure and physicochemical properties of starch. *Trends in Food Science & Technology*, 78, 234–242.

Zhu, F., & Cui, R. (2020). Comparison of physicochemical properties of oca (*Oxalis tuberosa*), potato, and maize starches. *International Journal of Biological Macromolecules*, 148, 601–607.

Zhu, F., & Liu, P. (2020). Starch gelatinization, retrogradation, and enzyme susceptibility of retrograded starch: Effect of amylopectin internal molecular structure. *Food Chemistry*, 316, 126036.

Zhu, F., & Xie, Q. (2018). Rheological and thermal properties in relation to molecular structure of New Zealand sweetpotato starch. *Food Hydrocolloids*, 83, 165–172.

Zhu, F., Corke, H., & Bertoft, E. (2011). Amylopectin internal molecular structure in relation to physical properties of sweetpotato starch. *Carbohydrate Polymers*, 84, 907–918.

Zhu, F., Bertoft, E., & Li, G. (2016). Morphological, thermal, and rheological properties of starches from maize mutants deficient in starch synthase III. *Journal of Agricultural and Food Chemistry*, 64, 6539–6545.

Zou, J., Xu, M., Wen, L., & Yang, B. (2020). Structure and physicochemical properties of native starch and resistant starch in Chinese yam (*Dioscorea opposita* Thunb.). *Carbohydrate Polymers*, 237, 116188.

Chapter 5

Starch Modifications for Controlled Digestion

Maribel Ovando-Martínez and Senay Simsek

CONTENTS

5.1 INTRODUCTION

Starch is a polysaccharide synthesized in plants. Cereals, roots, tubers, fruits, and legumes are major starch sources worldwide and are considered renewable, cheap, and abundant sources in nature (Ogunsona et al., 2018). Starch

DOI: 10.1201/9781003088929-5

is formed by amylose and amylopectin. Amylose is a linear molecule with glucose units linked with α-(1-4) bonds with a few branched points. On the other hand, amylopectin is a branched polymer of glucose units linked with α-(1-4) bonds and α-(1-6) bonds at the branch points (Bello-Perez et al., 2020; Buléon et al., 1998; Pérez & Bertoft, 2010). The organization of amylose and amylopectin polymers results in a granule structure like an onion where semi-crystalline growth rings are formed by alternate stacks of amorphous and crystalline lamellae. The crystalline lamellae are represented by the highly ordered structure of the amylopectin, while the amorphous lamellae are comprised of amylose (Zhong et al., 2020a; Baker et al., 2001; Tester et al., 2004; Jenkins & Donald, 1995; Gallant et al., 1997). The backbone structure of the starch granules and the amylose/amylopectin ratio has been reported to influence the physicochemical and digestibility properties in starchy foods (Zhong et al., 2020b).

In the food industry, starch has always been explored due to its physicochemical properties and application in different food systems (Fan & Picchioni, 2020). Starch has been classified based on amylose content as normal starch (20%–35% amylose), waxy starch (less than 15% amylose), and high-amylose or amylostarch starch (more than 40% amylose). There are clear differences in the physicochemical properties of starches based on their amylose content (Tester et al., 2004). For example, maize starch varying from 0.5% to 90% amylose content showed different gelatinization properties, swelling power and resistance to acid digestion. The amylose content of the starches impacts the formation of the double helices, the degree of crystallinity and amylose–lipid complexation. Thus these properties are related to the packing of the double helices within the starch granule and the highly ordered crystalline structure (Shi et al., 1998).

The amylose content of rice starch from different cultivars impacts the crystal structure pattern, crystallinity and the average chain length distribution of amylopectin. Thus, amylose content affects the pasting properties. Starch with high amylose content has less crystallinity and a high ratio of long-chain length in amylopectin compared to that with low amylose content. Further, higher amylose content presents lower viscosity and high gelatinization temperature. Pasting properties are of importance for rice, as they play a major a role in the taste and quality of rice and other rice-based foods (Park et al., 2020). On the other hand, wheat with a waxy phenotype requires more energy to disorganize the starch structure and gelatinize the starch due to its high crystallinity compared to native starch (Fujita et al., 1998). Native starch from different Indian lentils also presented differences in the physicochemical properties, indicating that amylose content significantly affects the functional and physicochemical properties of starch (Kaur et al., 2010).

Furthermore, from the nutritional point of view, starch digestibility, especially slowly digestible and resistant starch, has been of great interest due to the health-related benefits. Amylose content and starch structure play an important role in starch digestibility (Englyst & Englyst, 2005; Englyst et al., 1996; Englyst et al., 2007). Native maize starches with around 20%–30% amylose content have limited application and nutrition quality (Zhong et al., 2020b), because their physicochemical properties under certain pH and temperature influence the starch digestibility (Englyst et al., 2007). The size and crystalline nature of starch also have an impact on starch digestion due to the susceptibility of the starch to the digestion enzymes. Starches with B- and C-type X-ray diffraction patterns are more resistant to pancreatic amylase compared to those with an A-type pattern (Cummings & Englyst, 1995). It has been reported that starch with high crystallinity has shown lower digestibility, which indicates a close relationship between the fine structure of amylopectin (long-chain content) with the digestion rate (Bi et al., 2017). However, when starch is cooked, the granules swell, and its molecular organization is disrupted, causing loss of crystallinity and making it more susceptible to attack by digestive enzymes. The digestion rate will depend on the retrogradation rate of starch during cooling, size and morphology of the starch granule (fissures, pores, channels), the presence of amylose–lipid complexes, particle size, amylose/amylopectin ratio, enzyme specificity, and other properties (Cummings & Englyst, 1995; Kong et al., 2003). So, there are many factors that could be altered to improve the digestion properties of starch even after processing and cooking. Therefore, improving the nutraceutical or functional properties of starch through modification of its structure utilizing different methods such as chemical, physical, enzymatic, and genetic modification, or combination is vital.

The development of new technologies to improve starch digestibility is a big challenge for the food industry since native starch is mostly rapidly digestible, contributing to an increase in metabolic syndrome–related diseases such as obesity, stroke and diabetes. Among those technologies, chemical modification by introducing chemical groups into the starch structure and physical changes such as heat-moisture treatments (annealing, autoclaving) to reorder the amylopectin structure are widely used. Additionally, enzymatic modification, through debranching and reassociation of the amylopectin chains (Chi et al., 2019; Li et al., 2018a, 2019c; Lee et al., 2018; Zhang et al., 2016; Kim et al., 2013, 2014), and genetic modification by manipulating the genes involved in the starch biosynthesis to modify the ratio of amylopectin to amylose and their structures for specific applications may be utilized (Jobling, 2004; Edwards et al., 1999; Davis et al., 2003; Sharma et al., 2002; Burton et al., 2002; Dinges et al., 2003; Tanaka et al., 2004; Nakamura et al., 1992; Morell et al., 1997). However, innovative techniques such as the use of ozone, pulsed electric field (PEF) and ultrasound technology are environmentally friendly and have been used to create clean label starch with desirable functionalities (Maniglia et al., 2021).

5.2 NATIVE STARCH: ADVANTAGES AND DISADVANTAGES

Starch structure and functionality vary among starches isolated from different botanical sources or even from the same plant cultivar grown under different conditions. Such variations result in diverse properties and create disadvantages related to starchy food processing where uniformity of ingredient quality is key (Copeland et al., 2009). For this reason, it is crucial to understand how the advantages and disadvantages of native starch interfere with the food processing and nutritional properties, how they are related to the starch structure and morphology, and finally, how to improve or take advantage of them.

Starch is a semicrystalline biopolymer formed by amylose and amylopectin. Native starch structure is composed of single and double helices forming the crystalline and amorphous lamellae, both being part of the growth rings of starch granules (Imberty et al., 1991; Bertoft, 2017; Tester et al., 2004; Buléon et al., 1998; Vamadevan & Bertoft, 2015; Jane et al., 2003). The amorphous lamellae are made up of amylose and amylopectin in a chaotic organization, while the ordered double helices of amylopectin form the crystalline lamellae, which give different polymorphic A, B and C forms. Additionally, it is noteworthy that starch has a semicrystalline region formed by amylopectin branched points and amylose molecules alternated in a disordered manner (Pérez & Bertoft, 2010; Bertoft, 2017; Tester et al., 2004; Buléon et al., 1998). Therefore, the organization and ratio of these polymers can affect the physicochemical and functional properties of the starch. These properties are then responsible for the disadvantages of the native starch application because it presents undesirable properties such as low solubility in water and other solvents, low retrogradation, low shear stress resistance, and susceptibility to thermal decomposition (Ai & Jane, 2018; Jane et al., 2003; Srichuwong & Jane, 2007).

The unique characteristics of starches from different botanical sources are important and provide each starch source with unique properties, which allows the use of the starch in specific applications (advantages) (Olayemi et al., 2021; Vamadevan & Bertoft, 2015; Srichuwong & Jane, 2007; Srichuwong et al., 2017). However, these native starches have some disadvantages when used under certain conditions (temperature, pH, shear forces), primarily affecting the digestibility properties of starch. Table 5.1 indicates the advantages and disadvantages of such properties in different botanical sources of starch. As shown in Table 5.1, all native starches have disadvantages related to their granule size, amylose/amylopectin ratio, structural properties and organization. For example, Li et al. (2020) found that chain length distribution and amylose content play an important role in determining the physicochemical and digestibility properties of the starch. On the other hand, Jane et al. (2003) mentioned granule size, morphology and surface affect the physiochemical properties

TABLE 5.1   PHYSICOCHEMICAL AND DIGESTIBILITY PROPERTIES: ADVANTAGES AND DISADVANTAGES OF DIFFERENT NATIVE STARCH

Starch source	Advantage of native starch	Disadvantage of native starch	Reference
Legumes			
Pea Lentil Faba bean	They have a smooth surface with fissures and scratches, contain high amylose content and large amylopectin chains compared to cereal and tuber starches. Therefore, they have excellent gelling ability and high retrogradation tendency.	The high retrogradation tendency may cause syneresis during storage of food products elaborated with these starches. Additionally, they present low starch digestibility in the raw form, while after cooking, they show a similar glycemic response as cereal starches.	Ren et al. (2021)
Different legume starches: bean, lima bean, chickpea, lentil, cowpea, others	These starches are strongly attached between molecules, which gives them high gelatinization enthalpy values.	These starches present vulnerability to retrogradation and not good processing properties in extreme conditions.	Ashogbon et al. (2020)
Borlotti bean White kidney bean Chickpea	When starches are uncooked, they have a high content of resistant starch and low content of rapidly and slowly digestible starch fractions.	The rapidly digestible starch fraction increased after cooking, which can affect the glycemic response.	Güzel & Sayar (2010)
Black bean Pinto bean Smooth pea Lentil Wrinkled pea	Most of these starches present similar X-ray patterns, amylose content, and thermal properties.	Despite the similar physicochemical properties, legume starches presented differences in the rate and extent of hydrolysis by α-amylase.	Zhou, Hoover, & Liu (2004)

(Continued)

TABLE 5.1   (CONTINUED) PHYSICOCHEMICAL AND DIGESTIBILITY PROPERTIES: ADVANTAGES AND DISADVANTAGES OF DIFFERENT NATIVE STARCH

Starch source	Advantage of native starch	Disadvantage of native starch	Reference
Legumes			
Fruits			
Peach palm fruit	This starch could provide body or be used as a thickener agent because it does not form very strong gels.	The starch has nonpolar components, low amylose content, low proportion of long side chains from amylopectin, low enthalpy value, and low retrogradation rates.	Ferrari Felisberto et al. (2020)
Annona: sweetsop and soursop	The starch from both types of *Annona* presented weak gel characteristics.	Starch from sweetsop showed low pasting temperature and low setback, while soursop had low gelatinization temperature.	Nwokocha & Williams (2009)
Pumpkin: *Cucurbita máxima* Ambar type, *C. máxima* Amazonka type, and *C. máxima* Uchiki Kuri type	The starch from three types of pumpkin fruits formed hard gels with good cohesiveness, springiness, and chewiness compared to corn and potato starch.	Starch from three types of pumpkin fruits showed low thermal properties compared to corn and potato starch.	Przetaczek-Rożnowska (2017)

(Continued)

TABLE 5.1   (CONTINUED) PHYSICOCHEMICAL AND DIGESTIBILITY PROPERTIES: ADVANTAGES AND DISADVANTAGES OF DIFFERENT NATIVE STARCH

Starch source	Advantage of native starch	Disadvantage of native starch	Reference
Legumes			
Green banana: flesh and peel	Physicochemical properties indicated both starches have similar swelling power and enthalpy of gelatinization, but peel starch has better water solubility, gelatinization temperature, and higher resistance to porcine pancreatin α-amylase hydrolysis than flesh starch. In the raw form, both starches presented low content of rapidly digestible starch (~1.5%) and high levels of resistant starch (~95%).	When peel and flesh starch was gelatinized, rapidly digestible starch increased to levels around 73%, and resistant starch decreased to levels around 23%. This could be related to the high glycemic index after intake of these starches.	Li et al. (2018)
Kiwifruit: Gold3, Gold9, and Hort16A (core and outer pericarp)	Starch from the core pericarp presented higher amylose content than starch from the outer pericarp. However, starch from the outer pericarp showed better crystallinity and thermal parameters than core starch.	Raw starch from the outer and core pericarp showed lower enzyme susceptibility to porcine pancreatic α-amylase than corn starch but higher than potato starch. What could be the degree of hydrolysis if the starch undergoes cooking?	Li & Zhu (2017)

(Continued)

TABLE 5.1   (CONTINUED) PHYSICOCHEMICAL AND DIGESTIBILITY PROPERTIES: ADVANTAGES AND DISADVANTAGES OF DIFFERENT NATIVE STARCH

Starch source	Advantage of native starch	Disadvantage of native starch	Reference
Legumes			
Mango and banana	Starches from banana and mango showed good peak and final viscosity than corn starch.	Mango starch presented a lower retrogradation value than banana starch. According to the structural analysis, both starches presented more short chains than long chains. Structural properties of these starches can affect their digestibility properties, consequently an effect on the starch quality.	Espinosa-Solis et al. (2009)
Cereals			
Starch from cereals: white and red sorghum, millet, corn, wheat, quinoa, and amaranth	Starch from white and red sorghum, millet, and corn showed higher gelatinization temperature and retrogradation enthalpy than wheat quinoa and amaranth starches.	Amaranth, quinoa and wheat had low crystallinity values. Also, only amaranth and quinoa starch had low amylose content. It was noted these starches also showed high values of starch hydrolysis. It is important to note that structural properties and granule size play an important role in physiochemical properties, which sometimes can have negative effects on the starch application.	Srichuwong et al. (2017)

(Continued)

TABLE 5.1   (CONTINUED) PHYSICOCHEMICAL AND DIGESTIBILITY PROPERTIES: ADVANTAGES AND DISADVANTAGES OF DIFFERENT NATIVE STARCH

Starch source	Advantage of native starch	Disadvantage of native starch	Reference
Legumes			
Four rice cultivars: Baegokchal (waxy)/ Ilmi (normal) (*Japonica* type), Mimyeon (*Japonica* and *Indica*), Dodamssal (Japonica-type high amylose)	Dodamssal starch had high amylose content, low rapidly digestible starch, and high resistant starch. However, Baegokchal, Ilmi, and Mimyeon starches presented high crystallinity and better pasting properties.	Baegokchal starch had the lowest amylose content and the highest rapidly digestible starch, while Dodamssal starch did not present good pasting properties.	Park et al. (2020)
Tubers			
Three cultivars of Potato (Xisen 6, Hongmei, Heimeiren) and sweet potato (Yu15, Qin7, 10-6-5)	It was noted that potato starches presented high amylose content, hardness of the starch gel, chewiness, gumminess, peak viscosity and breakdown. Also, sweet potato starch presented a high temperature of gelatinization and crystallinity.	The minor granule size and low amylose content in sweet potato starches caused the low hardness of the starch gels.	Wang et al. (2020)
Roots			
Cassava: two varieties (TMS326 and TME419)	Starch from TMS326 showed high amylose content, crystallinity, peak viscosity and gelatinization temperature.	The low amylose content of starch from TME419 affected its physicochemical properties compared to starch from TMS326.	Oyeyinka et al. (2019)

of starch. According to the authors, B granules show a higher gelatinization temperature than A granules because of the granular and molecular structure of barley starch. In the case of corn starch, the presence of pinholes in maize, potato and sorghum increases the hydrolysis of the starch granule. Therefore, native starch needs to be modified to increase its area of application in the food industry, as well as to produce starch with improved digestibility properties, enhanced nutrition (more resistant starch and slowly digestible starch), and to decrease the connection to metabolic syndromes caused by the rapidly digestible starch intake (Luo et al., 2018).

5.3 STARCH MODIFICATION

Various technologies have been utilized to achieve better physicochemical, functional and nutritional properties to improve starch quality (Li et al., 2020). These technologies are used to improve starch insolubility in water and increase its application in the food industry (Haq et al., 2019). Still, they are also needed to produce starch with applications in drug delivery systems (wall material of encapsulated bioactive compounds or probiotics), clinical potential as regulators of blood glucose concentrations (Masina et al., 2017), and to produce resistant starch as a fiber, among others. With these technologies, amylose and amylopectin can be modified by introducing chemical components, changes in their structural organization, hydrolysis or elongation of their structure, or by the improvement of the starch biosynthesis to increase the production of amylose or amylopectin (Punia et al., 2021). All these methods and the changes in the starch digestibility properties will be described next.

5.3.1 Chemical Modification

Chemical modification is the most used technology to modify starch with a small amount of approved chemical reagents (Tomas & Atwell, 1999). Chemical modification introduces functional groups into the starch structure at the hydroxyl groups of the glucose units of amylose or amylopectin or the reducing end of C-type starch (Chen et al., 2015; Lawal, 2019). The chemical modification may affect the morphology or size distribution of the starch granules depending on the modification used. It has been reported that chemical modification causes changes in the starch behavior, gelatinization and retrogradation parameters, freeze–thaw stability, pasting properties, shear resistance, solubility, and digestibility. This type of modification is used to obtain desired characteristics needed for specific applications (Alcázar-Alay & Meireles, 2015; Punia et al., 2021). Degradation of the starch by hydrolysis and oxidation; substitution methods such as esterification by acetylation and succinylation;

etherification by carboxymethylation, hydroxypropylation and hydroxyethyl-ation; and cross-linking are classifications of chemical modification.

5.3.1.1 Degradation

Chemical degradation of starch includes hydrolysis and oxidation methods (Figure 5.1A a and b, respectively), which are the most widely used to obtain starch derivatives with low molecular weight (Lewicka et al., 2015; Lawal, 2019).

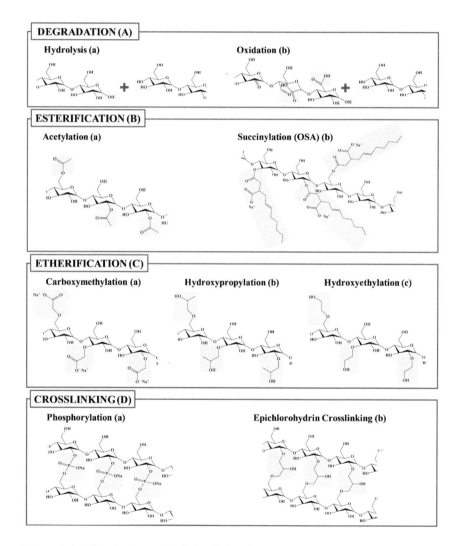

Figure 5.1   Chemically modified starch structures.

After modification processes with these methods, the starch solubility, water absorption, pasting and gelatinization properties are improved (Chen et al., 2015). Starch hydrolysis, mainly done with hydrochloric and sulfuric acids, break down the α-1-4 bonds in the starch structure to increase the proportion of the short chains. Thus, the starch generated can be used to develop healthy food because it is a good source of slowly digestible starch (Chen et al., 2015; Wang & Copeland, 2015). Espinosa-Solis et al. (2011) reported that acid-treated banana and plantain starches increased the slowly digestible starch content at low hydrolysis percentages (6%–35%) and decreased the resistant starch fraction at higher hydrolysis percentages (70%–80%), whereas acid treatment of mango starch did not result in any effect on the slowly digestible starch fraction. On the other hand, Miao et al. (2011) found that mild acid treatment of maize starch with hydrochloric acid increased rapidly digestible and resistant starch contents while decreasing the slowly digestible starch.

Starch oxidation transforms the hydroxyl groups of the C-2 and C-3 position of the glucose units of the starch into carbonyl and carboxyl groups through oxidation agents such as potassium permanganate, sodium hypochlorite, hydrogen peroxide, persulfate, and ozone (Vanier et al., 2017; Zhou et al., 2016; Chen et al., 2015). Due to the changes in the starch structure by the introduction of those functional groups, oxidized starches are used in the food industry as food additives, as well as ingredients for the development of low glycemic foods, among others (Lewicka et al., 2015; Vanier et al., 2017). Potato starch oxidized with different concentrations of active chlorine presented changes in digestibility. It was observed that the increase of the active chloride caused a decrease in the rapidly digestible starch, while the slowly digestible starch and resistant starch increased (Zhou et al., 2016). According to Zhou et al. (2016), there is a steric effect hindering the active site of the α-amylase due to the bulky functional groups at the C-2 position of the glucose unit in oxidized starches, which delays the hydrolysis of starch and, by consequence, changes the starch digestibility. Reyes et al. (2021) mentioned that those changes reduce rapidly and slowly digestible fractions of oxidized starches. Overall, hydrolysis or oxidative degradation of starches may be utilized to alter the starch structure resulting in changes to functionality and digestibility properties.

5.3.1.2 Esterification

Esterification involves the reaction of starch chains with inorganic and organic acids and anhydrides or chlorides of organic acids. During this process, acid groups are attached to the amylose and amylopectin molecules through ester bonds linked to the C-2, C3 and C-6 positions of the glucose units (Zdybel et al., 2020). Esterification is often used to improve starch hydrophobicity (Jianteng et al., 2020). Acetylation is considered the primary method to produce starch esters Figure 5.1B(a)], where the hydroxyl groups are changed to acetyl groups.

The greater the number of acetyl groups, the more significant the reduction of starch hydrophilicity (Lawal, 2019; Masina et al., 2017). A high degree of substitution has been observed in this type of modification when starch has low amylose content, as acetylation occurs in the amorphous region. Also, there is more interaction with the amylopectin branches to incorporate the carbonyl group (Masina et al., 2017). Recently, starches acylated with propionate and butyrate groups gained interest because of the capacity to deliver short-chain fatty acids to the colon and the relationship with treatment and prevention of bowel disorders (Xie et al., 2019). Wang et al. (2021) reported that the resistant starch content of rice starch increased after propionylation reaction. Starch modification with succinic anhydride (mainly done with octenylsuccinic anhydride) [Figure 5.1B(b)] has recently been the most popular esterification method. During starch modification with succinic anhydride in slightly basic conditions, the intermolecular bonding between the starch chains is reduced by replacing the hydroxyl groups for the octenylsuccinic groups (Bajaj et al., 2019).

Due to the amphiphilic characteristic granted by these modifications, starch esters have wide application in the pharmaceutical, plastic and food industries. Regarding the food industry, starch esters also showed a relationship with the degree of substitution and increased slowly digestible starch and resistant starch. In the case of octenylsuccinate starch modification, starches with varied amylose contents showed that starch hydrolysis was reduced independently of the amylose content (Lopez-Silva et al., 2020). Another study demonstrated that the introduction of octenylsuccinate groups in the starch structure decreased the rapidly digestible starch content and increased the slowly digestible starch in cereal (corn, rice and wheat), tuber (potato), and root (tapioca) succinate starches (Simsek et al., 2015). Thus, esterification is a valuable method to produce starches with unique functionality or digestibility.

5.3.1.3 Etherification

Contrary to esterification, the primary purpose of etherification is to increase the hydrophilic nature of starch by adding hydrophilic functional groups, increasing solubility and interaction with water (Jianteng et al., 2020). Commonly, starch etherification consists of the substitution of the hydroxyl groups with anionic carboxymethyl groups (carboxymethylation) [Figure 5.1C(a)], hydroxypropyl groups (hydroxypropylation) [Figure 5.1C(b)] or hydroxyethyl groups (hydroxyethylation) [Figure 5.1C(c)] with an alkaline catalyst (Masina et al., 2017). Regarding the nutritional properties, focusing on digestibility, Xiao et al. (2018) reported that carboxymethylated Chinese water chestnut starch had greater slowly digestible starch and resistant starch content than native, acetylated and succinate starches. In the case of hydroxypropylation, inhibition of the starch digestibility is expected because hydroxypropyl groups act as branch points. However, this is not always the case as some results indicated digestion of starch did not change, and others have shown the opposite results, as Fu et al. (2019) have described. An example of hydroxypropylated starch is starch citrate, modified with citric acid, a harmless reagent compared to others. Starch citrate presents a high resistance to hydrolysis by enzymes because of the citrate groups introduced to the starch molecules. Therefore, this starch can be used as an alternative to dietary fiber (Xie & Liu, 2004; Wepner et al., 1999). Various methods of etherification may be used to develop slowly digestible starch with improved water solubility.

5.3.1.4 Cross-Linking

Cross-linking consists of generating new linkages between the starch chains with another molecule, called cross-linker agents, such as sodium trimetaphosphate and epichlorohydrin (Haq et al., 2019; Lawal, 2019; Ding et al., 2022). The inter- and intramolecular cross-linking reactions occur between amylose and amylopectin and between amylopectin chains to produce the di-starch phosphate in the starch granule, as depicted in Figure 5.1D. Amylose chains do not cross-link to other amylose molecules because of their considerable distance from each other (Ding et al., 2022). The new linkages result in starch with thermal and mechanical stability and limited swelling power (Chen et al., 2015). Apart from other applications, cross-linked starch is categorized as type 4 resistant starch due to a restriction of hydrolysis by digestive enzymes as observed in potato starch (Ding et al., 2022). In addition, Sharma et al. (2020) reported that digestibility properties of cross-linked fava bean starch improved due to a reduction of the rapidly digestible starch fraction and an increase in the resistant starch with the higher concentration of the cross-linking reagent. Cross-linking is a helpful technique for producing resistant starches with good functional properties.

As was described, chemical modification affects the nutritional properties of the starch, causing changes in the starch fractions during the digestion process. Chemical reagents restricted the hydrolysis of the starch by the digestive enzymes. As a result, the resistant starch and slowly digestible starch contents increased, which could be reflected in a low glycemic response. However, other methods improve the digestibility properties of the starch without repeating the use of chemicals.

5.3.2 Physical Modification

Physical modification is an alternative technology to chemical modification, being economical and free of chemical reagents. It is the most accepted technology to prepare functional starch because it is considered safe for human consumption (Goel et al., 2020; Lawal, 2019; Ashogbon et al., 2020). This type of modification is carried out by thermal or mechanical force or physical field without breaking down the amylose and amylopectin polymers (Wang et al., 2020). However, some processes, such as pregelatinization, destroy the granular structure of the starch due to depolymerization and fragmentation of starch by heating (Ashogbon et al., 2020). Physical modification procedures are classified as either thermal or nonthermal methods. The thermal methods include pregelatinization, heat-moisture treatment and annealing; while among the nonthermal methods are ultrahigh pressure, ultrasound, microwave, pulsed electric field, dry heating, osmotic pressure, freezing and thawing (Wang et al., 2020; Ashogbon & Org, 2021; Wu & Zhou, 2018). Most of these methods enhance starch digestibility by improving levels of slowly digestible starch, increasing resistant starch content, and decreasing the rapidly digestible starch, which could have a relationship with the glycemic response and colonic fermentation.

5.3.2.1 Thermal Methods

The most used thermal methods are heat-moisture treatment and annealing, and both differ in the amount of water and temperature used (Wu & Zhou, 2018). Heat-moisture treatment consists of the treatment of starch granules with moisture levels no more than 35% (w/w) during a lapse time between 15 min to 72 h, using a temperature range above the glass transition temperature, typically between 80°C and 140°C (Goel et al., 2020; Bemiller & Huber, 2015). Heat-moisture treatment does not damage the granular morphology of rice, yam, taro and cassava starches. Nevertheless, some starches like potato and corn have shown damage or changes to granule morphology after heat-moisture treatment (Arns et al., 2015; Gunaratne and Hoover, 2002; Kawabata et al., 1994). After heat-moisture treatment of cereal, tuber and legume starches, it was observed that starch hydrolysis decreased compared to native starches, suggesting that this treatment makes crystalline regions less accessible to α-amylase. Thus, the botanical source influences the results after heat-moisture treatment because of the differences in amylose contents and the chain length distribution of the starches (Hoover & Vasanthan, 1994; Mathobo et al., 2021). In general, heat-moisture treatment alters the molecular weight, crystallinity and others through interactions of starch chains altering the crystalline order. Then, reassociation of the disrupted crystals and mobility of the amorphous regions occurs, causing a rearrangement of the double helices. These alterations affect the digestion properties of starch (Liu et al., 2019).

Regarding the annealing treatment, starch is heated with excess water (40%–60% water, w/w) at temperatures below the gelatinization temperature of starch and above the glass transition temperature (BeMiller & Huber, 2015; Ashogbon & Org, 2021; Mathobo et al., 2021). This physical modification increases the starch granule structure, stability, crystallinity, interactions between the amorphous and crystalline regions, and the emergence of double helices in starch granules (Mathobo et al., 2021). These alterations cause changes in the digestibility properties of starch (Zou et al., 2019). For example, annealed waxy corn and waxy barley starches showed higher enzyme susceptibility than waxy rice and waxy potato starches (Samarakoon et al., 2020). On the other hand, mung bean starch subjected to continuous annealing and repeated annealing showed that both treatments caused the increase of slowly digestible starch and resistant starch compared to native starch (Zou et al., 2019).

As mentioned, both heat-moisture and annealing treatments affect the digestibility properties, but this effect depends on the botanical sources and the thermal treatment applied. Heat-moisture treatments have shown better improvement of the digestion properties of starch than annealing, as was reported by Chung et al. (2009) in corn, pea and lentil starches subjected to heat-moisture and annealing treatments. The authors reported that heat-moisture treatments presented a higher impact on resistant starch formation than the annealing treatments (Zavareze & Dias 2011; Chung et al., 2009).

5.3.2.2 Nonthermal Methods

Nonthermal methods do not use temperature to modify starch. Instead, nonthermal physical modification methods involve the application of techniques such as the use of irradiation, high hydrostatic pressure, ultrasound treatment, microwaves and electric pulses. The use of these technologies implies the destruction of most pathogenic or spoilage microorganisms. Also, they can inactivate enzymes and preserve the taste, color, texture, nutrients and heat-labile functional characteristics of starch (Lawal, 2019; Wang et al., 2020). However, it is important to know if these nonthermal methods affect starch digestibility. For example, gamma irradiation can increase or decrease the resistant starch fraction depending on the botanical source and irradiation conditions (Kong, 2018). In the case of gamma irradiation of potato and bean starches, an increase in resistant starch fraction, a decrease of the slowly digestible starch fraction and no effect on the rapidly digestible starch was observed (Chung & Liu, 2010). Structural changes and chain cleavage, or an increase in the proportion of β-bonded starch due to transglucosidation or an increase in carboxyl groups that could inhibit the enzyme attack could be possible causes of these changes in starch digestibility (Chung & Liu, 2010; Kong, 2018). Hence, Sujatha et al. (2018) suggest that gamma irradiation can be used as a technology to improve the starch digestibility of selected milled grains.

Another nonthermal treatment used to modify starch's physicochemical and digestibility properties is pulsed electric field. It has been reported that japonica rice starches subjected to this method had increased rapidly digestible starch, decreased slowly digestible starch and no change in the resistant starch content (Wu et al., 2019). Wu et al. (2019) speculated that damage to the granule surface allowed the digestive enzymes to reach the inner part of the starch granule and alter its structure, making it less compact, then causing faster hydrolysis of the starch and decreasing the slowly digestible starch. These results are not beneficial for the nutritional properties of starch because they could indicate that this treatment enhances the starch hydrolysis and negatively impact glycemic response. Abduh et al. (2019) reported that pulsed electric field treatment did not significantly change the digestibility of potato starch.

Other nonthermal treatments, such as microwave treatment, could enhance the slowly digestible starch fraction and reduce the digestion rate of starch (Li et al., 2020). Reduced digestion rates were also observed in normal maize and potato starches subjected to low and high microwave power. According to Li et al. (2019, 2020), microwave treatment promoted the formation of more order and periodic amorphous-crystalline structures, which were reassembled as starch chain packing domains with an intermediate bulk density to generate slowly digestible domains. Here, Xu et al. (2019) found that the slowly and rapidly digestible starch fractions increased with the microwave power, while a reduction of the resistant starch fraction was observed in both normal maize and potato starches. Such results indicated microwave treatment could damage the morphological structure, crystalline structure and helical structure of starches having stronger destructive effects on the B-type and A-type granules of potato and normal maize starch, respectively. The presence of pinholes in normal maize or hollows in potato starch also determined the diffusion of enzymes into the starch granule. Normal maize presented an inside-out digestion pattern (crystalline and helical structures were crucial factors affecting digestibility). In contrast, potato starch showed an exo-pitting digestion pattern where the morphological structure was the first barrier to enzymatic hydrolysis. Still, the presence of hollows after high-power microwave treatment increases the enzyme diffusion into the granule interior (Xu et al., 2019).

In general, it seems like most of the nonthermal treatments improve the physicochemical properties of starches but not their digestibility. Therefore, these methods should be combined with other technologies or thermal processes to improve the digestion properties of starch, as was reported by Liu et al. (2021). The use of more than one technology is named dual modification, and it is used to improve some undesirable functional properties or improve the digestibility of starch. Dual technology denotes the combination of two modifications between chemical and physical treatments or chemical and enzymatic modification, among others combinations (Punia, 2020).

5.3.3 Enzymatic Modification

Enzymatic modification has been used to partly replace the physical or chemical modifications, mainly because this technology is environmentally clean and safe for consumers (Punia, 2020). The main advantage of this modification is related to the mild reaction conditions, high selectivity and the reduction of undesirable by-products as those generated in the chemical modification (Wu & Zhou, 2018). The enzymes used to modify starch are α-amylase, β-amylase, pullulanase, glucoamylase, debranching enzymes, cyclodextrins glycosyltransferase and glucose isomerase (Wang et al., 2020). Enzymatic modification represents an alternative to modifying the physicochemical properties of starch but also obtaining desirable starch digestibility. For example, amylosucrase has been used to modify waxy starches to increase the content of slowly digestible starch and resistant starch (Zhang et al., 2019). The slowly digestible and resistant starch contents are increased by elongating nonreducing ends of amylose and amylopectin (Kim et al., 2016). The development of B-type crystallinity is responsible for the decrease in enzymatic susceptibility (Rolland-Sabaté et al., 2004). Another possible explanation may be given by the molecular model with α-amylase showing that the hydrolytic site of the enzyme could not fit inside the large and stiff fragments. Thus, the resistance to digestion was controlled by the structure of the double helices and not by the relative crystallinity (Zhang et al., 2017). Additionally, the same amylosucrase enzyme used to modify starch indicates this treatment favors the formation of resistant starch (Zhang et al., 2019) because the double-helical structure controls the rate of starch digestion (Zhang et al., 2017).

Another enzyme that promotes the formation of resistant starch is pullulanase, which produces short amylose and amylopectin molecules by debranching starch. This action causes molecular alignment and aggregation opportunities (Ma et al., 2018; Li et al., 2020). According to Li et al. (2020), pullulanase treatment can be combined with autoclaving to improve the physicochemical properties of rice starch; however, the sequential modification of autoclaving, followed by β-amylase, transglucosidase, and pullulanase, increased the slowly digestible starch and resistant starch fractions, and also reduced the glycemic index. The autoclaving and sequential triple enzymatic modification of starch showed those results due to the significant increase of linear short chains inside the modified starch granules. Such short linear segments have fewer enzymatic attack sites than the long straight chains, so the structure formed is less susceptible to enzymatic hydrolysis during digestion.

Dual enzyme modification of oat starches using β-amylase and transglucosidase resulted in a decrease in the amount of glucose released during simulated gastrointestinal tract conditions and increased the resistant starch content. These results can be due to the following reasons: (a) β-amylase catalyzed the breaking of α-1,4 linkages of amylose and amylopectin while transglucosidase transferred D-glucosyl residues of maltose to the primary hydroxyl groups on C6 of suitable substrate, which increased the number of α-1,6 glycosidic bonds and shorter proportions of amylose and amylopectin chains; (b) formation of short linear chains and trimming of external chains of amylopectin that enhances the amylose–amylose interactions and RS3 crystallites; and (c) dual enzymatic modification increased the amylose–lipid complex, which is resistant to digestive enzymes (Shah et al., 2018). Another dual enzymatic modification to improve the digestibility of starch was reported by Li et al. (2019), who treated corn starch with α-D-glucan branching enzyme and cyclodextrin glucosyltransferase. These authors reported that dual enzyme modification improved the amount of slowly digestible starch and resistant starch. These results are due to the increase of the short side chains and branching degree. Also, they mention that cyclodextrins formed with cyclodextrin glucosyltransferase treatment formed complexes with the branched amylopectin clusters by hydrogen bonds and other intermolecular forces, slowing the digestion of starch. On the other hand, modifying corn, waxy corn, potato and tapioca starches with α-D-glucan branching enzyme decreased the starch digestion rate and increased the slowly digestible starch depending on the starch botanical source. Among these starches, tapioca starch had better digestibility properties after enzymatic treatment. Such behavior was attributed to the shortening of the average chain length, the decrease of the crystallinity and the increase of the α-1,6 linkages (Li et al., 2018).

Overall, single or dual enzymatic modification improved the digestibility properties of starch and the glycemic index. According to this, authors indicate this type of modification is an alternative to produce functional ingredients to develop starch-based foods for controlling the glucose release due to the resistant starch content acting as dietary fiber in enzymatically modified starches. However, it should be taken into account enzymatic modification is complex and time consuming.

5.3.4 Biological Modification

Biological modification of plants helps create novel starches with new functional properties similar to those of chemically or physically modified starches while maintaining some important qualities of the native starches (Visser & Jacobsen, 1993). It is well known that normal starch has an amylose/amylopectin ratio around 1:3 to 1:5, or around 25% amylose and 75% amylopectin. This characteristic is responsible for the physicochemical properties of the starch. For the biosynthesis of amylose and amylopectin, enzymes such as starch synthase (S.S.), starch branching enzymes (SBE) and debranching enzymes, and granule bound starch synthases (GBSS or waxy proteins) are involved. The alteration of these enzymes can generate novel starches rich in amylose or amylopectin, both presenting unique physicochemical properties. Therefore, the mechanisms controlling the starch biosynthesis can be identified to manipulate the starch composition through alteration of the genes encoding enzymes involved in the starch synthesis (Lafiandra et al., 2008). In addition, the use of biological modifications of starch could have better environmental benefits compared to chemical modifications (Visser & Jacobsen, 1993).

Biological modification has been done for some time, but the production of biologically modified starch coming from these types of crops faces high regulatory costs. Examples of this type of modification are the production of waxy and high amylose starches mostly from cereals (rice, wheat, barley, corn), the primary starch sources worldwide. Between waxy and high amylose starches, high amylose starches show better nutritional benefits due to their high resistant starch content. For this reason, high amylose starches can bring benefits to the consumers through a decrease in the incidence of lifestyle diseases and substantial health costs associated with these conditions (Regina et al., 2014). According to Li et al. (2019) there are several biological modifications to enhance the amylose content in starch: (a) suppression of SBE, (b) suppression of soluble S.S. and (c) overexpression of GBSS. These biological alternatives alter the biosynthesis enzymes, increasing the proportion of long chains with a certain degree of polymerization and decreasing the proportions of very short chains and short chains (Li et al., 2019). For example, suppression of SBE in maize increased the amylose content from 27% to 57%–71%, but changed the amylopectin molecular structure by decreasing very short chains with a degree of polymerization between 6 and 10 and short chains with a degree of polymerization between 11 and 20, while increasing the long chains with a degree of polymerization higher than 20 (Shi et al., 1998).

The granule size, porosity, amylopectin chain length distribution, and crystallinity influence the digestibility of normal and waxy rice starches. Knowledge about such characteristics can help choose waxy rice starches with better starch digestibility properties. The starch nutritional fractions and estimated glycemic index of six waxy rice starches were determined. These starches presented high levels of rapidly digestible starch, followed by slowly digestible starch and, finally, a minor amount of resistant starch. Regarding the estimated glycemic index, high values were observed. Therefore, the proportion of short branch chains of amylopectin and the crystallinity are factors involved in controlling the starch digestibility of waxy rice starches (You et al., 2014). In the case of waxy maize starch (Xu et al., 2020b), the digestibility properties differed because the rapidly digestible starch showed lower values than those reported in waxy rice starches. Additionally, waxy maize starches presented high values of slowly digestible starch and resistant starch compared to waxy rice starches. So, the genetically modified starch will deliver different digestibility depending on the botanical source the starch backbone structure, and crystallinity. For this reason, this type of starch has been subjected to dual modification (with enzymatic, chemical or physical modifications) because of the interest to increase its application in starch-based foods with evidence-based health benefits.

5.4 INNOVATIVE METHODS FOR STARCH MODIFICATION

Due to environmental concerns and the need to produce starch labeled as generally recognized as safe (GRAS), innovative technologies have been developed to modify starch. These innovative technologies target increased starch application in the food industry. Due to the poor solubility of starch in water and other solvents (Chen et al., 2018), strong solvents and hazardous chemicals are used to increase the degree of substitution of the modified starch. However, new technologies have been developed to replace the traditional reagents/solvents to dissolve native starch and increase the degree of substitution. For example, ionic liquid or supercritical carbon dioxide can be used with the starch modification methods mentioned above (Fan & Picchioni, 2020).

5.4.1 Ionic Liquids

Ionic liquids are a group of salts with a delocalized charge, composed of a combination of organic cation and smaller organic and inorganic anion. Also, they are low volatile, non-flammable, and have thermal and chemical stability (Fan & Picchioni, 2020; Chen et al., 2018). Ionic liquids can be considered environmentally friendly compared to typical solvents used to modify starch based on these characteristics. Examples of ionic liquids are imidazolium, pyridinium, ammonium, others (Fan & Picchioni, 2020). Ionic liquids are expected to be used to increase the degree of substitution of starches, development of starch-based composite films, fabrication of biopolymers, starch nanoparticles, and drug carriers (Ren et al., 2020). Still, there is no information about the application of ionic liquids to develop starch-based foods, mainly in using this technology to produce starch with appropriate digestion properties.

5.4.2 Supercritical CO_2

Supercritical fluids are substances held at or above their critical point at a determined temperature and pressure, which results in the disappearance of distinct liquid and gas phases. Supercritical fluids have properties from both gas and liquid, including low viscosity, high diffusion coefficient, and good permeability and dissolving capacity. Supercritical CO_2 is the most common supercritical fluid used (Fan & Picchioni, 2020; Muljana et al., 2011; Villegas et al., 2020). Properties of supercritical fluids make them a good option for starch modification through chemical or physical technologies. The use of this fluid as a solvent in starch modification helps to increase the degree of substitution but also changes the physical structure of the amylose and amylopectin by the presence of CO_2 (Muljana et al., 2009).

In general, most of the research using these types of innovative methods to modify starch, combined with chemical or physical modification, was done to improve the starch quality for applications in the elaboration of starch-based materials in an efficient and green way (Fan & Picchioni, 2020). However, more research needs to be done to use these technologies to produce starch with better digestibility properties.

5.5 HOW STARCH MODIFICATION AFFECTS STARCH DIGESTIBILITY

Starch plays a vital role in regulating the glycemic load response via digestion and the absorption rate. Starch is classified into three fractions according to its digestibility (rapidly digestible starch, slowly digestible starch and resistant starch). The proportion of these fractions will be related to the glycemic index, which is the postprandial glucose rise in blood compared to a reference standard glucose. Starches with a high proportion of rapidly digestible starch are related to a high glycemic index, while starches with high levels of slowly digestible starch and resistant starch are important because they have been related to a low glycemic index (Cai et al., 2020; Sorndech et al., 2018). In addition, resistant starch and very slowly digestible starch can reach the colon and act as a prebiotic fermented by the microbiota to produce short-chain fatty acids. These relevant characteristics in both fractions of starch are helpful because they can aid in the management of obesity, metabolic syndrome and other conditions (Sorndech et al., 2018; Ross, 2019).

The modification of starch is an essential tool to improve the nutrition and energy values of starch-based foods. As was mentioned earlier, the chemical, physical (by thermal treatments), enzymatic and dual modifications are the technologies impacting the levels of starch fractions. In the case of enzymatic modification using a combination of starch-active enzymes in the glucosyltransferase and glucanohydrolase families, slowly digestible and nondigestible glucans are reported to be produced. The slowly digestible products are gradually degraded and absorbed over an extended period in the small intestine, maintaining blood glucose levels and decreasing the glycemic index. On the other hand, the nondigestible products become fermented by microbiota in the colon, promoting gut health (Sorndech et al., 2018; Ross, 2019).

Much research combining starch modifications has been done to improve the physicochemical properties of native starch. Still, the effect of these modifications on the starch nutritional properties, especially on the starch fractions and the glycemic response, needs to be investigated. Additional information is required to increase the application of these starches in starch-based foods for specialized diets to treat obesity, metabolic syndrome and other related diseases.

5.6 CONCLUDING REMARKS AND FUTURE TRENDS

Native starches are modified to improve their physicochemical properties, functionality and digestibility properties. Currently, starch is explicitly modified to develop starch-based foods for specific applications to treat obesity, metabolic syndrome and other diseases related to the high intake of glucose.

- Because chemical modifications generate non-environmentally friendly residues, it is vital to implement effective green technologies to produce chemically modified starches.
- Metabolomic and molecular biology studies to determine the production of short-chain fatty acids and microbiota composition, respectively, are necessary to investigate the effect of improving slowly digestible starch and resistant starch fractions on the lower gastrointestinal tract and its relationship with health benefits in different target tissues.
- The development of starch-based foods directed to people with specific dietary requirements needs attention because of obesity, metabolic syndrome and cancers cases worldwide.

REFERENCES

Abduh, S. B. M., Leong, S. Y., Agyei, D., & Oey, I. (2019). Understanding the properties of starch in potatoes (Solanum Tuberosum Var. Agria) after being treated with pulsed electric field processing. *Foods*, 8(5), Multidisciplinary Digital Publishing Institute: 159. doi:10.3390/FOODS8050159.

Ai, Y., & Jane, J. L. (2018). Understanding starch structure and functionality. In *Starch in Food: Structure, Function and Applications: Second Edition, January* (pp. 151–178). Woodhead Publishing. doi:10.1016/B978-0-08-100868-3.00003-2.

Alcázar-Alay, S. C., & Almeida Meireles, M. A.. (2015). Physicochemical properties, modifications and applications of starches from different botanical sources. *Food Science and Technology*, 35(2), Sociedade Brasileira de Ciência e Tecnologia de Alimentos: 215–236. doi:10.1590/1678-457X.6749.

Arns, B., Bartz, J., Radunz, M., do Evangelho, J. A., Pinto, V. Z., da Rosa Zavareze, E., & Guerra Dias, A. R.. (2015). Impact of heat-moisture treatment on rice starch, applied directly in grain paddy rice or in isolated starch. *LWT: Food Science and Technology*, 60(2), Academic Press: 708–713. doi:10.1016/J.LWT.2014.10.059.

Ashogbon, A. O. (2021). Current research addressing physical modification of starch from various botanical sources. Accessed November 29. https://www.researchgate.net/publication/325877081.

Ashogbon, A. O., Akintayo, E. T., Oladebeye, A. O., Oluwafemi, A. D., Akinsola, A. F., & Evans Imanah, O. (2020). Developments in the isolation, composition, and physicochemical properties of legume starches. 61(17), *Critical Reviews in Food Science and Nutrition*, Taylor & Francis: 2938–2959. doi:10.1080/10408398.2020.1791048.

Bajaj, R., Singh, N., & Kaur, A.. (2019). Properties of octenyl succinic anhydride (OSA) modified starches and their application in low fat mayonnaise. *International Journal of Biological Macromolecules*, 131(June). Elsevier: 147–157. doi:10.1016/J. IJBIOMAC.2019.03.054.

Baker, A. A., Miles, M. J., & Helbert, W.. (2001). Internal structure of the starch granule revealed by AFM. *Carbohydrate Research*, 330(2), Elsevier: 249–256. doi:10.1016/ S0008-6215(00)00275-5.

Bello-Perez, L. A., Flores-Silva, P. C., Agama-Acevedo, E., & Tovar, J.. (2020). Starch digestibility: Past, present, and future. *Journal of the Science of Food and Agriculture*, John Wiley and Sons Ltd. 100: 5009–5016. doi:10.1002/jsfa.8955.

Bemiller, J. N, & Huber, K. C. (2015). Physical modification of food starch functionalities. *The Annual Review of Food Science and Technology*, 6, 19–69. doi:10.1146/ annurev-food-022814-015552.

Bertoft, E. (2017). Understanding starch structure: Recent progress. *Agronomy*, 7(3), Multidisciplinary Digital Publishing Institute: 56. doi:10.3390/ AGRONOMY7030056.

Bi, Y., Zhang, Y., Jiang, H., Hong, Y., Gu, Z., Cheng, L., Li, Z., & Li, C.. (2017). Molecular structure and digestibility of banana flour and starch. *Food Hydrocolloids*, 72(November), Elsevier B.V.: 219–227. doi:10.1016/j.foodhyd.2017.06.003.

Buléon, A., Colonna, P., Planchot, V., & Ball, S.. (1998). Starch granules: Structure and biosynthesis. *International Journal of Biological Macromolecules*, 23(2), Elsevier: 85–112. doi:10.1016/S0141-8130(98)00040-3.

Burton, R. A., Jenner, H., Carrangis, L., et al. (2002). Starch granule initiation and growth are altered in barley mutants that lack isoamylase activity. *The Plant Journal*, 31(1), John Wiley & Sons, Ltd: 97–112. doi:10.1046/J.1365-313X.2002.01339.X.

Cai, C., Tian, Y., Zhiwei, Y., Sun, C., & Jin, Z.. (2020). In vitro digestibility and predicted glycemic index of chemically modified rice starch by one-step reactive extrusion. *Starch/Stärke*, 72(1–2). John Wiley & Sons, Ltd: 1900012. doi:10.1002/ STAR.201900012.

Chen, J., Xie, F., Li, X., & Chen, L.. (2018). Ionic liquids for the preparation of biopolymer materials for drug/gene delivery: A review. *Green Chemistry*, 20(18), The Royal Society of Chemistry: 4169–4200. doi:10.1039/C8GC01120F.

Chen, Q., Haojie, Y., Wang, L., et al. (2015). Recent progress in chemical modification of starch and its applications. *RSC Advances*, 5(83). The Royal Society of Chemistry: 67459–67474. doi:10.1039/C5RA10849G.

Chi, C., Li, X., Lu, P., Miao, S., Zhang, Y., & Chen, L.. (2019). Dry heating and annealing treatment synergistically modulate starch structure and digestibility. *International Journal of Biological Macromolecules*, 137(September), Elsevier B.V.: 554–561. doi:10.1016/j.ijbiomac.2019.06.137.

Chung, H. J., & Liu, Q.. (2010). Molecular structure and physicochemical properties of potato and bean starches as affected by gamma-irradiation. *International Journal of Biological Macromolecules*, 47(2), Elsevier: 214–222. doi:10.1016/J. IJBIOMAC.2010.04.019.

Chung, H. J., Liu, Q., & Hoover, R.. (2009). Impact of annealing and heat-moisture treatment on rapidly digestible, slowly digestible and resistant starch levels in native and gelatinized corn, pea and lentil starches. *Carbohydrate Polymers*, 75(3), Elsevier: 436–447. doi:10.1016/J.CARBPOL.2008.08.006.

Copeland, Les, J. B., Salman, H., & Tang, M. C.. (2009). Form and functionality of starch. *Food Hydrocolloids*, 23(6), Elsevier: 1527–1534. doi:10.1016/J. FOODHYD.2008.09.016.

Cummings, J. H., & Englyst, H. N.. (1995). Gastrointestinal effects of food carbohydrate. *The American Journal of Clinical Nutrition*, 61(4, Supplement), Oxford Academic: 938S–945S. doi:10.1093/AJCN/61.4.938S.

Davis, J. P., Supatcharee, N., Khandelwal, R. L., & Chibbar, R. N. (2003). Synthesis of Novel Starches in Planta: Opportunities and Challenges. *Starch/Stärke*, 55(3–4), 107–120. John Wiley & Sons, Ltd:. doi:10.1002/STAR.200390036.

De Ambrogio E., Lafiandra D., Saliola Bucelli A., Sestili F., Silvestri M. Genetic modification of starch composition in wheat. In: Molina-Cano J.L. (ed.), Christou P. (ed.), Graner A. (ed.), Hammer K. (ed.), Jouve N. (ed.), Keller B. (ed.), Lasa J.M. (ed.), Powell W. (ed.), Royo C. (ed.), Shewry P. (ed.), Stanca A.M. (ed.). *Cereal Science and Technology for feeding ten billion people: genomics era and beyond.* Zaragoza: CIHEAM/IRTA, 2008, pp. 206–270 (Options Mediterraneennes: Serie A. Seminaires Mediterraneens, n. 81).

Ding, L., Huang, Q., Xiang, W., Fu, X., Zhang, B., & Wu, J.-Y. (2022). Chemical cross-linking reduces in vitro starch digestibility of cooked potato parenchyma cells. *Food Hydrocolloids*, 124(March). Elsevier: 107297. doi:10.1016/J.FOODHYD.2021 .107297.

Dinges, J.R., Colleoni, C., James, M. G., & Myers, A. M. (2003). Mutational analysis of the pullulanase-type debranching enzyme of maize indicates multiple functions in starch metabolism. *The Plant Cell*, 15(3). Oxford Academic: 666–680. doi:10.1105/ TPC.007575.

Edwards, A., Fulton, D. C., Hylton, C. M., Jobling, S. A., Gidley, M., Rössner, U., Martin, C., & Smith, A. M. (1999). A combined reduction in activity of starch synthases II and III of potato has novel effects on the starch of tubers. *The Plant Journal*, 17(3), John Wiley & Sons, Ltd: 251–261. doi:10.1046/J.1365-313X.1999.00371.X.

Englyst, K. N., & Englyst, H. N.. (2005). Carbohydrate bioavailability. *British Journal of Nutrition*, 94(1). Cambridge University Press: 1–11. doi:10.1079/BJN20051457.

Englyst, H. N., Kingman, S. M., Hudson, G. J., & Cummings, J. H. (1996). Measurement of resistant starch in vitro and in vivo. *British Journal of Nutrition*, 75(5), Cambridge University Press: 749–755. doi:10.1079/BJN19960178.

Englyst, K. N., Liu, S., & Englyst, H. N.. (2007). Nutritional characterization and measurement of dietary carbohydrates. *European Journal of Clinical Nutrition*, 61(1). Nature Publishing Group: S19–S39. doi:10.1038/sj.ejcn.1602937.

Espinosa-Solis, V., Jane, J.-L., & Bello-Perez, L. A. (2009). Physicochemical characteristics of starches from unripe fruits of mango and banana. *Starch/Stärke*, 61(5). John Wiley & Sons, Ltd: 291–299. doi:10.1002/STAR.200800103.

Espinosa-Solis, V., Sanchez-Ambriz, S. L., Hamaker, B. R., & Bello-Pérez, L. A. (2011). Fine structural characteristics related to digestion properties of acid-treated fruit starches. *Starch/Stärke*, 63(11). John Wiley & Sons, Ltd: 717–727. doi:10.1002/ STAR.201100050.

Fan, Y., & Picchioni, F. (2020). Modification of starch: A review on the application of 'green' solvents and controlled functionalization. *Carbohydrate Polymers*, 241(August). Elsevier: 116350. doi:10.1016/J.CARBPOL.2020.116350.

Felisberto, M. H. F., Costa, M. S., Boas, F.V., Leivas, C. L., Franco, C. M. L., de Souza, S. M., Silva Clerici, M. T. P., & Cordeiro, L. M. C. (2020). Characterization and technological properties of peach palm (Bactris Gasipaes Var. Gasipaes) fruit starch. *Food Research International*, 136(October). Elsevier: 109569. doi:10.1016/J. FOODRES.2020.109569.

Fu, Z., Zhang, L., Ren, M. H., & BeMiller, J. N. (2019). Developments in hydroxypropylation of starch: A review. *Starch/Stärke*, 71(1–2). John Wiley & Sons, Ltd: 1800167. doi:10.1002/STAR.201800167.

Fujita, S., Yamamoto, H., Sugimoto, Y., Morita, N., & Yamamori, M. (1998). Thermal and crystalline properties of waxy wheat (Triticum Aestivum L.) starch. *Journal of Cereal Science*, 27(1). Academic Press: 1–5. doi:10.1006/JCRS.1997.0152.

Gallant, D. J., Bouchet, B., & Baldwin, P. M.. (1997). Microscopy of starch: Evidence of a new level of granule organization. *Carbohydrate Polymers*, 32(3–4). Elsevier: 177–191. doi:10.1016/S0144-8617(97)00008-8.

Goel, Charu, Semwal, A. D., Ayub, K., Kumar, S., & Sharma, G. K. (2020). Physical modification of starch: Changes in glycemic index, starch fractions, physicochemical and functional properties of heat-moisture treated buckwheat starch. *Journal of Food Science and Technology*, 57(8). Springer: 2941–2948. doi:10.1007/S13197-020-04326-4/TABLES/3.

Gunaratne, A., & Hoover, R.. (2002). Effect of heat–moisture treatment on the structure and physicochemical properties of tuber and root starches. *Carbohydrate Polymers*, 49(4). Elsevier: 425–437. doi:10.1016/S0144-8617(01)00354-X.

Güzel, F., & Sayar, S. (2010). Digestion profiles and some physicochemical properties of native and modified borlotti bean, chickpea and white kidney bean starches. *Food Research International*, 43(8). Elsevier: 2132–2137. doi:10.1016/J. FOODRES.2010.07.025.

Haq, F., Haojie, Y., Wang, L., Teng, L., Haroon, M., Khan, R. U., Mehmood, S., et al. (2019). Advances in chemical modifications of starches and their applications. *Carbohydrate Research*, 476(April). Elsevier: 12–35. doi:10.1016/J. CARRES.2019.02.007.

Hoover, R., & Vasanthan, T. (1994). Effect of heat-moisture treatment on the structure and physicochemical properties of cereal, legume, and tuber starches. *Carbohydrate Research*, 252(C). Elsevier: 33–53. doi:10.1016/0008-6215(94)90004-3.

Imberty, Anne, Alain Buléon, Vinh Tran, & Serge Péerez. (1991). Recent advances in knowledge of starch structure. *Starch/Stärke*, 43(10). John Wiley & Sons, Ltd: 375–384. doi:10.1002/STAR.19910431002.

Jane, Jay-Lin, Aol, Z., Duvick, S. A., Wiklund, M., Yoo, S.-H., Wong, K.-S., and Candice Gardner. (2003). Structures of amylopectin and starch granules: How are they synthesized? *Journal of Applied Glycoscience*, 50(2). The Japanese Society of Applied Glycoscience: 167–172. doi:10.5458/JAG.50.167.

Jenkins, P. J., & Donald, A. M. (1995). The influence of amylose on starch granule structure. *International Journal of Biological Macromolecules*, 17(6). Elsevier: 315–321. doi:10.1016/0141-8130(96)81838-1.

Jobling, S. (2004). Improving starch for food and industrial applications. *Current Opinion in Plant Biology*, 7(2). Elsevier Current Trends: 210–218. doi:10.1016/J. PBI.2003.12.001.

Kaur, M., Sandhu, K.S., & Taik Lim, S. (2010). Microstructure, physicochemical properties and in vitro digestibility of starches from different indian lentil (lens culinaris) cultivars. *Carbohydrate Polymers*, 79(2). Elsevier: 349–355. doi:10.1016/J. CARBPOL.2009.08.017.

Kawabata, A., Takase, N., Miyoshi, E., Sawayama, S., Kimura, T., & Kudo, K.. (1994). Microscopic observation and X-ray diffractometry of heat/moisture-treated starch granules. *Starch/Stärke*, 46(12). John Wiley & Sons, Ltd: 463–469. doi:10.1002/STAR.19940461204.

Kim, B. K., Kim, H. I., Moon, T. W., & Choi, S. J. (2014). Branch chain elongation by amylosucrase: Production of waxy corn starch with a slow digestion property. *Food Chemistry*, 152(June). Elsevier: 113–20. doi:10.1016/J.FOODCHEM.2013.11.145.

Kim, B. S., Kim, H. S., Hong, J. S., K. C. Huber, J. H. Shim, & Yoo, S. H. (2013). Effects of amylosucrase treatment on molecular structure and digestion resistance of pre-gelatinised rice and barley starches. *Food Chemistry*, 138(2–3). Elsevier: 966–975. doi:10.1016/J.FOODCHEM.2012.11.028.

Kim, J. H., Kim, H. R., Choi, S. J., Park, C. S., & Moon, T. W.. (2016). Production of an in vitro low-digestible starch via hydrothermal treatment of amylosucrase-modified normal and waxy rice starches and its structural properties. *Journal of Agricultural and Food Chemistry*, 64(24). American Chemical Society: 5045–5052. doi:10.1021/ACS.JAFC.6B01055/SUPPL_FILE/JF6B01055_SI_001.PDF.

Kong, B. W., Kim, J. I., Kim, M. J., & Kim, J. C.. (2003). Porcine pancreatic α-amylase hydrolysis of native starch granules as a function of granule surface area. *Biotechnology Progress*, 19(4). American Chemical Society (ACS): 1162–1166. doi:10.1021/BP034005M.

Kong, X. (2018). Gamma Irradiation of Starch. In: Sui, Z., Kong, X. (eds). *Physical Modifications of Starch*, (January). Springer, Singapore. https://doi.org/10.1007/978-981-13-0725-6_5.

Lawal, M. V. (2019). Modified starches as direct compression excipients: Effect of physical and chemical modifications on tablet properties: A review. *Starch/Stärke*, 71(1–2). John Wiley & Sons, Ltd: 1800040. doi:10.1002/STAR.201800040.

Lee, E. S., Lee, B. H., Shin, D. U., Lim, M. Y., Chung, W. H., Park, C. S., Baik, M. Y., Do Nam, Y., & Seo, D. H. (2018). Amelioration of obesity in high-fat diet-fed mice by chestnut starch modified by amylosucrase from deinococcus geothermalis. *Food Hydrocolloids*, 75(February). Elsevier: 22–32. doi:10.1016/J.FOODHYD.2017.09.019.

Lewicka, K., Siemion, P., & Kurcok, P. (2015). Chemical Modifications of Starch: Microwave Effect. *International Journal of Polymer Science*, vol. 2015, Article ID 867697, 10 pages, 2015. https://doi.org/10.1155/2015/867697.

Li, C., Wu, A., Wenwen, Y., Hu, Y., Li, E., Zhang, C., & Liu, Q. (2020a). Parameterizing starch chain-length distributions for structure-property relations. *Carbohydrate Polymers*, 241(August). Elsevier: 116390. doi:10.1016/J.CARBPOL.2020.116390.

Li, D., & Zhu, F. (2017). Physicochemical properties of kiwifruit starch. *Food Chemistry*, 220(April). Elsevier: 129–136. doi:10.1016/J.FOODCHEM.2016.09.192.

Li, H., Gidley, M. J., & Dhital, S. (2019a). High-amylose starches to bridge the 'fiber gap': Development, structure, and nutritional functionality. *Comprehensive Reviews in Food Science and Food Safety*, 18(2), John Wiley & Sons, Ltd: 362–379. doi:10.1111/1541-4337.12416.

Li, H., Gui, Y., Li, J., Zhu, Y., Cui, B., & Guo, L. (2020b). Modification of rice starch using a combination of autoclaving and triple enzyme treatment: Structural, physicochemical and digestibility properties. *International Journal of*

Biological Macromolecules, 144(February). Elsevier: 500–508. doi:10.1016/J. IJBIOMAC.2019.12.112.

Li, N., Cai, Z., Guo, Y., Xu, T., Qiao, D., Zhang, B., Zhao, S., et al. (2019b). Hierarchical structure and slowly digestible features of rice starch following microwave cooking with storage. *Food Chemistry*, 295(October). Elsevier: 475–483. doi:10.1016/J. FOODCHEM.2019.05.151.

Li, N., Wang, L., Zhao, S., Qiao, D., Jia, C., Niu, M., Lin, Q., & Zhang, B. (2020c). An insight into starch slowly digestible features enhanced by microwave treatment. *Food Hydrocolloids*, 103(June). Elsevier: 105690. doi:10.1016/J.FOODHYD.2020.105690.

Li, Y., Ren, J., Jia, L., Sun, L., Wang, Y., Liu, B., Li, C., & Li, Z. (2018a). Modification by α-D-glucan branching enzyme lowers the in vitro digestibility of starch from different sources. *International Journal of Biological Macromolecules*, 107(February). Elsevier: 1758–1764. doi:10.1016/J.IJBIOMAC.2017.10.049.

Li, Y., Li, C., Gu, Z., Cheng, L., Hong, Y., & Li, Z. (2019c). Digestion properties of corn starch modified by α-D-glucan branching enzyme and cyclodextrin glycosyltransferase. *Food Hydrocolloids*, 89(April). Elsevier: 534–541. doi:10.1016/J. FOODHYD.2018.11.025.

Li, Z., Guo, K., Lin, L., He, W., Zhang, L., & Wei, C. (2018b). Comparison of physicochemical properties of starches from flesh and peel of green banana fruit. *Molecules*, 23(9). Multidisciplinary Digital Publishing Institute: 2312. doi:10.3390/ MOLECULES23092312.

Liu, K., Zhang, B., Chen, L., Li, X., & Zheng, B. (2019). Hierarchical structure and physicochemical properties of highland barley starch following heat moisture treatment. *Food Chemistry*, 271(January). Elsevier: 102–108. doi:10.1016/J. FOODCHEM.2018.07.193.

Liu, Q., Jiao, A., Yang, Y., Wang, Y., Li, J., Xu, E., Yang, G., & Jin, Z. (2021). The combined effects of extrusion and recrystallization treatments on the structural and physicochemical properties and digestibility of corn and potato starch. *LWT*, 151(November), Academic Press: 112238. doi:10.1016/J.LWT.2021.112238.

Lopez-Silva, M., Bello-Perez, L. A., Castillo-Rodriguez, V. M., Agama-Acevedo, E., & Alvarez-Ramirez, J. (2020). In vitro digestibility characteristics of octenyl succinic acid (OSA) modified starch with different amylose content. *Food Chemistry*, 304(January), Elsevier: 125434. doi:10.1016/J.FOODCHEM.2019.125434.

Luo, K., Wang, X., & Zhang, G. (2018). The anti-obesity effect of starch in a whole grain-like structural form. *Food & Function*, 9(7), The Royal Society of Chemistry: 3755–3763. doi:10.1039/C8FO00602D.

Ma, Z., Yin, X., Chang, D., Hu, X., & Boye, J. I. (2018). Long- and short-range structural characteristics of pea starch modified by autoclaving, α-amylolysis, and pullulanase debranching. *International Journal of Biological Macromolecules*, 120(December), Elsevier: 650–656. doi:10.1016/J.IJBIOMAC.2018.08.132.

Maniglia, B. C., Castanha, N., Le-Bail, P., Le-Bail, A., & Augusto, P. E. D. (2021). Starch modification through environmentally friendly alternatives: A review. *Critical Reviews in Food Science and Nutrition*, 61:15, 2482–2505. doi: 10.1080/10408398.2020.1778633.

Masina, N., Choonara, Y. E., Kumar, P., du Toit, L. C., Govender, M., Indermun, S., & Pillay, V. (2017). A review of the chemical modification techniques of starch. *Carbohydrate Polymers*, 157(February), Elsevier: 1226–1236. doi:10.1016/J.CARBPOL.2016.09.094.

Mathobo, V.M., Silungwe, H., Ramashia, S.E. et al. (2021). Effects of heat-moisture treatment on the thermal, functional properties and composition of cereal, legume and tuber starches—a review. *J Food Sci Technol* 58, 412–426. https://doi.org/10.1007/s13197-020-04520-4.

Miao, M., Jiang, B., Zhang, T., Jin, Z., & Mu, W. (2011). Impact of mild acid hydrolysis on structure and digestion properties of waxy maize starch. *Food Chemistry*, 126(2), Elsevier: 506–513. doi:10.1016/J.FOODCHEM.2010.11.031.

Morell, M. K., Blennow, A., Kosar-Hashemi, B., & Samuel, M. S. (1997). Differential expression and properties of starch branching enzyme isoforms in developing wheat endosperm. *Plant Physiology*, 113(1), Oxford Academic: 201–208. doi:10.1104/PP.113.1.201.

Muljana, H., Picchioni, F., Heeres, H. J., & Janssen, L.P.B.M. (2009). Supercritical carbon dioxide (ScCO$_2$) induced gelatinization of potato starch. *Carbohydrate Polymers*, 78(3). Elsevier: 511–519. doi:10.1016/J.CARBPOL.2009.05.005.

Muljana, H., Picchioni, F., Knez, Z., Heeres, H. J., & Janssen, L. P. B. M. (2011). Insights in starch acetylation in sub- and supercritical CO2. *Carbohydrate Research*, 346(10), Elsevier: 1224–1231. doi:10.1016/J.CARRES.2011.04.002.

Nakamura, Y., Takeichi, T., Kawaguchi, K., & Yamanouchi, H. (1992). Purification of two forms of starch branching enzyme (Q-enzyme) from developing rice endosperm. *Physiologia Plantarum*, 84(3), John Wiley & Sons, Ltd: 329–335. doi:10.1111/J.1399-3054.1992.TB04672.X.

Nwokocha, L. M., & Williams, P. A. (2009). New starches: Physicochemical properties of sweetsop (Annona Squamosa) and Soursop (Anonna Muricata) starches. *Carbohydrate Polymers*,78(3) Elsevier: 462–468. doi:10.1016/J.CARBPOL.2009.05.003.

Ogunsona, E., Ojogbo, E., & Mekonnen, T. (2018). Advanced material applications of starch and its derivatives. *European Polymer Journal*. Elsevier Ltd. doi:10.1016/j.eurpolymj.2018.09.039.

Olayemi O, Adetunji O, Isimi C. (2021). Physicochemical, structural characterization and pasting properties of pre-gelatinized Neorautanenia mitis starch. *Polimery w Medycynie*. Jan-Jun;51(1):7–16. doi: 10.17219/pim/138964. PMID: 34180609.

Oyeyinka, S.A., Adeloye, A.A., Smith, S.A., Adesina, B.O., & Akinwande, F.F. (2019). Physicochemical properties of flour and starch from two cassava varieties. *Agrosearch*, 19(1), African Journals Online (AJOL): 28–45. doi: 10.4314/agrosh.v19i1.3.

Park, J., Oh, S. K., Chung, H. J., & Park, H. J. (2020). Structural and physicochemical properties of native starches and non-digestible starch residues from Korean Rice cultivars with different amylose contents. *Food Hydrocolloids*, 102(May). Elsevier: 105544. doi:10.1016/J.FOODHYD.2019.105544.

Pérez, S., & Bertoft, E. (2010). The molecular structures of starch components and their contribution to the architecture of starch granules: A comprehensive review. *Starch/Stärke*, 62(8), John Wiley & Sons, Ltd: 389–420. doi:10.1002/STAR.201000013.

Przetaczek-Rożnowska, I. (2017). Physicochemical properties of starches isolated from pumpkin compared with potato and corn starches. *International Journal of Biological Macromolecules*, 101(August), Elsevier: 536–542. doi:10.1016/J.IJBIOMAC.2017.03.092.

Punia, S. (2020). Barley starch modifications: Physical, chemical and enzymatic: A review. *International Journal of Biological Macromolecules* 144(February), Elsevier: 578–585. doi:10.1016/J.IJBIOMAC.2019.12.088.

Punia, S., Kumar, M., Siroha, A. K., Kennedy, J. F., Dhull, S. B., & Whiteside, W. S. (2021). Pearl millet grain as an emerging source of starch: A review on its structure, physicochemical properties, functionalization, and industrial applications. *Carbohydrate Polymers*, 260(May), Elsevier: 117776. doi:10.1016/J.CARBPOL.2021.117776.

Regina, A., Li, Z., Morell, M.K., & Jobling, S.A. (2014). Genetically modified starch: State of art and perspectives. In Peter J. Halley & Luc Avérous (Eds.) *Starch Polymers*, Elsevier, 13–29. https://doi.org/10.1016/B978-0-444-53730-0.00019-1.

Ren, F., Wang, J., Xie, F., Zan, K., Wang, S., & Wang, S. (2020). Applications of ionic liquids in starch chemistry: A review. *Green Chemistry* 22(7), The Royal Society of Chemistry: 2162–2183. doi:10.1039/C9GC03738A.

Ren, Y., Yuan, T.Z., Chigwedere, C.M., & Ai, Y. F. (2021). A current review of structure, functional properties, and industrial applications of pulse starches for value-added utilization. *Comprehensive Reviews in Food Science and Food Safety* 20(3), Blackwell Publishing Inc.: 3061–3092. doi:10.1111/1541-4337.12735.

Reyes, I., Hernandez-Jaimes, C., Vernon-Carter, E. J., Bello-Perez, L. A., Alvarez-Ramirez, J., Reyes, I., Hernandez-Jaimes, C, Vernon-Carter, E. J., Alvarez-Ramirez, J., & Bello-Perez, L. A. CEPROBI Instituto Politécnico Nacional Yautepec Morelos. (2021). Air oxidation of corn starch: Effect of heating temperature on physicochemical properties and in vitro digestibility. *Starch/Stärke* 73(3–4), John Wiley & Sons, Ltd: 2000237. doi:10.1002/STAR.202000237.

Rolland-Sabaté, A., Colonna, P., Potocki-Véronèse, G., Monsan, P., & Planchot, V. (2004). Elongation and insolubilisation of α-Glucans by the action of Neisseria Polysaccharea Amylosucrase. *Journal of Cereal Science* 40(1), Academic Press: 17–30. doi:10.1016/J.JCS.2004.04.001.

Ross, S. M. (2019). Vive La Resistance!: Resistant starch supports blood sugar and weight maintenance. *Holistic Nursing Practice* 33(3), Lippincott Williams and Wilkins: 187–190. doi:10.1097/HNP.0000000000000329.

Samarakoon, E. R. J., Waduge, R., Liu, Q., Shahidi, F., & Banoub, J. H.. (2020). Impact of annealing on the hierarchical structure and physicochemical properties of waxy starches of different botanical origins. *Food Chemistry* 303(January), Elsevier: 125344. doi:10.1016/J.FOODCHEM.2019.125344.

Shah, A., Masoodi, F. A., Gani, A., & Ashwar, B. (2018). Dual enzyme modified oat starch: Structural characterisation, rheological properties, and digestibility in simulated GI tract. *International Journal of Biological Macromolecules*, 106(January). Elsevier: 140–147. doi:10.1016/J.IJBIOMAC.2017.08.013.

Sharma, R., Sissons, M. J., Rathjen, A. J., & Jenner, C. F.. (2002). The null-4A Allele at the waxy locus in durum wheat affects pasta cooking quality. *Journal of Cereal Science*, 35(3), Academic Press: 287–297. doi:10.1006/JCRS.2001.0423.

Sharma, V., Kaur, M., Sandhu, K. S., & Godara, S. K. (2020). Effect of cross-linking on physico-chemical, thermal, pasting, in vitro digestibility and film forming properties of faba bean (Vicia Faba L.) starch. *International Journal of Biological Macromolecules*, 159(September), Elsevier: 243–249. doi:10.1016/J.IJBIOMAC.2020.05.014.

Shi, Y. C., Capitani, T., Trzasko, P., & Jeffcoat, R. (1998). Molecular structure of a low-amylopectin starch and other high-amylose maize starches. *Journal of Cereal Science*, 27(3), Academic Press: 289–299. doi:10.1006/JCRS.1997.9998.

Simsek, S., Ovando-Martinez, M., Marefati, A., Sj, M., & Rayner, M. (2015). Chemical composition, digestibility and emulsification properties of octenyl succinic esters of various starches. *Food Research International*, 75, 41–49. doi:10.1016/j.foodres.2015.05.034.

Sorndech, W., Tongta, S., & Blennow, A. (2018). Slowly digestible- and non-digestible α-glucans: An enzymatic approach to starch modification and nutritional effects. *Starch/Stärke*, 70(9–10), John Wiley & Sons, Ltd: 1700145. doi:10.1002/STAR.201700145.

Srichuwong, S., & Jane, J. I. (2007). Physicochemical properties of starch affected by molecular composition and structures: A review. *Food Science and Biotechnology*, 16(5), 663–674. https://www.koreascience.or.kr/article/JAKO200735822355808.page.

Srichuwong, S., Curti, D., Austin, S., King, R., Lamothe, L., & Gloria-Hernandez, H. (2017). Physicochemical properties and starch digestibility of whole grain sorghums, millet, quinoa and amaranth flours, as affected by starch and non-starch constituents. *Food Chemistry*, 233(October), Elsevier: 1–10. doi:10.1016/J.FOODCHEM.2017.04.019.

Sujatha, M., Hymavathi, T.V., Uma, D. K., Pradeepa, R. T., & Praveen, K. D. (2018). Effect of heat treatment and gamma irradiation on in vitro starch digestibility of selected millet grains. *The Pharma Innovation Journal*, 7(12), 409–413. www.thepharmajournal.com.

Tanaka, N., Fujita, N., Nishi, A., Satoh, H., Hosaka, Y., Ugaki, M., Kawasaki, S., & Nakamura, Y. (2004). The structure of starch can be manipulated by changing the expression levels of starch branching enzyme IIb in rice endosperm. *Plant Biotechnology Journal*, 2(6), John Wiley & Sons, Ltd: 507–516. doi:10.1111/J.1467-7652.2004.00097.X.

Tester, R. F., Karkalas, J., & Qi, X. (2004). Starch: Composition, fine structure and architecture. *Journal of Cereal Science*, 39(2), Academic Press: 151–165. doi:10.1016/J.JCS.2003.12.001.

Tomas, D. J., & Atwell, W. A. (1999). Starch modifications. In *Starches*, Eagan Press, St. Paul, U.S.A., pp. 31–48.

Vamadevan, V., & Bertoft, E. (2015). Structure-Function Relationships of Starch Components. *Starch/Stärke*, 67(1–2), John Wiley & Sons, Ltd: 55–68. doi:10.1002/STAR.201400188.

Vanier, N. L., El Halal, S. L. M., Guerra Dias, A. R., & da Rosa Zavareze, E. (2017). Molecular structure, functionality and applications of oxidized starches: A review. *Food Chemistry*, 221(April), Elsevier: 1546–1559. doi:10.1016/J.FOODCHEM.2016.10.138.

Villegas, M. E., Aredo, V., Asevedo, K. J. E., Lourenço, R. V., Camino Bazito, R., & Oliveira, A. L. (2020). Commercial starch behavior when impregnated with food additives by moderate temperature supercritical CO2 processing. *Starch/Stärke*, 72(11–12), John Wiley & Sons, Ltd: 1900231. doi:10.1002/STAR.201900231.

Visser, R.G.F., & Jacobsen, E. (1993). Towards modifying plants for altered starch content and composition. *Trends in Biotechnology*, 11(2), Elsevier Current Trends: 63–68. doi:10.1016/0167-7799(93)90124-R.

Wang, H., Yang, Q., Gao, L., Gong, X., Qu, Y., & Feng, B. (2020). Functional and physicochemical properties of flours and starches from different tuber crops. *International Journal of Biological Macromolecules*, 148(April), Elsevier: 324–332. doi:10.1016/J.IJBIOMAC.2020.01.146.

Wang, R., Li, M., Liu, J., Wang, F., Wang, J., & Zhou, Z. (2021). Dual modification manipulates rice starch characteristics following debranching and propionate esterification. *Food Hydrocolloids*, 119(October), Elsevier: 106833. doi:10.1016/J. FOODHYD.2021.106833.

Wang, S., & Copeland, L. (2015). Effect of acid hydrolysis on starch structure and functionality: A review. 55(8), *Critical Reviews in Food Science and Nutrition*, Taylor & Francis: 1081–1097. doi:10.1080/10408398.2012.684551.

Wang, S., Wang, J., Liu, Y., & Liu, X. (2020). Starch modification and application. In *Starch Structure, Functionality and Application in Foods* (pp. 131–149). Springer. doi:10.1007/978-981-15-0622-2_8.

Wepner, B., Berghofer, E., Miesenberger, E., Tiefenbacher, K., & Ng, P N. K. (1999). Introduction. *Starch/Starke*, 51(10), 354–361.

Wu, C., & Zhou, X.. (2018). The overview of functional starch. *Functional Starch and Applications in Food* (pp. 1–26). Springer. doi:10.1007/978-981-13-1077-5_1.

Wu, C., Wu, Q. Y., Wu, M., Jiang, W., Qian, J. Y., Rao, S.Q., Zhang, L., Li, Q., & Zhang, C.. (2019). Effect of pulsed electric field on properties and multi-scale structure of japonica rice starch. *LWT*, 116(December), Academic Press: 108515. doi:10.1016/J. LWT.2019.108515.

Xiao, L., Chen, J., Wang, X., Bai, R., Chen, D., & Liu, J.. (2018). Structural and physicochemical properties of chemically modified Chinese water chestnut [Eleocharis Dulcis (Burm. f.) Trin. Ex Hensch] Starches. *International Journal of Biological Macromolecules*, 120(December), Elsevier: 547–556. doi:10.1016/J.IJBIOMAC.2018.08.161.

Xie, X., & Liu, Q. (2004). Development and physicochemical characterization of new resistant citrate starch from different corn starches. *Starch/Stärke*, 56(8), John Wiley & Sons, Ltd.: 364–370. doi:10.1002/STAR.200300261.

Xie, Z., Wang, S., Wang, Z., Fu, X., Huang, Q., Yuan, Y., Wang, K., & Zhang, B.. (2019). In vitro fecal fermentation of propionylated high-amylose maize starch and its impact on gut microbiota. *Carbohydrate Polymers*, 223(November), Elsevier: 115069. doi:10.1016/J.CARBPOL.2019.115069.

Xu, J., Andrews, T. D., & Shi, Y. C. (2020a). Recent advances in the preparation and characterization of intermediately to highly esterified and etherified starches: A review. *Starch/Stärke*, 72(3–4), John Wiley & Sons, Ltd.: 1900238. doi:10.1002/ STAR.201900238.

Xu, J., Chen, L., Guo, X., Liang, Y., & Xie, F. (2020b). Understanding the multi-scale structure and digestibility of different waxy maize starches. *International Journal of Biological Macromolecules*, 144(February), Elsevier: 252–258. doi:10.1016/J. IJBIOMAC.2019.12.110.

Xu, X., Chen, Y., Luo, Z., & Lu, X.. (2019). Different variations in structures of a- and b-type starches subjected to microwave treatment and their relationships with digestibility. *LWT*, 99(January), Academic Press: 179–187. doi:10.1016/J. LWT.2018.09.072.

You, S. Y., Taik Lim, S., Lee, J. H., & Chung, H. J. (2014). Impact of molecular and crystalline structures on in vitro digestibility of waxy rice starches. *Carbohydrate Polymers*, 112(November), Elsevier: 729–735. doi:10.1016/J.CARBPOL.2014.06.065.

Zavareze, E. D. R., & Guerra Dias, A. R. (2011). Impact of heat-moisture treatment and annealing in starches: A review. *Carbohydrate Polymers*, 83(2). Elsevier: 317–328. doi:10.1016/J.CARBPOL.2010.08.064.

Zdybel, E., Zieba, T., Tomaszewska-Ciosk, E., & Rymowicz, W. (2020). Effect of the esterification of starch with a mixture of carboxylic acids from yarrowia lipolitica fermentation broth on its selected properties. *Polymers*, 12(6), Multidisciplinary Digital Publishing Institute: 1383. doi:10.3390/POLYM12061383.

Zhang, H., Zhou, X., Wang, T., Luo, X., Wang, L., Li, Y., Wang, R., & Chen, Z. (2016). New insights into the action mode of amylosucrase on amylopectin. *International Journal of Biological Macromolecules*, 88(July), Elsevier: 380–384. doi:10.1016/J. IJBIOMAC.2016.04.007.

Zhang, H., Zhou, X., Wang, T., He, J., Yue, M., Luo, X., Wang, L., Wang, R., & Chen, Z. (2017). Enzymatically modified waxy corn starch with amylosucrase: The effect of branch chain elongation on structural and physicochemical properties. *Food Hydrocolloids*, 63(February), Elsevier: 518–524. doi:10.1016/J. FOODHYD.2016.09.043.

Zhang, H., Wang, R., Chen, Z., & Zhong, Q. (2019). Enzymatically modified starch with low digestibility produced from amylopectin by sequential amylosucrase and pullulanase treatments. *Food Hydrocolloids*, 95(October), Elsevier: 195–202. doi:10.1016/J.FOODHYD.2019.04.036.

Zhong, Y., Bertoft, E., Li, Z., Blennow, A., & Liu, X. (2020a). Amylopectin starch granule lamellar structure as deduced from unit chain length data. *Food Hydrocolloids*, 108(November), Elsevier B.V.: 106053. doi:10.1016/j.foodhyd.2020.106053.

Zhong, Y., Liu, L., Qu, J., Blennow, A., Hansen, A. R., Wu, Y., Guo, D., & Liu, X. (2020b). Amylose content and specific fine structures affect lamellar structure and digestibility of maize starches. *Food Hydrocolloids*, 108(November), Elsevier B.V.: 105994 doi:10.1016/j.foodhyd.2020.105994.

Zhou, F., Liu, Q., Zhang, H., Chen, Q., & Kong, B. (2016). Potato starch oxidation induced by sodium hypochlorite and its effect on functional properties and digestibility. *International Journal of Biological Macromolecules*, 84(March), Elsevier: 410–417. doi:10.1016/J.IJBIOMAC.2015.12.050.

Zhou, Y., Hoover, R., & Liu, Q. (2004). Relationship between α-amylase degradation and the structure and physicochemical properties of legume starches. *Carbohydrate Polymers*, 57(3), Elsevier: 299–317. doi:10.1016/J.CARBPOL.2004.05.010.

Zou, J., Xu, M., Wang, R., & Li, W. (2019). Structural and physicochemical properties of mung bean starch as affected by repeated and continuous annealing and their in vitro digestibility. 22(1), *International Journal of Food Properties*, Taylor & Francis: 898–910. doi:10.1080/10942912.2019.1611601.

Chapter 6

Starch: Nutraceutical Properties

Luis A. Bello-Pérez, Daniel E. Garcia-Valle
and Roselis Carmona-Garcia

CONTENTS

6.1 INTRODUCTION

Starch digestibility has become an important topic in the last 30 years ago when it was found that not all starch present in foods was completely hydrolyzed by digestive enzymes and consequently absorbed in the small intestine; then, that fraction, named resistant starch, reached the colon, where it was fermented by the microbiota (Asp, 1992). In the same decade, it was also reported that starch in starchy foods was hydrolyzed at different rates after ingestion through *in vitro* analysis with physiological mechanisms. This classification produced three fractions: rapidly digestible starch (RDS), slowly digestible starch (SDS) and resistant starch (RS) based on the time at which starch was hydrolyzed by digestive enzymes. RDS is the fraction digested within 20 minutes after ingestion, SDS is the fraction digested between 20 and 120 minutes after ingestion, and RS is the fraction that remained after 120 minutes of hydrolysis (Englyst et al., 1992).

The study of starch digestion led to taking into consideration the inclusion of RS as a fraction of the dietary fiber, because in the traditional method to determine the total dietary fiber (DF), heating of the sample is included with the change of the RS to digestible starch. Heating does not alter the content of nonstarch polysaccharides such as cellulose. The foods "as eaten" should be analyzed without additional heating to avoid underestimating the DF content (McCleary, 2009, 2015). All of them are due to the health benefits associated with the consumption of undigestible carbohydrates, such as reductions in weight, obesity, diabetes, colon cancer and cardiovascular diseases (Champ et al., 2003).

The glycemic response associated with the starch hydrolysis rate is important when suggesting the type of foods to be consumed by groups of people with specific caloric requirements and glucose contents. To determine the starch hydrolysis rate through in vitro tests, the use of mathematical models has been reported. Models such as Goñi's method (Goñi et al., 1997), the logarithm of slope (LOS) approach (Butterworth et al., 2012; Edwards et al., 2014) and multiscale hydrolysis kinetics (Bello-Perez et al., 2019) have been used to fit the experimental data in vitro to obtain rapid information, as starch hydrolysis is achieved in vivo without the use of volunteers. While the mathematical model is more complex, the prediction is to the in vivo results. It has been recognized that in vitro methods to determine starch digestion in the small intestine cannot predict the amount of nonhydrolyzed starch that reaches the large intestine, a matter that depends on host factors and the characteristics of the starch or food (Chanvrier et al., 2007b). The use of in vitro methods is reinforced because of some drawbacks of the in vivo methods, which require ethical consent, specialized nutritional or clinical studies, the use of humans, a well-controlled diet, and high expense (Gidley, 2013). Also, the structural changes of the food matrix during the digestion time course are difficult to understand and play an important role in the starch digestion (Zhang et al., 2015). In this sense, in vitro methods should consider the structure of digestive enzymes as well as the interactions between both digestive enzymes (α-amylase and amyloglucosidase) that participate in starch digestion because heterogeneous reactions are involved. Additionally, the interpretation of the kinetic data with the use of Michaelis–Menten kinetics and first-order kinetics is a matter that should be considered (Goñi et al., 1997; Bello-Perez et al., 2019; Meraz et al., 2020).

Associated with the starch hydrolysis fractions and the beneficial effect on health, the concept of "nutraceutical starch" was suggested to include the two starch digestion fractions (SDS and RS) (Magallanes-Cruz et al., 2017). The RS fraction has been reviewed, including its production (Raigond et al.,

2015), health benefits (Jenkins et al., 1998; Lockyer & Nugent, 2017), potential as a prebiotic (Zaman & Sarbini, 2015), management of metabolic diseases (Bindels et al., 2015) and colonic fermentation (Lee et al., 2013). The structural basis for digestion properties for slow digestion (Zhang et al., 2006a, b) and the role of SDS associated with health benefits have been reviewed (Lehmann & Robin, 2007; Zhang & Hamaker, 2009), and the importance of the location of its digestion has been determined (Lee et al., 2013; Strader & Woods, 2005).

6.2 STARCH DIGESTION FRACTIONS

Starch digestion was a topic of interest 30 years ago when it was reported that not all starch consumed in the diet is digested at the same rate or is not digested due to intrinsic (starch type, degree of gelatinization, food matrix) and extrinsic factors (transit through the bowel, chewing, amylase concentration) (Englyst et al., 1992). Starch digestion in humans is achieved in two phases: (1) intraluminal and (2) brush border. Starch digestion starts in the mouth when salivary α-amylase hydrolyzes part of the starch to maltose and dextrins. At this point, chewing is important because it supports saliva production and reduces the size of the food, which increases the surface area. Only approximately 5% of starch present in the food is hydrolyzed in the mouth (Lee et al., 2013; Insel & Turner, 2007). The pH (3.0) of the stomach slows the activity of salivary α-amylase, but some activity remains after passing the stomach to the proximal section of the small intestine (duodenum). The α-amylase in the lumen of the small intestine is secreted via the pancreatic duct. The α-amylase produces maltose, maltotriose and α-limit dextrin, but the generation of glucose is quite low.

Englyst et al. (1992) proposed an in vitro test with the use of porcine α-amylase and fungal glucoamylase instead of human α-amylase and mucosal α-glucosidases to simulate gastrointestinal starch digestion. The first fraction of the starch digestion was named RDS. This fraction is present in starchy foods with completely gelatinized starch and no interactions with other components, such as proteins, lipids, nonstarch polysaccharides and polyphenols, that restrict the hydrolysis of starch. White bread is a food with starch that is rapidly hydrolyzed. This fraction (RDS) produces a glycemic peak in the blood that provokes a rapid insulinemic response to metabolize glucose. The frequent consumption of foods with RDS can produce insulin resistance, which can result in type 2 diabetes. For this reason, it is important to decrease the consumption of foods where starch is rapidly hydrolyzed by digestive enzymes. The consumption of foods with the other two starch digestion fractions (SDS and RS) is associated with beneficial effects on human health; thus, they were named "nutraceutical starch" (Magallanes-Cruz et al., 2017). These two fractions are discussed in this chapter.

6.3 SLOWLY DIGESTIBLE STARCH (SDS)

There are several in vitro methods reported for SDS quantification that are not official. Englyst's method was developed to determine SDS. SDS is a nutritional concept that depends on the physiological response. RDS is correlated with the glycemic index (GI), but an in vivo method that reflects the SDS content does not exist (Zhang & Hamaker, 2009). The importance of both starch digestion fractions (RDS and SDS) associated with their nutritional relevance and health benefits suggested a validation of Englyst's method where the samples were analyzed "as eaten" with a particle size that is produced during chewing of the food. The results of the study in different laboratories showed repeatability and reproducibility, and the SDS content determined in different foods indicates foods with slow starch hydrolysis that can be associated with health benefits (Englyst et al., 2018).

Low GI foods are related to slow and prolonged release of glucose in the small intestine; however, low GI foods are not always slowly digested, e.g., when RDS is partially replaced by a nonstarch filler as protein or insoluble nonstarch polysaccharides. Conversely, foods with SDS do not always have low GI because the area under the curve may produce a relatively high GI value. The beneficial effects on human health associated with the consumption of starchy food with low GI, which presents a slow and sustained glucose release, fit the definition of SDS (Zhang & Hamaker, 2009). Among the health benefits of SDS is controlling obesity, diabetes and cardiovascular diseases (Miao et al., 2015).

In a review, the concept of extended GI (EGI) was introduced to specify the amount of glucose release over a long time. The EGI was based on in vivo studies performed on SDS digestion (Venkatachalam et al., 2006) and was proposed as an in vivo method that can be used to understand the concept of SDS and its physiological effects; additionally, the EGI can be used to evaluate the qualities of different starchy foods with SDS (Zhang & Hamaker, 2009).

The SDS fraction is digested between 20 and 120 minutes after starchy food ingestion. This is considered a physiological concept because after 120 minutes, the insulinemic response produces a regulation of the blood glucose concentration, and it is not possible to know if more glucose is liberated by starch hydrolysis and adsorbed in the small intestine. One of the two phases involved in starch digestion is in the lumen of the small intestine, where α-amylase is secreted via the pancreatic duct to produce small oligosaccharides. Subsequently, two protein complexes (maltase–glucoamylase and sucrase–isomaltase) present in the small intestine brush border hydrolyze the oligosaccharides to produce glucose (Lin et al., 2012).

6.3.1 SDS Production

There is interest in developing SDS ingredients because the amount in common foods is low (Björck et al., 2000). The preparation of SDS can be performed in pure starch from diverse botanical sources (Table 6.1).

The effect of different botanical sources was evaluated on SDS production. Native lentil starch with a higher SDS content than starch of other legumes (pea) and corn showed a higher SDS content after heat-moisture treatment (HMT) but a similar SDS content to pea starch after annealing treatment (Chung et al., 2009b). The effect of amylose content on SDS production was analyzed after double thermal treatment: autoclaving and hydrothermal treatment at low temperature, and autoclaving and HMT. Corn starch with 50% amylose content showed the highest SDS content with both dual treatments and decreased when amylose content decreased in the starch granule. The SDS value in corn starch with 70% amylose content was not clear because it was lower than that in corn starch with 50% amylose (Agama-Acevedo et al., 2018a). Normal maize starch treated with different salts ($CaCl_2$, $CaCO_3$ and $Ca(OH)_2$) simulated the effect of those salts on starch digestibility during nixtamalization; maize starch treated with $CaCl_2$ and $CaCO_3$ showed the highest SDS content. The authors mentioned that salt treatment produced calcium-mediated cross-linking; however, it was reported that the mechanisms responsible for the changes in starch digestibility by the effect of the salts were not fully elucidated (Roldan-Cruz et al., 2020).

The use of starchy flours with high nonstarch polysaccharides, proteins and polyphenols is considered in Chapter 8.

Pretreatment methods such as ultrasonication (Sujka, 2017; Zhu, 2015; Lou et al., 2008), high-pressure (Li et al., 2020b; Liu et al., 2016; Wang et al., 2008) and pulsed electric field (Han et al., 2009, 2012) technologies have been explored in recent years with the aim of increasing the yield of SDS content. Han and BeMiller (2007) performed diverse chemical modifications (esterification, cross-linking, acetylation, hydroxypropylation and combinations of two modifications) in waxy corn starch. Dual modification and cross-linking produced high amounts of SDS (approximately 20%). The combination of esterification and heating at diverse temperatures showed the highest SDS content (approximately 40%). This pattern was related to that during heating of esterified starch the carboxyl group of the octenylsuccinate half-ester group produced cross-linking with a decrease in digestibility. This matter is important because the heat treatment of esterified starch produces starch that can be consumed in

TABLE 6.1 SLOWLY DIGESTIBLE STARCH PRODUCTION FROM DIVERSE BOTANICAL SOURCES AND METHODS

Source	Process	Starch	Native	Annealing	HMT			Reference
				Slowly digestible starch (g/100 g)				
Starch	Annealing (ANN) Heat-moisture treatment (HMT)	Corn	65.7	48.6	24.2			Chung et al. (2009b)
		Pea	62.9	59.8	25.9			
		Lentil	70.1	60.4	30.4			
Normal corn starch	Annealing (ANN) Heat-moisture treatment (HMT)		Native	ANN	HMT	ANN–HMT	HMT–ANN	Chung et al. (2009a)
			56.9	54.8	48.9	49.7	51.2	
Corn starches:	Autoclaving-cooling treatment (AC) Hydrothermal treatment to low temperature-moisture (LMT) Heat-moisture treatment (HMT)	Starch	Native	AC	HMT	AC–HMT	AC–LMT	Agama-Acevedo et al. (2018b)
Waxy		Waxy	—	—	—	2.3	3.2	
Normal		Normal	—	—	—	13.7	15.8	
Amylomaize V		Amylomaize V	17.1	13.0	13.9	27.0	26.4	
Amylomaize VII		Amylomaize VII	6.5	11.1	10.7	21.5	21.6	
Normal maize starch	Acid modification (AM) Hydrothermal treatment (HT) Extrusion (E)		Native	AM	HT	HT–AM	AM–E	Hasjin & Jane (2009)
			7.4	3.2	11.8	11.1	9.7	

(Continued)

TABLE 6.1 (CONTINUED) SLOWLY DIGESTIBLE STARCH PRODUCTION FROM DIVERSE BOTANICAL SOURCES AND METHODS

Source	Process	Slowly digestible starch (g/100 g)						Reference
Native maize starch	Calcium-treated $CaCO_3$ $CaCl_2$ $Ca(OH)_2$	Native		$CaCl_2$	$CaCO_3$	$Ca(OH)_2$		Roldan-Cruz et al. (2020)
		64		75	74	30		
Waxy corn starch	Modified with amylosucrase		Native			Modified		Kim et al. (2017)
			63.01			42.48		
High-amylose maize starch	Starch/lauric acid complexes		Native			Modified		Lu et al. (2020)
			6.5			14.7		
Mung bean starches: Low amylose Medium amylose High amylose	Citric acid (CA) Annealing (ANN) Heat-moisture treatment (HMT)	Starch		Native	CA–HMT		CA–ANN	Duyen et al. (2020)
		Low amylose		14.5	27.4		22.6	
		Medium amylose		12.8	25.2		24.2	
		High amylose		11.0	25.5		18.4	
Corn and sorghum starch	Citric acid (CA) Lactic acid (LA) Heat-moisture treatment (HMT)	Starch	Native	CA	CA–HMT	LA	LA–HMT	Shaikh et al. (2019)
		Corn	12.74	4.41	5.41	10.02	6.18	
		Sorghum	16.67	9.34	7.20	6.34	4.07	

(Continued)

TABLE 6.1   (CONTINUED) SLOWLY DIGESTIBLE STARCH PRODUCTION FROM DIVERSE BOTANICAL SOURCES AND METHODS

Source	Process	Slowly digestible starch (g/100 g)				Reference
Sweet potato starch	OSA modified; Heat-moisture treatment (HMT)	Native: 1.89	OSA: 3.16	HMT–OSA: 4.03		Lv et al. (2018)
Sweet potato starches: Yellow, White, Purple	Annealing (ANN); Heat-moisture treatment (HMT)	Starch: Yellow / White / Purple	Native: 1.0 / 0.7 / 1.0	HMT: 3.7 / 4.0 / 4.9	ANN: 3.2 / 4.7 / 2.7	Trung et al. (2017)
Sweet potato starch	Heat-moisture treatment (HMT)	Native: 41.5	HMT: 52.9			Na et al. (2020)
Indica rice starch; Waxy rice starch	Microwave (M); Heat-moisture treatment (HMT)	Starch: Indica / Waxy	Native: 14.07 / 32.21	M–HMT: 19.0 / 37.38		Gou et al. (2019)
Rice starch	Heat-moisture treatment (HMT)	Native: 56.4	HMT: 31.6	Steamed–HMT: 20.7	Cooked–HMT: 11.5	Xie et al. (2020)
Rice starch	Citric acid esterified; Extrusion	Native: 15.6	Modified: 4.5			Ye et al. (2019)
Wheat starch; A-type wheat starch granules; B-type wheat starch granules	Citric acid treated (CA); Heat-moisture treatment (HMT)	Starch: Wheat / A-type / B-type	Native: 22.38 / 25.86 / 20.07	CA: 2.37 / 4.79 / 1.74	CA–HMT: 2.84 / 1.37 / 1.62	Li et al. (2019b)

(Continued)

TABLE 6.1   (CONTINUED) SLOWLY DIGESTIBLE STARCH PRODUCTION FROM DIVERSE BOTANICAL SOURCES AND METHODS

Source	Process	Starch	Slowly digestible starch (g/100 g) Native	Modified	Reference
Potato starch	Citric acid treated (CA)	Potato	10.5	21.2	Remya et al. (2018)
Cassava starch		Cassava	20.3	22.4	
Sweet potato starch		Sweet potato	14.2	18.3	
Lentil starch		Lentil	0.8	19.1	
Banana starch		Banana	15.6	22.1	
Maize starches: Waxy	Esterified with acetic anhydride (EAA); Pulsed electric fields (PEF)	Waxy	Native 14.6	EAA 28.0; PEF 2.4	Hong et al. (2018)
Normal		Normal	12.4	EAA 23.6; PEF 28.2	
Gelose 50		Gelose 50	8.1	EAA 8.6; PEF 16.5	
Gelose 80		Gelose 80	4.8	EAA 5.7; PEF 6.5	
Banana starches: Dominico harton (*Musa aab Simmonds*)	OSA modified	Dominico harton		Modified 18.69	Quintero-Castaño et al. (2020)
Gros Michel (*Musa acuminate AAA*)		Gros Michel		14.89	
FHIA 21 (hybrid)		FHIA 21		14.42	
Faba bean starch	Cross-linked (CL)		Native 35.8	Modified 35.6	Sharma et al. (2020)

(Continued)

TABLE 6.1   (CONTINUED) SLOWLY DIGESTIBLE STARCH PRODUCTION FROM DIVERSE BOTANICAL SOURCES AND METHODS

Source	Process	Slowly digestible starch (g/100 g)						Reference
Starches: Plantain (*Musa paradisiaca* L.), Mango	β-amylase (AM) Transglucosidase (TG)	Starch	Native	AM	AM–TG			Casarrubias-Castillo et al. (2012a)
		Plantain	10.96	7.15	18.53			
		Mango	6.37	22.32	11.78			
Waxy potato starch	Amylosucrase	Native		Modified				Kim et al. (2020)
		4.64		16.4				
Banana (*Musa paradisiaca* L.) starch	Cross-linked (CL) Esterification (ES)	Native	CL	ES	ES–CL	CL–ES		Carlos-Amaya et al. (2011)
		10.79	16.79	11.63	13.72	12.77		
Hylon VII	Autoclaving-cooling treatment (AC) Oven-dried (OD) Freeze-dried (FR) Hydrothermal treatment (HT)	Native	AC	OD	FR	OD–HT	FR–HT	Agama-Acevedo et al. (2018b)
		7	19	14	18	32	17	

foods without additional cooking (e.g., yogurt, smoothies, fruits); this issue deserves more investigation.

The production of SDS ingredients involves changes in the starch structure; the native starch structure (crystalline arrangement, chain length distribution of amylopectin) plays an important role in the starch structure after modification and consequently in the digestibility of the modified starch (SDS). Physical modification, such as HMT, has been used to modulate starch digestion. HMT involves the use of high temperature and low moisture, where the starch components (amylose and amylopectin) are mobilized within the amorphous and crystalline areas of starch granules without destruction. SDS is related to short crystallites and dense-branched regions in the lamellar structure (Zhang et al., 2006).

Foods with SDS produce a low GI (Lehmann & Robin, 2007). The consumption of low GI foods is related to a reduction in the risk of some chronic diet-related metabolic diseases (e.g., overweight, obesity, type 2 diabetes), which are related to SDS ingesta. In association with the beneficial health effects in the human body due to the consumption of SDS, there is interest in developing starchy ingredients with high SDS content. Diverse botanical sources and processes have been evaluated to produce SDS ingredients for possible use in foods (Table 6.1). Physical and enzymatic methods are preferred over chemical methods due to the pollution problems associated with the final deposition of reagents used in chemical methods. It is important to mention that the physical methods in which heat treatments are involved can produce SDS ingredients, which are suggested in products such as yogurt, smoothies and fruits, where no cooking is necessary because an appreciable reduction in the SDS content was observed. High-amylose starch (70% amylose) was autoclaved, dried and subjected to hydrothermal treatment (HT). Two drying methods were evaluated (oven-drying and freeze-drying) after autoclaving and before HT. The native starch showed 7% and 19% autoclaving SDS, but the HT after freeze-drying did not change the SDS content (18%), and a marked increase was observed when HT was used after oven-drying (32%) (Agama-Acevedo et al., 2018b). High-amylose starches (50% and 70% amylose) were modified by thermal treatment (pregelatinization and HMT at low and high temperatures); HMT at low (4°C) and high (100°C) temperatures after pregelatinization of high-amylose starch (50 g amylose/100 g starch) produced the highest SDS content (27%) and was lower in the starch with the highest amylose content (70 g amylose/100 g starch) starch (21%). Both HMT treatments in waxy and normal maize starches produced low SDS contents, and higher SDS contents were found in normal maize starches than in waxy maize starches (Agama-Acevedo et al., 2018a).

The starch source and amylose/amylopectin ratio, as well as the combination of physical methods (drying, HT, etc.), are important in the development of SDS.

Structural changes in starch after enzymatic methods modify the SDS content. Unconventional starch sources such as those from unripe fruits (mango and plantain) are suggested to produce ingredients with reduced digestibility. Cooked (gelatinized) native and enzyme-treated mango and plantain starches showed a decrease in RDS content; mango starch showed an increase in SDS content with both enzymatic treatments (β-amylase and β-amylase-transglucosidase), but plantain starch treated with β-amylase showed a decrease compared with its native counterpart. The starch source (starch structure) plays an important role in enzymatic modification to modify its hydrolysis by digestive enzymes (Casarrubias-Castillo et al., 2012a).

In summary, the use of different botanical sources and methods produces SDS contents in the range between 3 and 60 g/100 g; however, analyses of yield, cost and functionality in food where SDS-rich powder can be added are mandatory before suggesting its production at an industrial level.

6.4 RESISTANT STARCH (RS)

RS is the fraction that resists hydrolysis by digestive enzymes in the small intestine passing into the colon where it is largely fermented by the microbiota to short-chain fatty acids (acetate, propionate and butyrate) and gases (H, CO_2 and CH_4). Currently, RS is classified as part of dietary fiber and has functionality similar to fermentable nonstarch polysaccharides. Fermentation with RS produces diverse effects, such as decreasing colon pH, relieving constipation, and increasing stool weight and bulking effects (Jenkins et al., 1998). It is well known that fermentation of RS by microbiota produces high butyrate content with the stimulation of beneficial bacteria with the concomitant reduction of harmful microorganisms. The preventive effect of butyrate in colorectal cancer has been reported. The potential mechanisms associated with the prevention of colorectal oncogenesis by butyrate are not completely understood; in vitro studies showed that high butyrate levels in the colon produce a decrease in pH that shows strong antitumorigenic properties. Additionally, studies in cell lines showed that butyrate inhibits the proliferation and induces the apoptosis of colorectal cancer (Fung et al., 2012). RS is classified into five categories that have been reported (Bello-Perez et al., 2020); this classification has been used to explain the resistance of polysaccharides to enzymatic hydrolysis and to suggest processing methods to produce RS-rich ingredients. Current trends in the study of RS are related to the search for alternative or nonconventional starchy sources that, after heat treatment (cooking), maintain resistance to hydrolysis by digestive enzymes. This matter can be considered from two points of view: (1) the isolation of starch, in which physical and enzymatic modification methods are preferred because they are environmentally friendly to produce high resistance to hydrolysis; and (2) starchy sources are modified by the aforementioned methods, where the development of a matrix and interactions with other components (e.g., polyphenols) produce resistance to enzymatic hydrolysis. In recent years, the structure of RS that provides resistance to enzymatic hydrolysis has been a topic of interest with the objective of developing tailor-made RS. The effects of RS products on gut microbiota species and host metabolism are a matter of interest that has been studied. It was reported that the microbiota species were different when different RS types were fermented in the colon (Birt et al., 2013). Changes in the microbiota have been related to obesity (Zhang et al., 2015). Obesity is a strong risk factor for diverse chronic degenerative diseases, such as cardiovascular disease and diabetes. The gut microbiota is distinct in healthy and obese persons (Ley, 2010). This pattern is related to the fact that bacterial species in the colon can extract calories from food that humans cannot digest, and when excess calories (persons with excess weight) are present, more microbes can grow because the gut microbiota of lean persons is distinct from that of obese persons (Bik, 2015).

6.4.1 RS Production

The health benefits associated with the consumption of foods with low calories (high dietary fiber) are well known. Currently, there is interest in developing foods fortified with dietary fiber such as bakery products, pasta, snacks and breakfast flakes. The addition of dietary fiber from fruit peel, bran and other sources decreases the sensory characteristics of products in a manner that is not preferred by consumers. RS-rich powder is an alternative to improve the sensory characteristics of foods and increase dietary fiber consumption (Tian & Sun, 2020). The market demand for RS-rich ingredients has motivated the development of technological innovations to achieve the production of RS at the industrial level. Different methods, such as chemical, physical, enzymatic and combination methods, as well as different botanical starch sources (Table 6.2), have been suggested to produce RS-rich ingredients (Jian et al., 2020).

Enzymatic treatment with amylosucrase modifies the starch structure by transferring monomers to a growing chain; the modified starch structure cannot be recognized by digestive enzymes. The effects of the starch source and the amylose content were analyzed. Native normal maize starch (uncooked) showed a lower RS (8.8 g/100 g) content than potato starch (29.9 g/100 g). Starch from rice varieties showed low RS content in the native state (between 8.4 and 11.0 g/100 g), and an appreciable increase in RS was found after enzymatic treatment with amylosucrase (between 63.5 and 69.2 g/100 g). After enzymatic modification, an important increase in RS content was found in normal maize starch (60.9 g/100 g) compared with potato starch (47.2 g/100 g). Native waxy maize starch showed the lowest RS (5.6 g/100 g), and after modification, it showed the highest RS content (68.7 g/100 g) (Park et al., 2019). In our opinion, the effect of starch source is not fully clear, and more studies should be performed.

The raw (ungelatinized) starches are resistant to enzymatic hydrolysis. It is important to mention that raw cereal starches (maize, wheat, rice) showed higher SDS contents than tuber starch (potato) (Zhang et al., 2006). One first strategy to produce RS-rich ingredients was the use of high-amylose maize starch obtained by genetic modification. Li et al. (2019a) reported that in the 1940s, the first breeding program was initiated to produce high-amylose crops, and in 1958, the first hybrid of maize with high amylose content was commercialized. Alterations in the amylose content in the crop may affect the yield of starch and the starch structure at different levels, and both can impact the functional and nutritional characteristics.

Although the amylose component is associated with starch retrogradation, the high-amylose maize starch granule shows restricted swelling that produces low gelatinization with low hydrolysis by digestive enzymes. Commercial products such as Hi-Maize® 260 are recommended as ingredients to prepare foods such as bakery products, snacks, and pasta with high undigestible carbohydrate contents even after thermal treatments (cooking).

RS production includes chemical modification of starch from diverse botanical sources, such as plantain starch (Carlos-Amaya et al., 2011); however, the safety and pollution due to the final deposition of the reagents is considered a critical point in the preparation of this kind of RS. Additionally, the use of chemically modified starch in foods and pharmaceutical products is restricted by the US Food and Drug Administration (FDA) because the RS content in those starches depends on the substitution degree. Chemical treatments such as cross-linking, acetylation, hydroxypropylation and their combination produced RS contents between 12.9% and 25.6% (Han & BeMiller, 2007); these RS values were limited by the reagent concentration approved by the FDA, but it is important to keep in mind that the level of chemically modified starches used in foods is low. Unconventional starch sources, such as plantain starch, that is isolated from unripe fruits are considered an important source of RS in their raw or uncooked state (Faisant et al., 1995). From this point of view, plantain starch was chemically modified with the aim of maintaining a high RS content after cooking. Plantain starch was modified by esterification, cross-linking and dual modification (esterification–cross-linking and cross-linking–esterification). The native (unmodified) starch and those chemically modified without cooking showed RS contents between 74% and 85%, and no appreciable increase in the RS content was observed due to chemical modification. After cooking, the unmodified plantain starch showed 21.5% RS, which increased to 29.1% in the cross-linked esterified starch (Carlos-Amaya et al., 2011). The increase in RS due to chemical modifications is not appreciable; for this reason,

TABLE 6.2 RESISTANT STARCH PRODUCTION FROM DIVERSE BOTANICAL SOURCES AND METHODS

Source	Process	Resistant starch (g/100 g)	Reference
Waxy potato starch	Amylosucrase	Native 29.9; Modified 47.2	Kim et al. (2020)
Corn starches: Amylomaize VII	Amylosucrase of *Neisseria subflava*	Starch: Amylomaize VII — Native 32.9, Modified 55.0	Park et al. (2019)
Normal		Native 8.8, Modified 60.9	
Waxy		5.6, 68.7	
Rice starches: Goamy		9.3, 63.5	
Dongjin		10.0, 64.7	
Manmi		11.0, 65.0	
Sinsunch		8.4, 69.2	
Hylon VII	Autoclaving-cooling treatment (AC), Oven-dried (OD), Freeze-dried (FR), Hydrothermal treatment (HT)	Native 78; AC 58; OD 70; FR 48; OD–HT 76; FR–HT 81	Agama-Acevedo et al. (2018b)

(Continued)

TABLE 6.2 (CONTINUED) RESISTANT STARCH PRODUCTION FROM DIVERSE BOTANICAL SOURCES AND METHODS

Source	Process		Resistant starch (g/100 g)				Reference
		Starch	Native	HC	P	AM	
Corn starch	Treatment	Starch	Native	HC	P	AM	Simons et al. (2018)
Bean starches:	Heat-cold (HC)	Corn	0.12	8.3	27.25	14.76	
Great northern	Pullulanase (P)	Great northern	40.83	18.33	28.35	27.71	
Pinto	α-amylase (AM)	Pinto	43.86	17.13	28.86	29.65	
Black		Black	30.52	14.86	25.60	29.58	
Lima		Lima	60.28	14.88	27.64	23.16	
Corn starches:	Esterification	Starch	Native		Modified		Lopez-Silva et al. (2020)
Waxy	OSA modified	Waxy	27.22		36.26		
Normal		Normal	34.66		43.65		
High amylose		High amylose	74.84		87.27		
Corn starches	Autoclaved	Starch	Native		Modified		Soler et al. (2020)
Normal		Normal	2.1		5.5		
High amylose		High amylose	12.7		30.2		
Native maize starch	Calcium-treated CaCO$_3$ CaCl$_2$ Ca(OH)$_2$	Native 2.7	CaCl$_2$ 6.7	CaCO$_3$ 6.8	Ca(OH)$_2$ 4.5		Roldan-Cruz et al. (2020)

(Continued)

TABLE 6.2 (CONTINUED) RESISTANT STARCH PRODUCTION FROM DIVERSE BOTANICAL SOURCES AND METHODS

Source	Process	Type	Resistant starch (g/100 g)					Reference
Waxy corn starch	Modified with Amylosucrase		Native 15.22		Modified 39.54			Kim et al. (2017)
Corn and sorghum starch	Citric acid (CA) Lactic acid (LA) Heat-moisture treatment (HMT)	Starch	Native	CA	CA-HMT	LA	LA-HMT	Shaikh et al. (2019)
		Corn	64.66	87.96	90.19	77.95	85.71	
		Sorghum	68.81	85.26	87.66	81.50	87.93	
Sweet potato starch	OSA modified Heat-moisture treatment (HMT)	Native	66.10	OSA 76.96		HMT-OSA 77.86		Lv et al. (2018)
Sweet potato starches:	Annealing (ANN) Heat-moisture treatment (HMT)	Starch	Native	HMT		ANN		Trung et al. (2017)
Yellow			24.1	30.6		28.8		
White			24.0	39.3		29.2		
Purple			25.3	35.4		32.0		
Sweet potato starch	Heat-moisture treatment (HMT)	Native	49.9		HMT 52.0			Na et al. (2020)
Indica rice starch	Microwave (M) Heat-moisture treatment (HMT)	Starch	Native		M-HMT			Gou et al. (2019)
		Indica	82.31		79.60			
Waxy rice starch		Waxy	57.96		56.78			
Rice starch	Heat-moisture treatment (HMT)	Native	13.0	HMT 15.9	Steamed-HMT 13.7	Cooked-HMT 11.8		Xie et al. (2020)

(*Continued*)

TABLE 6.2	(CONTINUED) RESISTANT STARCH PRODUCTION FROM DIVERSE BOTANICAL SOURCES AND METHODS					
Source	**Process**	**Resistant starch (g/100 g)**				**Reference**
Banana starches: Dominico harton (*Musa aab Simmonds*) Gros Michel (*Musa acuminate AAA*) FHIA 21 (hybrid)	OSA modified	Starch Dominico harton Gros Michel FHIA 21		Modified 75.10 77.98 80.20		Quintero-Castaño et al. (2020)
Faba bean starch	Cross-linked (CL)	Native 49.8	Native	Modified 61.1		Sharma et al. (2020)
Three rice lines starches: 1563 K3 HRS	Hydroxypropylation (HP) Cross-linked (CL)	Starch 1563 K3 HRS	Native 0.11 0.99 7.14	CL 0.08 1.23 5.70	HP 0.13 0.76 6.39	Shen et al. (2019)
Maize starches: Waxy Normal Gelose 50 Gelose 80	Esterified with acetic anhydride (EAA) Pulsed electric fields (PEF)	Starch Waxy Normal Gelose 50 Gelose 80	Native 16.7 18.7 30.5 50.7	EAA 32.1 33.6 39.3 56.7	PEF 24.8 21.2 28.0 54.1	Hong et al. (2018)

(Continued)

TABLE 6.2 (CONTINUED) RESISTANT STARCH PRODUCTION FROM DIVERSE BOTANICAL SOURCES AND METHODS

Source	Process	Resistant starch (g/100 g)						Reference
Potato starch Cassava starch Sweet potato starch Lentil starch Banana starch	Citric acid treated (CA)	Starch		Native		Modified		Remya et al. (2018)
		Potato		4.0		41.1		
		Cassava		1.6		39.6		
		Sweet potato		3.6		39.5		
		Lentil		5.3		48.8		
		Banana		3.01		47.9		
Wheat starch A-type wheat starch granules B-type wheat starch granules	Citric acid treated (CA) Heat-moisture treatment (HMT)	Starch		Native	CA	CA–HMT		Li et al. (2019b)
		Wheat		49.48	77.32	85.54		
		A-type		49.42	77.29	78.89		
		B-type		53.97	76.72	86.46		
Rice starch	Citric acid esterified Extrusion	Native			Modified			Ye et al. (2019)
		37.9			91.0			
Waxy maize starch	Debranching treatment (DB) Esterification with citrate (EC)	Native		DB	EC	DB–EC		Liu et al. (2020)
		17.90		44.74	78.96	83.78		
Potato starch	Microwave-toughening treatment	Native			Modified			Li et al. (2018a)
		11.54			27.09			

(*Continued*)

TABLE 6.2   (CONTINUED) RESISTANT STARCH PRODUCTION FROM DIVERSE BOTANICAL SOURCES AND METHODS

Source	Process	Resistant starch (g/100 g)						Reference
High amylose maize starch	Starch/lauric acid complexes	Native	Modified					Lu et al. (2020)
		18.8	29.1					
Mung bean starches:	Citric acid (CA) Annealing (ANN) Heat-moisture treatment (HMT)	Starch	Native	CA–HMT	CA–ANN			Duyen et al. (2020)
Low amylose		Low amylose	10.1	25.9	27.1			
Medium amylose		Medium amylose	13.4	32.3	37.7			
High amylose		High amylose	15.6	35.6	41.1			
Corn starches	Autoclaving–cooling treatment (AC)	Starch	Native	AC	HMT	AC–HMT	AC–LMT	Agama-Acevedo et al. (2018b)
Waxy	Hydrothermal treatment to low temperature moisture (LMT)	Waxy	–	–	–	10.2	7.8	
Normal		Normal	–	–	–	12.4	11.9	
Amylomaize V		Amylomaize V	69.1	21.1	54.9	36.2	31.5	
Amylomaize VII	Heat-moisture treatment (HMT)	Amylomaize VII	81.2	30.6	55.8	47.5	37.5	
Pea starch	Autoclaving–cooling treatment (AC) Partial acid hydrolysis (PAH) Pullulanase (P)	Native	AC	PAH–AC	AC–PAH	P–AC	AC–P	Polesi et al. (2011)
		39.9	38.5	54.6	53.6	38.8	42.3	

(Continued)

TABLE 6.2 (CONTINUED) RESISTANT STARCH PRODUCTION FROM DIVERSE BOTANICAL SOURCES AND METHODS

Source	Process			Native	AM	AM–TG	Reference
Starches:	β-amylase (AM)	Starch		Native	AM	AM–TG	Casarrubias-Castillo et al. (2012a)
Plantain (*Musa paradisiaca* L.)	Transglucosidase (TG)			14.09	32.74	18.05	
Mango				19.34	16.84	32.92	
Barley starches:	Micronization	Starch			Modified	Modified	Emami et al. (2012)
Waxy					2.1		
Normal					14.8		
High amylose					13.4		
Bean starches:	Steam-cooking	Starch		Native		Modified	Tovar & Melito (1996)
Black				2.2		18.9	
Red				1.7		21.2	
Lima				2.0		30.7	
Starches:	Annealing (ANN)	Starch		Native	ANN	HMT	Chung et al. (2009b)
Corn	Heat-moisture treatment (HMT)			4.6	8.7	12.3	
Pea				10.0	11.2	14.5	
Lentil				9.1	11.5	14.7	
Normal corn starch	Annealing (ANN)	Native	ANN	HMT	ANN–HMT	HMT–ANN	Chung et al. (2009a)
	Heat-moisture treatment (HMT)	19.7	18.3	16.9	17.3	19.7	

(Continued)

TABLE 6.2   (CONTINUED) RESISTANT STARCH PRODUCTION FROM DIVERSE BOTANICAL SOURCES AND METHODS

Source	Process	Resistant starch (g/100 g)						Reference
Taro (*Colocasia esculenta* L. Schott)	Heating, autoclaving, storage at room temperature	Native			Modified		Modified	Simsek & El (2012)
		2.2			35.1			
Hylon V	Acid modification	Starch			Native		Native	Ozturk et al. (2011)
Hylon VII		Hylon V			43		42.9	
		Hylon VII			53		53.7	
Starches:	Extrusion	Starch			Extruded			González-Soto et al. (2006)
Corn		Corn			2.05			
Mango		Mango			4.05			
Banana		Banana			4.98			
Banana (*Musa paradisiaca* L.) starch	Cross-linked (CL) Esterification (ES)	Native	CL	ES	ES–CL		CL–ES	Carlos-Amaya et al. (2011)
		85.07	74.01	82.61	80.66		80.63	
Normal maize starch	Acid modification (AM) Hydrothermal treatment (HT) Extrusion (E)	Native	AM	HT	HT–AM		AM–E	Hasjin & Jane (2009)
		11.0	12.2	32.9	32.7		29.8	

(*Continued*)

TABLE 6.2   (CONTINUED) RESISTANT STARCH PRODUCTION FROM DIVERSE BOTANICAL SOURCES AND METHODS

Source	Process	Resistant starch (g/100 g)					Reference
		Native	DB	PHA	PHA (1 day)–DB	PHA (7 day)–DB	
Smooth pea starch	Partial acid hydrolysis (PHA) Debranched (DB)	Native	DB	PHA	PHA (1 day)–DB	PHA (7 day)–DB	Lehmann et al. (2003)
		21.4	42.4	51.3	37.5	19.7	
Starches Potato	Partial acid hydrolysis (PAH)	Starch	AC	AC–HMT	PAH–CA	PAH–A C–HMT	Shin et al. (2004)
Sweet potato	Autoclaving-cooling treatment (AC) Heat-moisture treatment (HMT)	Potato	5.5	5.15	8.3	14.2	
	Heat-moisture treatment (HMT)	Sweet potato	5.6	8.15	19.6	22.7	
Hylon VII	Partial acid hydrolysis (PAH) Annealing (ANN) Heat-moisture treatment (HMT)	Native		PAH	PAH-ANN	PAH-HMT	Brumovsky & Thompson (2001)
		78.7		79.2	74	78.1	

other treatments to modify starch digestibility, such as physical and enzymatic treatments, have been proposed.

Physical treatments, such as annealing, have been suggested to produce RS due to their high moisture content (up to 35%) and gelatinization temperatures. The high water content in the annealed samples increases the mobility of the amylopectin chains that are present in the crystalline areas of the starch granule, which are reorganized in densely packed regions. Additionally, other combinations of thermal treatments and conditions have been used to produce RS-rich powder. Resistance to hydrolysis by digestive enzymes of high-amylose starches is well known. The combination of pregelatinization and thermal treatment at low (4°C for 24 h) and high (100°C for 5 h) temperatures and subsequent drying at 40°C in an air oven produced a higher RS content at high temperatures in starch with 70% amylose (47.5%) than its counterpart at low temperatures (37.5%). Additionally, the amylose content had an effect on the RS content because the highest values were in starch with 70% amylose (Agama-Acevedo et al., 2018b). The effect of drying after pregelatinization (oven-drying and freeze-drying) on RS production was evaluated. Oven-drying of the pregelatinized high-amylose starch (70%) produced a higher RS content than freeze-drying due to the mechanisms involved in water removal; additionally, at the longest thermal treatment time, the RS content increased. The combination of the drying method and hydrothermal treatment can produce RS-rich powder with the objective of increasing the undigestible carbohydrate content (dietary fiber) of foods (Agama-Acevedo et al., 2018a).

The last classification of RS types included RS5, which is attributed to amylose–lipid complexes (Xia et al., 2018); this RS type has been rarely studied and represents a topic that deserves more investigation. Panyoo and Emmambux (2017) reported the production of amylose–lipid complexes and the potential health benefits. The health benefits of RS as part of dietary fiber are widely recognized; however, the formation of amylose–lipid complexes in food could decrease the lipid content and consequently the caloric value of the product. Amylose–lipid complexes self-assemble during and after gelatinization and stabilize the amylose helix, producing a V-type X-ray diffraction pattern. The change in the starch structure also modifies the physicochemical and functional characteristics of starch (Zheng et al., 2020a). Amylose is the main component that participates in short-term retrogradation, and amylopectin participates in long-term retrogradation. However, amylose–lipid complexes affect both short- and long-term retrogradation (Zheng et al., 2020b). The linear chains of starch with a degree of polymerization (DP) of up to 100 glucose monomers can produce left-handed single helices with a hydrophobic center or channel that accommodates guest molecules (Tan & Kong, 2020). This hydrophobic center can complex with hydrophobic molecules such as fatty acids (Okumus et al., 2018). The amylose–lipid complexes are stable to temperature and cannot be hydrolyzed by digestive enzymes because they are not recognized in the active center (catalytic center). In this sense, high-amylose starches and starches with long-chain amylopectin (e.g., potato) can be used to produce this kind of RS after heating, cooling and incubation with isoamylase to produce linear chains and complexes with fatty acids (palmitic acid, oleic, stearic, etc.). This strategy to produce RS was achieved with waxy corn starch debranched with pullulanase.

The fatty acid type (i.e., the carbon number in the chain and the unsaturation number) is important in the properties of the amylose–lipid complex. For example, palmitic acid and maize amylose retard the retrogradation of amylopectin upon long-term storage (Mariscal et al., 2018). The study of different fatty acids and starches can provide more information on the formation of this type of RS. The RS content in the powder was 53% (determined with the official method of dietary fiber); the high dietary fiber content was attributed to the presence of both retrograded starch (due to reorganization of linear chains) and amylose–lipid complexes (Hasjim et al., 2010).

6.5 USE OF NUTRACEUTICAL STARCH

The use of nutraceutical starch as an ingredient or the use of starchy flours with starch with slow and resistant digestion is a matter of interest by health agencies, health ministries and the food industry. Health agencies recommend a daily average consumption of dietary fiber of approximately 35 g to maintain health and wellness. Starch powders including both SDS and RS fractions are suggested by the industry because they have no taste, odor, or appearance and do not modify the texture of foods. RS powder as a dietary fiber source is preferred because fermentation products, mainly butyric acid, are related to the prevention of colon cancer (Brouns et al., 2002). Commercial products sold as RS sources are available in the food market, and they are recommended to increase the dietary fiber content in bakery products, pasta, snacks, battered fried products and breakfast cereals (Haralampu, 2000; Homayouni et al., 2014; Ashwar et al., 2016; Jiang et al., 2020). The use of RS-rich powder has been suggested as a wall material in the microencapsulation of probiotics, as RS-rich powder is a substrate for beneficial bacteria (Shafiei et al., 2012). SDS application in foods has been achieved in experimental studies. Table 6.3 shows the use of RS and SDS powders to produce different foods with the objective of increasing the dietary fiber content.

Gluten-free cookies were prepared with the addition of 50 g of Hi-Maize 260/100 g of flours; these cookies presented approximately 20 g of dietary fiber/100 g of product with a low hydrolysis percentage and glycemic response (43) (Garcia-Solis et al., 2018). Gluten-free cookies with high dietary fiber

TABLE 6.3   SLOWLY DIGESTIBLE STARCH (SDS) AND RESISTANT STARCH (RS) (NUTRACEUTICAL STARCH) IN FOODS

| Product | Ingredient | Starch fractions | | Reference |
		SDS (%)	RS (%)	
Gluten-free cookies	Hi-Maize 260	16.1	13.2	García-Solis et al. (2018)
Gluten-free cookies	Debranched waxy rice starch	17.3	13.3	Giuberti et al. (2017)
	Annealed waxy rice starch	13.0	3.0	
	Acid–heat-moisture	18.0	8.0	
Bread crumb	Extruded banana starch	7.76	1.89	Roman et al. (2019)
Cookies	Resistant starch-rich powder (lintnerized banana starch)	—	8.42	Aparicio-Saguilán et al. (2007)
Cookies	High-amylose maize starch	49.0	10.7	Giuberti et al. (2016)
Noodles	Pea starch	7.76	20.38	Ge et al. (2014)
	Mung bean starch	8.53	21.51	
	Sweet potato starch	8.50	19.66	
	Potato starch	10.05	17.59	
	Kudzu starch	13.01	11.76	
	Fernery starch	12.98	11.92	
Custard pastes	Corn starch	9.23	11.89	Alimi et al. (2017)
	Native banana–corn starch	14.28	5.48	
	Annealed banana–corn starch	10.20	13.47	
	HMT banana–corn starch	10.94	13.38	
Noodles	Plantain starch	—	4.6	Osorio-Díaz et al. (2008)
Noodles	NUTRIOSE®	21.65	52.11	Menon et al. (2015)
Noodles	Acetylated corn starch	21.82	27.98	Lin et al. (2019)
Noodles	Mung bean starch	—	27.98	Saito et al. (2019)
	Potato starch	—	10.4	

(Continued)

TABLE 6.3   (CONTINUED) SLOWLY DIGESTIBLE STARCH (SDS) AND RESISTANT STARCH (RS) (NUTRACEUTICAL STARCH) IN FOODS

Product	Ingredient	Starch fractions SDS (%)	RS (%)	Reference
Noodles	Modified starches			Hong et al. (2020)
	Waxy (~0.5% amylose contents)	17.6	43.3	
	Normal (23% amylose contents)	10.9	44.1	
	Gelose 50 (50% amylose contents)	8.8	52.5	
	Gelose 80 (80% amylose contents)	9.6	46.7	
Pasta	Banana starch (*Musa paradisiaca* L.)	—	10.33	Hernández-Nava et al. (2009)
Pasta	Hi-Maize 260	85.5	1.2	García-Valle et al. (2020)
Pasta	RS II (Hi-Maize 260)	—	9.82	Aravind et al. (2013)
	RS III (Novelose 330)	—	11.85	
Pasta	Hi-Maize 260	—	79.4	Foschia et al. (2016)
Pasta	Hi-Maize 260	1.83	10.20	Agama-Acevedo et al. (2019)
Cake batters	Hi-Maize	—	18.43	Magallanes-Cruz et al. (2020)
Muffins	Hi-Maize 260	—	9.76	Sanz et al. (2009)
	Novelose 240	—	8.37	
	Novelose 330	—	12.02	
	C*Actistar	—	13.50	

content could be an alternative to battle overweight and obesity problems. The addition of Hi-Maize 260 was also studied in cakes. The addition of Hi-Maize 260 to different flours (plantain, wheat and rice) to produce a cake showed increases in the dietary fiber content ranging between 11.6% and 18.4% without appreciable effects on the panelists' perceptions. Cakes elaborated with conventional or nonconventional flour sources (e.g., gluten-free) can be an alternative to include Hi-Maize 260 and increase the dietary fiber content (Magallanes-Cruz et al., 2020).

Pasta is a food that can be enriched with different ingredients. Pasta was formulated with commercial RS-rich powders (RS2 and RS4). Pasta with the highest substitution level (10 g RS/100 g of wheat flour) showed RS values of 11.3 and 6.4 g/100 g for RS2 and RS4, respectively. The RS type has an important role in the compact structure of pasta, restricting the disorganization of starch during cooking and consequently its hydrolysis, which can increase the RS content (Bustos et al., 2011, 2013).

RS-rich powder (RS5) was prepared and included in bread. Commercial high-amylose starch from Cargill (HA7) with 70% amylose content was gelatinized, cooled and debranched with isoamylase. The objective of debranching is the production of linear chains that can complex with fatty acids (palmitic acid). The formation of RS was assessed as total dietary fiber (53%) and was due to starch retrogradation and amylose–lipid complexes. The bread was evaluated in healthy nonsmoker male humans (19–38 years old). After consumption of RS-rich bread, the postprandial plasma glucose and insulin response decreased by 50% compared with the control bread. It was concluded that RS-rich bread could be used in people with obesity, insulin resistance and diabetes (Hasjim et al., 2010).

The production of commercial RS-rich powder with conventional starches (maize, wheat, potato, rice) can involve a combination of methods with the objective of increasing the yield and lowering the cost. The processes used gelatinization (heating) and cooling cycles, chemical treatments (mainly acid hydrolysis), extrusion cooking, and hydrothermal treatments with "normal" maize starch (Hasjim & Jane, 2009). However, structural features such as the amylose/amylopectin ratio, chain length distribution, molar mass and gyration radius do not allow the production of RS-rich ingredients with the same conditions during processing and should be evaluated for each starch source. Additionally, the end use of RS-rich ingredients needs to be considered with caution (food type, process, shelf life, etc.).

6.6 FUTURE TRENDS

Consumers are increasingly interested in the consumption of healthy foods that prevent some diseases, such as overweight, obesity, diabetes and cardiovascular problems. Foods with high dietary fiber (undigestible carbohydrates) are a group of interest because, in general, they supply low calorie content. It is well known that the worldwide consumption of dietary fiber is low, and there is a growing interest in the production of food with high dietary fiber content. The development of starchy ingredients such as those with high SDS and RS contents is an area of opportunity because such ingredients decrease the glucose peak and function as a substrate for the microbiota because some undesirable bacterial groups have been associated with diverse body disorders. There is opportunity to develop new RS-rich ingredients from unconventional starches (unripe fruits, tubers, rhizomes, etc.) that can produce those ingredients with better functionalities, e.g., low digestibility, improvements in the texture and sensory characteristics of foods, an environmentally friendly design process, reduced cost of production, and increased yield. The application of RS-rich powders prepared with different methods to produce RS types such as RS3 (retrograded starch by physical methods) and RS5 (amylose–lipid complexes) in different foods is a challenge for the food industry. The effects of RS on microbiota should be evaluated due to their potential relationship with some health problems.

6.7 CONCLUSIONS

The health benefits associated with the consumption of SDS and RS (nutraceutical starch) have resulted in increasing interest from the food industry and consumers to produce ingredients and foods with high contents of both starch digestion fractions. More studies are necessary to develop methods with conventional or unconventional processes used in the food industry, such as extrusion and nonthermal treatments (high pressure, ultrasonication, plasma technologies, etc.) that produce nutraceutical starch. Another challenge is the end use of nutraceutical starch, which does not modify the "traditional" sensory and texture characteristics of foods such as bakery products, pasta, snacks and breakfast cereals. Studies on the nutraceutical starch–structure relationship deserve more investigation.

ACKNOWLEDGMENTS

The authors thank the SIP-IPN, COFAA-IPN and EDI-IPN for support. DEGV also acknowledges the scholarship from CONACYT-Mexico.

REFERENCES

Asp, N. G. (1992). Resistant starch. Proceeding from the second plenary meeting of EURESTA: European FLAIR concerted action no. 11 on physiological implications of the consumption of resistant starch in man. *European Journal of Clinical Nutrition*, 2, S1–148.
Agama-Acevedo, E., Islas-Hernández, J. J., Pacheco-Vargas, G., Osorio-Díaz, P., & Bello-Pérez, L. A. (2012). Starch digestibility and glycemic index of cookies partially substituted with unripe banana flour. *LWT- Food Science and Technology*, 46, 177–182.

Agama-Acevedo, E., Bello-Perez, L. A., Lim, J., Lee, B.-H., & Hamaker, B. R. (2018a). Pregelatinized starches enriched in slowly digestible and resistant fractions. *LWT- Food Science and Technology*, 97, 187–192.

Agama-Acevedo, E., Pacheco-Vargas, G., Bello-Perez, L. A., & Alvarez-Ramirez, J. (2018b). Effect of drying method and hydrothermal treatment of pregelatinized Hylon VII starch on resistant starch content. *Food Hydrocolloids*, 77, 817–824.

Agama-Acevedo, E., Bello-Pérez, L. A., Pacheco-Vargas, G., Tovar, J., & Sáyago-Ayerdi, S. G. (2019). Unripe plantain flour as a dietary fiber source in gluten-free spaghetti with moderate glycemic index. *Journal of Food Processing and Preservation*, 43, e14012.

Alimi, A. A., Workneh, T., & Oyeyinka, S. A. (2017). Structural, rheological and *in vitro* digestibility properties of composite corn-banana starch custard paste. *LWT- Food Science and Technology*, 79, 84–91.

Aparicio-Saguilan, A., Sayago-Ayerdi, S. G., Vargas-Torres, A., Tovar, J., Ascencio-Otero, T. E., & Bello-Pérez, L. A. (2007). Slowly digestible cookies prepared from resistant starch-rich lintnerized banana starch. *Journal of Food Composition and Analysis*, 20, 175–181.

Aravind, N., Sissons, M., Fellows, C. M., Blazek, J., & Gilbert, E. P. (2013). Optimisation of resistant starch II and III levels in durum wheat pasta to reduce *in vitro* digestibility while maintaining processing and sensory characteristics. *Food Chemistry*, 136, 1100–1109.

Ashwar, B.A, Gani, A., Shah, A., Wani, I. A., & Masoodi, F. A. (2016). Preparation, health benefits and applications of resistant starch: A review. *Starch/Stärke*, 68, 287–301.

Bello-Pérez, L. A., Agama-Acevedo, E., Garcia-Valle, D. E., & Alvarez-Ramirez, J. (2019). A multiscale kinetics model for the analysis of starch amylolysis. *International Journal of Biological Macromolecules*, 122, 405–409.

Bello-Perez, L. A., Flores-Silva, P. C., Agama-Acevedo, E., & Tovar, J. (2020). Starch digestibility: Past, present, and future. *Journal of The Science of Food and Agriculture*, 100, 5009–5016.

Bik, E. M. (2015). You lose some, you win some: Weight loss induces microbiota and metabolite shifts. *EbioMedicine*, 2, 806–807.

Bindels, L. B., Walter, J., & Ramer-Tait, A. E. (2015). Resistant starches for the management of metabolic. *Current Opinion in Clinical Nutrition & Metabolic Care*, 18, 559–565.

Birt, D. F., Boylston, T., Hendrich, S., Jane, J-L., Hollis, J., Li, L., McClelland, J., Moore, S., Phillips, G. J., Rowling, M., Schalinske, K., Scott, M. P., & Whitley, E. M. (2013). Resistant starch: Promise for improving human health. *Advances in Nutrition*, 4, 587–601.

Björck, I., Liljeberg, H., & Östman, E. (2000). Low glycaemic-index food. *British Journal Nutrition*, 83, S149–S155.

Brouns, F., Kettlitz, B., & Arrigoni, E. (2002). Resistant starch and "the butyrate revolution". *Trends in Food Science & Technology*, 13, 251–261.

Brumovsky, J., & Thompson, D. B. (2011). Production of boiling- stable granular resistant starch by partial acid hydrolysis and hydrothermal treatments of high-amylose maize starch. *Cereal Chemistry*, 78, 680–689.

Bustos, M. C., Pérez, G. T., & León, A. E. (2011). Effect of four types of dietary fibre on the technological quality of pasta. *Food Science and Technology International*, 17, 213–219.

Bustos, M. C., Pérez, G. T., & León, A. E. (2013). Combination of resistant starches types II and IV with minimal amounts of oat bran yields good quality, low glycaemic index pasta. *International Journal of Foods Science & Technology*, 48, 309–315.

Butterworth, P. J., Warren, F. J., Grassby, T., Patel, H., & Ellis, P. R. (2012). Analysis of starch amylolysis using plots for first-order kinetics. *Carbohydrate Polymers*, 87, 2189–2197.

Cahyana, Y., Wijaya, E., Halimah, T. S., Marta, H., Suryadi, E., & Kurniati, D. (2019). The effect of different thermal modifications on slowly digestible starch and physicochemical properties of Green banana flour (*Musa acuminata colla*). *Food Chemistry*, 274, 274–280.

Carlos-Amaya, F., Osorio-Díaz, P., Agama-Acevedo, E., Yee-Madeira, H., & Bello-Pérez, L. A. (2011). Physicochemical and digestibility properties of double-modified banana (*Musa paradisiaca* L.) starches. *Journal of Agriculture and Food Chemistry*, 59, 1376–1382.

Casarrubias-Castillo, M. G., Hamaker, B. R., Rodrgiuez-Ambriz, S. L., & Bello-Pérez, L. A. (2012a). Physicochemical, structural, and digestibility properties of enzymatic modified plantain and mango starches. *Starch/Stärke*, 64, 304–312.

Casarrubias-Castillo, M. G., Méndez-Montealvo, G., Rodríguez-Ambriz, S. L., Sánchez-Rivera, M. M., & Bello-Pérez, L. A. (2012b). Structural and rheological differences between fruit and cereal starches. *Agrociencia*, 46, 455–466.

Champ, M., Langkilde, A.-N., Brouns, F., Kettlitz, B., & Le Bail-Collet, Y. (2003). Advances in dietary fiber characterization. 2. Consumption, chemistry, physiology and measurement of resistant starch; implications for health and food labelling. *Nutrition Research Reviews*, 16, 143–161.

Chanvrier, H., Appelqvist, I. A. M., Bird, A. R., Gilbert, E., Htoon, A., Li, Z., Lillford, P. J., Lopez-Rubio, A., Morell, M. K., & Topping, D. L. (2007a). Processing of novel elevated amylose wheats: Functional properties and starch digestibility of extruded products. *Journal of Agricultural and Food Chemistry*, 55, 10248–10257.

Chanvrier, H., Uthayakumaran, S., Appelqvist, I. A. M., Gidley, M. J., Gilbert, E. P., & López-Rubio, A. (2007b). Influence of storage conditions on the structure, thermal behavior, and formation of enzyme-resistant starch in extruded starches. *Journal of Agricultural and Food Chemistry*, 55, 9883–9890.

Chung, H-J., Hoover, R., & Liu, Q. (2009a). The impact of single and dual hydrothermal modifications on the molecular structure and physicochemical properties of normal corn starch. *International Journal of Biological Macromolecules*, 44, 203–210.

Chung, H-J., Liu, Q., & Hoover, R. (2009b). Impact of annealing and heat-moisture treatment on rapidly digestible, slowly digestible and resistant starch levels in native and gelatinized corn, pea and lentil starches. *Carbohydrate Polymers*, 75, 436–447.

Duyen, T. T. M., Houng, N. T. M., Phi, N. T. L., & Hung, P. V. (2020). Physicochemical properties and *in vitro* digestibility of mung-bean starches varying amylose contents under citric acid and hydrothermal treatments. *International Journal of Biological Macromolecules*, 164, 651–658.

Edwards, C. H., Warren, F. J., Milligan, P. J., Butterworth, P. J., & Ellis, P. R. (2014). A novel method for classifying starch digestion by modelling the amylolysis of plant foods using first-order enzyme kinetic principles. *Food & Function*, 5, 2751–2758.

Emami, S., Perera, A., Meda, V., & Tyler, R. T. (2012). Effect of microwave treatment on starch digestibility and physic-chemical properties of three barley types. *Food and Bioprocess Technology*, 5, 2266–2274.

Englyst, H. N., Kingman, S. M., & Cummings, J. H. (1992). Classification and measurement of nutritionally important starch fractions. *European Journal of Clinical Nutrition*, 46(Supplement 2), S33–S50.

Englyst, K., Goux, A., Meynier, A., Quigley, M., Englyst, H., Brack, O., & Vinoy, S. (2018). Inter-laboratory validation of the starch digestibility method for determination of rapidly digestible and slowly digestible starch. *Food Chemistry*, 245, 1183–1189.

Faisant, N., Buléon, A., Colonna, P., Molis, C., Lartigue, S., Galmiche, J. P., & Champ, M. (1995). Digestion of raw banana starch in the small intestine of healthy humans: Structural features of resistant starch. *British Journal of Nutrition*, 73, 111–123.

Flores-Silva, P. C., Rodriguez-Ambriz, S. L., & Bello-Pérez, L. A. (2015). Gluten-free snacks using plantain-chiickpea and maize blend: Chemical composition, starch digestibility, and predicted glycemic index. *Journal of Food Science*, 80, C961–C966.

Foschia, M., Beraldo, P., & Peressini, D. (2016). Evaluation of the physicochemical properties of gluten-free pasta enriched with resistant starch. *Journal of the Science of Food and Agriculture*, 97, 572–577.

Fung, K. Y. C., Cosgrove L., Lockett, T., Head, R., & Topping D. L. (2012). A review of the potential mechanisms for the lowering of colorectal oncogenesis by butyrate. *British Journal of Nutrition*, 108, 820–831.

García-Solís, S. E., Bello-Pérez, L. A., Agama-Acevedo, E., & Flores-Silva, P. C. (2018). Plantain flour: A potential nutraceutical ingredient to increase fiber and reduce starch digestibility of gluten-free cookies. *Starch/Stärke*, 70, 1700107.

Garcia-Valle, D. E., Agama-Acevedo, E., Alvarez-Ramirez, J., & Bello-Perez, L. A. (2020). Semolina pasta replaced with whole unripe plantain flour: Chemical, cooking quality, texture, and starch digestibility. *Starch/Stärke*, 72, 1900097.

Ge, P-Z., Fan, D-C., Ding, M., Wang, D., & Zhou, C-Q. (2014). Characterization and nutritional quality evaluation of several starch noodles. *Starch/Stärke*, 66, 880–886.

Gidley, M. J. (2013). Hydrocolloids in the digestive tract and related health implications. *Current Opinion in Colloid & Interface Science*, 18, 371–378.

Giuberti, G., Gallo, A., Fortunati, P., & Rossi, F. (2016). Influence of high-amylose maize starch addition on *in vitro* starch digestibility and sensory characteristics of cookies. *Starch/Stärke*, 68, 469–475.

Giuberti, G., Marti, A., Fortunati, P., & Gallo, A. (2017). Gluten free rice cookies with resistant starch ingredients from modified waxy rice starches: Nutritional aspect and textural characteristics. *Journal of Cereal Science*, 76, 157–164.

Goñi, I., Garcia-Alonso, A., & Saura-Calixto, F. (1997). A starch hydrolysis procedure to estimate glycemic index. *Nutrition Research*, 17, 427–437.

González-Soto, R. A., Agama-Acevedo, E., Solorza-Feria, J., Rendón-Villalobos, R., & Bello-Pérez, L. A. (2004). Resistant starch made from banana starch by autoclaving and debranching. *Starch/Stärke*, 56, 495–499.

González-Soto, R. A., Sánchez-Hernández, L., Solorza-Feria, J., Núñez-Santiago, C., Flores-Huicochea, E., & Bello-Pérez, L. A. (2006). Resistant starch production from non-conventional starch sources by extrusion. *Food Science and Technology International*, 12, 5–11.

Gou, Y., Xu, T., Li, N., Cheng, Q., Qiao, D., Zhang, B., Zhao, S., Huang, Q., & Lin, Q. (2019). Supramolecular structure and pasting/ digestion behaviors of rice starches following concurrent microwave and heat moisture treatment. *International Journal of Biological Macromolecules*, 135, 437–444.

Han, J. A., & BeMiller, J. N. (2007). Preparation and physical characteristics of slowly digesting modified food starches. *Carbohydrate Polymers*, 67, 366–374.

Han, Z., Zeng, X. A., Yu, S. J., Zhang, B. S., & Chen, X. D. (2009). Effects of pulsed electric fields (PEF) treatment on physicochemical properties of potato starch. *Innovative Food Science and Emerging Technologies*, 10, 481–485.

Han, Z., Zeng, X. A., Fu, N., Yu, S. J., Chen, X. D., & Kennedy, J. F. (2012). Effects of pulsed electric field treatments on some properties of tapioca starch. *Carbohydrate Polymers*, 89, 1012–1017.

Haralampu, S. (2000). Resistant starch: A review of the physical properties and biological impact of RS3. *Carbohydrate Polymers*, 41, 285–292.

Hasjim, J., & Jane, J. L. (2009). Production of resistant starch by extrusion cooking of acid-modified normal-maize starch. *Journal of Food Science*, 74, C556–C562.

Hasjim, J., Lee, S-O., Hendrich, S., Setiawan, S., Ai, Y., & Jane, J. L. (2010). Characterization of a novel resistant-starch and its effects on postprandial plasma-glucose and insulin responses. *Cereal Chemistry*, 87, 257–262.

Hernández-Nava, R. G., Berrios, J. D. J., Pan, J., Osorio-Díaz, P., & Bello-Pérez, L. A. (2009). Development and characterization of spaghetti with high resistant starch content supplemented with banana starch. *Food Science and Technology International*, 15, 73–78.

Homayouni, A., Amini, A., Keshtiban, A. K., Mohammad, A., Mortazavian, A. M., Esazadeh, K., & Pourmoradian, S. (2014). Resistant starch in food industry: A changing outlook for consumer and producer. *Starch/Stärke*, 66, 102–114.

Hong, J., An, D., Liu, C., Li, L., Han, Z., Guan, E., Xu, B., Zheng, X., & Bian, K. (2020). Rheological, textural, and digestible properties of fresh noodles: Influence of starch esterified by conventional and pulsed electric field-assisted dual technique with full range of amylose content. *Journal of Food Processing and Preservation*, 44, e14567.

Hong, J., Zeng, X-A., Buckow, R., & Han, Z. (2018). Structural, thermodynamic and digestible properties of maize starches esterified by conventional and dual methods: Differentiation of amylose contents. *Food Hydrocolloids*, 83, 419–429.

Insel, P. M., & Turner, R. E. (2007). *Nutrition*. Jones & Bartlett Learning.

Jenkins, D. J. A., Vuksan, V., Kendall, C. W. C., Würsch, P., Jeffcoat, R., Waring, S., Mehling, C. C., Vidgen, E., Augustin, L. S. A., & Wong, E. (1998). Physiological effects of resistant starches on fecal bulk, short chain fatty acids, blood lipids and glycemic index. *Journal of the American College of Nutrition*, 17, 609–616.

Jiang, F., Du, C., Jiang, W., Wang, L., & Du, S.-L. (2020). The preparation, formation, fermentability, and applications of resistant starch. *International Journal of Biological Macromolecules*, 150, 1155–1161.

Kim, H. M., Choi, S. J., Choi, H-D., Park, C-S., & Moon, T. W. (2020). Amylosucrase-modified waxy potato starches recrystallized with amylose: The role of amylopectin chain length on formation of low-digestible fractions. *Food Chemistry*, 320, 126490.

Kim, H. R., Choi, S. J., Park, C.-S., & Moon, T. W. (2017). Kinetic studies of *in vitro* digestion of amylosucrase-modified waxy corn starches based on branch chain length distributions. *Food Hydrocolloids*, 65, 46–56.

Lee, B. H., Bello-Pérez, L. A., Lin, A. H. M., Kim, C. Y., & Hamaker, B. R. (2013). Importance of location of digestion and colonic fermentation of starch related to its quality. *Cereal Chemistry*, 90, 335–343.

Lehmann, U., & Robin, F. (2007). Slowly digestible starch- its structure and health implications: A review. *Trends in Food Science & Technology*, 18, 346–355.

Lehmann, U., Rössler, C., Schmiedl, D., & Jacobasch, G. (2003). Production and physicochemical characterization of resistant starch type III derived from pea starch. *Molecular Nutrition Food Research*, 47, 60–63.

Ley, R. E. (2010). Obesity and the human microbiome. *Current Opinion Gastroenterology*, 26, 5–11.

Li, H., Gidley, M. J., & Dhital, S. (2019a). High-Amylose starches to bridge the "fiber gap": Development, structure, and nutritional functionality. *Comprehensive Reviews in Food Science and Food Safety*, 18, 362–379.

Li, H., Yan, S., Ji, J., Xu, M., Mao, H., Wen, Y., Wang, J., & Sun, B. (2020). Insights into maize starch degradation by high pressure homogenization treatment from molecular structure aspect. *International Journal of Biological Macromolecules*, 161, 72–77.

Li, M-N., Xie, Y., Chen, H-Q., & Zhang, B. (2019b). Effects of heat-moisture treatment after citric acid esterification on structural properties and digestibility of wheat starch, A- and B-type starch granules. *Food Chemistry*, 272, 523–529.

Li, Y-D., Xu, T-C., Xiao, J-X., Zong, A-Z., Qiu, B., Jia, M., Liu, L-N., & Liu, W. (2018). Efficacy of potato resistant starch prepared by microwave-toughening treatment. *Carbohydrate Polymers*, 192, 299–307.

Lin, A. H.-M., Hamaker, B. R., & Nichols, B. L. J. (2012). Direct starch digestion by sucrose-isomaltase and maltase-glucoamylase. *Journal of Pediatric Gastroenterology and Nutrition*, 55(Supplement 2), S43–S45.

Lin, D., Zhou, W., Yang, Z., Zhong, Y., Xing, B., Wu, Z., Chen, H., Wu, D., Zhang, Q., Qin, W., & Li, S. (2019). Study on physicochemical, digestive properties and application of acetylated starch in noodles. *International Journal of Biological Macromolecules*, 128, 948–956.

Liu, H., Wang, L., Cao, R., Fan, H., & Wang, M. (2016). *In vitro* digestibility and changes in physicochemical and structural properties of common buckwheat starch affected by high hydrostatic pressure. *Carbohydrate Polymers*, 144, 1–8.

Liu, Y., Liu, J., Kong, J., Wang, R., Liu, M., Strappe, P., Blanchard, C., & Zhou, Z. (2020). Citrate esterification of debranched waxy maize starch: Structural, physicochemical and amylolysis properties. *Food Hydrocolloids*, 104, 105704.

Lockyer, S., & Nugent, A. P. (2017). Health effects of resistant starch. *Nutrition Bulletin*, 42, 10–41.

Long, J., Zhang, B., Li, X., Zhan, X., Xu, X., Xie, Z., & Jin, Z. (2018). Effective production of resistant starch using pullulanase immobilized onto magnetic chitosan/ Fe3O4 nanoparticles. *Food Chemistry*, 239, 276–286.

López-Silva, M., Bello-Pérez, L. A., Castillo-Rodríguez, V. M., Agama-Acevedo E., & Alvarez-Ramirez, J. (2020). *In vitro* digestibility characteristics of octenyl succinic acid (OSA) modified starch with different amylose content. *Food Chemistry*, 304, 125434.

Lu, X., Liu, H., & Huang, Q. (2020). Fabrication and characterization of resistant starch stabilized Pickering emulsions. *Food Hydrocolloids*, 103, 105703.

Luo, Z., Fu, X., He, X., Luo, F., Gao, Q., & Yu, S. (2008). Effect of ultrasonic treatment on the physicochemical properties of maize starches differing in amylose content. *Starch/ Stärke*, 60, 646–653.

Lv, Q-Q., Li, G-Y., Xie, Q-T., Zhan, B., Li, X-M., Pan, Y., & Chen, H-Q. (2018). Evaluation studies on the combined effect of hydrothermal treatment and octenyl succinylation on the physic-chemical, structural and digestibility characteristics of sweet potato starch. *Food Chemistry*, 256, 413–418.

Magallanes-Cruz, P. A., Flores-Silva, P. C., & Bello-Perez, L. A. (2017). Starch structure influences its digestibility: A review. *Journal of Food Science*, 82, 2016–2023.

Magallanes-Cruz, P. A., Bello-Pérez, L. A., Agama-Acevedo, E., Tovar, J., & Carmona-García, R. (2020). Effect of the addition of thermostable and non-thermostable type 2 resistant starch (RS2) in cake batters. *LWT: Food Science and Technology*, 118, 108834.

Mariscal-Morenos, R. M., Figueroa-Cárdenas, J. D., Santiago-Ramos, D., & Rayas-Duarte, P. (2018). Amylose lipid complexes formation as an alternative to reduce amylopectin retrogradation and staling of stored tortillas. *International Journal of Food Science Technology*, 54, 1651–1657.

McCleary, B. V., Mills, C., Draga, A. (2009). Development and evaluation of an integrated method for the measurement of total dietary fibre. *Quality Assurance and Safety of Crops & Foods*, 4, 213–224.

McCleary, B. V., Sloane, N, Draga, A. (2015). Determination of total dietary fibre and available carbohydrates: A rapid integrated procedure that simulates in vivo digestion. *Starch/Starke*, 67, 860–883.

Menon, R., Padmaja, G., & Sajeev, M. S. (2015). Cooking behavior and starch digestibility of NUTRIOSE® (resistant starch) enriched noodles from sweet potato flour and starch. *Food Chemistry*, 182, 217–223.

Meraz, M., Alvarez-Ramirez, J., Vernon-Carter, E. J., Reyes, I., Hernandez-Jaimes, C., & Martinez-Martinez, F. (2020). A two competing substrates Michaelis-Menten Kinetics scheme for the analysis of *in vitro* starch digestograms. *Starch/Stärke*, 72, 1900170.

Miao, M., Jiang, B., Cui, S. W., Zhang, T., & Jin, Z. (2015). Slowly digestible starch-A review. *Critical Reviews in Food Science and Nutrition*, 55, 1642–1657.

Na, J. H., Kim, H. R., Kim, Y., Lee, J. S., Park, H. J., Moon, T. W., & Lee, C. J. (2020). Structural characteristics of low-digestible sweet potato starch prepared by heat-moisture treatment. *International Journal of Biological Macromolecules*, 151, 1049–1057.

Okumus, B. N., Tacer-Caba, Z., Kahraman, K., & Nilufer-Erdil, D. (2018). Resistant starch type V formation in brown lentil (*Lens culinaris* Medikus) starch with different lipids/ fatty acids. *Food Chemistry*, 240, 550–558.

Osorio-Díaz, P., Aguilar-Sandoval, A., Agama-Acevedo, E., Rendón-Villalobos, R., Tovar, J., & Bello-Pérez, L. A. (2008). Composite durum wheat flour/ plantain starch White salted noodles: Proximal composition, starch digestibility, and indigestible fraction content. *Cereal Chemistry*, 85, 339–343.

Ovando-Martinez, M., Sáyago-Ayerdi, S., Agama-Acevedo, E., Goñi, I., & Bello-Pérez, L. A. (2009). Unripe banana flour as an ingredient to increase the undigestible carbohydrates of pasta. *Food Chemistry*, 113, 121–126.

Ozturk, S., Köksel, H., & Ng, P. K. W. (2011). Production of resistant starch from acid-modified amylotype starches with enhanced functional properties. *Journal of Food Engineering*, 103, 156–164.

Panyoo, A. E., & Emmambux, M. N. (2017). Amylose–lipid complex production and potential health benefits: A mini-review. *Starch/Stärke*, 69, 1–24.

Park, M.-O., Chandrasekaran, M., & Yoo, S.-H. (2019). Production and characterization of low-calorie turanose and digestion-resistant starch by an amylosucrase from *Neisseria subflava*. *Food Chemistry*, 300, 125225.

Polesi, L. F., Sarmento, S. B. S., & Franco, C. M. L. (2011). Production and physicochemical properties of resistant starch from hydrolysed wrinkled pea starch. *International Journal of Food Science& Technology*, 46, 2257–2265.

Quintero-Castaño, V. D., Castellanos-Galeano, F. J., Álvarez-Barreto, C. L., Bello-Pérez, L. A., & Alvarez-Ramirez, J. (2020). *In vitro* digestibility of octenyl succinic anhydride-starch from the fruit of three Colombian Musa. *Food Hydrocolloids*, 101, 105566.

Raigond, P., Ezekiel, R., & Raigond, B. (2015). Resistant starch in food: A review. *Journal of the Science of Food and Agriculture*, 95, 1968–1978.

Roldan-Cruz, C. R., García-Dias, S., Garcia-Hernandez, A., Alvarez-Ramirez, J., Vernon-Carter, E. J. (2020). Microstructural changes and *in vitro* digestibility of maize starch treated with different calcium compounds used in nixtamalization processes. *Starch/Starke*, 72, 1900303.

Roman, L., Gomez, M., Hamaker, B. R., & Martinez, M. M. (2019). Banana starch and molecular shear fragmentation dramatically increase structurally drive slowly digestible starch in fully gelatinized bread crumb. *Food Chemistry*, 274, 664–671.

Remya, R., Jyothi, A. N., & Sreekumar, J. (2018). Effect of chemical modified with critic acid on the physicochemical properties and resistant starch formation in different starches. *Carbohydrate Polymers*, 202, 29–38.

Saito, H., Tamura, M., & Ogawa, Y. (2019). Starch digestibility of various Japanese commercial noodles made from different starch sources. *Food Chemistry*, 283, 390–396.

Sanz, T., Salvador, A., Baixauli, R., & Fiszman, S. M. (2009). Evaluation of four types of resistant starch in muffins. II. Effects in texture, colour and consumer response. *European Food Research and Technology*, 229, 197–204.

Shafiei, Y., Razavilar, V., Javadi, A., & Mirzaei, H. (2012). Survivability of free and micro-encapsulated Lactobacillus plantarum with alginate and resistant starch in simulated gastrointestinal conditions. *Journal of Food Agriculture and Environment*, 10, 207–212.

Shaikh, F., Ali, T. A., Mustafa, G., & Hasnain, A. (2019). Comparative study on effects of citric and lactic acid treatment on morphological, functional, resistant starch fraction and glycemic index of corn and sorghum starches. *International Journal of Biological Macromolecules*, 135, 314–327.

Sharma, V., Kaur, M., Sandlhu, K.S and Godara, S. K. (2020). Effect of cross-linking on physicochemical, thermal, pasting, *in vitro* digestibility and film forming properties of faba bean (*Vicia faba* L.) starch. *International Journal of Biological Macromolecules*, 159, 243–249.

Shen, Y., Zhang, N., Xu, Y., Huang, J., Yuan, M., Wu, D., & Shu, X. (2019). Physicochemical properties of hydroxypropylated and cross-linked rice starches differential in amylose content. *International Journal of Biological Macromolecules*, 128, 775–781.

Shin, S. I., Byun, J., Park, K. H., & Moon, T. W. (2004). Effect of partial acid hydrolysis and heat-moisture treatment on formation of resistant tuber starch. *Cereal Chemistry*, 81, 194–198.

Simons, C. W., Hall, C., & Vatansever, S. (2018). Production of resistant starch (RS3) from edible beans starches. *Journal of Processing and Preservation*, 42, e13587.

Simsek, S., & El, S. N. (2012). Production of resistant starch from taro (*Colocasia esculenta* L. Schott) corn and determination of its effects on health by *in vitro* methods. *Carbohydrate Polymers*, 90, 1204–1209.

Sofi, S. A., Singh, J., Mir, S. A., & Dar, B. N. (2020). In vitro starch digestibility, cooking quality, rheology and sensory properties of gluten-free pregelatinized rice noodle enriched with germinated chickpea flour. *LWT- Food Science and Technology*, 133, 110090.

Soler, A., Méndez-Montealvo, G., Velázquez-Castillo, R., Hernández-Gama, R., Osorio-Díaz, P., & Velázquez, G. (2020). Effect of crystalline and double helical structures on the resistant fraction of autoclaved corn starch with different amylose content. *Starch/Starke*, 72, 1900306.

Strader, A. D., & Woods, S. C. (2005). Gastrointestinal hormones and food intake. *Gastroenterology*, 128, 175–191.

Sujka, M. (2017). Ultrasonic modification of starch – Impact on granules porosity. *Ultrasonics Sonochemistry*, 37, 424–429.

Tan, L., & Kong, L. (2020). Starch-guest inclusion complexes: Formation, structure, and enzymatic digestion. *Critical Reviews in Food Science and Nutrition*, 60, 780–790.

Tian, S., & Sun, Y. (2020). Influencing factor of resistant starch formation and application in cereal products: A review. *International Journal of Biological Macromolecules*, 149, 424–431.

Tovar, J., & Melito, C. (1996). Steam-cooking and dry heating produce resistant starch in legumes. *Journal of Agricultural and Food Chemistry*, 44, 2642–2645.

Trung, P. T. B., Ngoc, L. B. B., Hoa, P. N., Tien, N. N. T. T., & Hung, P. V. (2017). Impact of heat-moisture and annealing treatments on physicochemical properties and digestibility of starches from different colored sweet potato varieties. *International Journal of Biological Macromolecules*, 105, 1071–1078.

Venkatachalam, M., Zhang, G., & Hamaker, B. R. (2006). Use polymer encapsulated starches as a novel method to make low glycemic foods. In World Grain Summit: Food and Beverages (AACC International), Abstract P-285, Sept. 17–20, San Francisco, CA.

Wang, B., Li, D., Wang, L., Chiu, Y. L., Chen, X. D., & Mao, Z. (2008). Effect of high-pressure homogenization on the structure and thermal properties of maize starch. *Journal of Food Engineering*, 87, 436–444.

Xia, J., Zhu, D., Wang, R., Cui, Y., & Yan, Y. (2018). Crop resistant starch and genetic improvement: A review of recent advances. *Theoretical and Applied Genetics*, 131, 2495–2511.

Xie, X., Qi, L., Xu, C., Shen, Y., Wang, H., & Zhang, H. (2020). Understanding how the cooking methods affected structures and digestibility of native and heat-moisture treated rice starches. *Journal of Cereal Science*, 95, 103085.

Ye, J., Lou, S., Huang, A., Chen, J., Liu, C., & McClements, D. J. (2019). Synthesis and characterization of citric acid esterified rice starch by reactive extrusion: A new method of producing resistant starch. *Food Hydrocolloids*, 92, 135–142.

Zaman, S. A., & Sarbini, S. R. (2015). The potential of resistant starch as prebiotic. *Critical Reviews in Biotechnology*, 36, 578–584.

Zhang, B., Dhital, S., & Gidley, M. J. (2015). Densely packed matrices as rate determining features in starch hydrolysis. *Trends in Food Science & Technology*, 43, 18–31.

Zhang, C., Yin, A., Li, H., Wang, R., Wu, G., Shen, J., Zhang, M., Wang, L., Hou, Y., Ouyang, H., Zhang, Y., Zheng, Y., Wang, J., Lv, X., Wang, Y., Zhang, F., Zeng, B., Li, W., Yan, F., Zhao, Y., Pang, X., Zhang, X., Fu, H., Chen, F., Zhao, N., Hamaker, B. R., Bridgewater, L. C., Weinkove, D., Clement, K., Dore, J., Holmes, E., Xiao, H., Zhao, G., Yang, S., Bork, P., Nicholson, K. J., Wei, H., Tang, H., Zhang, X., & Zhao, L. (2015). Dietary modulation of gut microbiota contributes to alleviation of both genetic and simple obesity in children. *EbioMedicine*, 2, 968–984.

Zhang, G., & Hamaker, B. R. (2009). Slowly digestible starch: Concept mechanism, and proposed extended glycemic index. *Critical Reviews in Food Science and Nutrition*, 49, 852–867.

Zhang, G., Ao, Z., & Hamaker, B. R. (2006a). Slow digestion property of native cereal starches. *Biomacromolecules*, 7, 3252–3258.

Zhang, G., Venkatachalam, M., Hamaker, B. R. (2006b). Structural basis for the slow digestion properties of native cereal starch. *Biomacromolecules*, 7, 3259–3266.

Zhang, X., Chen, Y., Zhang, R., Zhong, Y., Luo, Y., Xu, S., Liu, J., Xue, J., & Guo, D. (2016). Effects of extrusion treatment on physicochemical properties and *in vitro* digestion of pregelatinized high amylose maize flour. *Journal of Cereal Science*, 68, 108–115.

Zheng, Y., Ou, Y., Zhang, Y., Zheng, B., Zeng, H., & Zeng, S. (2020a). Physicochemical properties and *in vitro* digestibility of lotus seed starch-lecithin complexes prepared by dynamic high pressure homogenization. *International Journal of Biological Macromolecules*, 156, 196–203.

Zheng, Y., Wang, B., Gou, Z., Zhang, Y., Zheng, B., Zeng, S., & Zeng, H. (2020b). Properties of lotus seed starch-glycerin monostearin V-complexes after long-term retrogradation. *Food Chemistry*, 311, 125887.

Zhu, F. (2015). Impact of ultrasound on structure, physicochemical properties, modifications, and applications of starch. *Trends in Food Science and Technology*, 43, 1–17.

Chapter 7

Resistant Starch

The Dietary Fiber of the Starch World

Darrell W. Cockburn

CONTENTS

7.1 INTRODUCTION

Starch makes up about 25% of the calories in the typical human diet, making it one of the most important carbohydrates that we consume. As it enters the digestive tract it is first subjected to mastication by the teeth, breaking apart materials surrounding starch and making it accessible to the enzymes in our saliva, particularly salivary amylase, which can initiate some starch break-down. It then passes through to the stomach, where the acids present may further hydrolyze some of the glycosidic bonds that join the starch together before passing through to the small intestine. This is where the vast majority of starch digestion occurs, driven by pancreatic amylase, efficiently reducing the starch to oligosaccharides, which can be further hydrolyzed to glucose by the

DOI: 10.1201/9781003088929-7

intestinal brush border enzymes sucrase–isomaltase and maltase–glucoamylase before being absorbed (Mackie & Rigby, 2015; Brownlee et al., 2018). This is where the typical story of starch digestion ends. However, depending on a number of factors, some may escape the small intestine intact to the colon where the dense microbial population can make use of it as an energy source (Cockburn & Koropatkin, 2016; DeMartino & Cockburn, 2019). Microbes earlier in the digestive tract likely have some ability to break down and utilize starch, however, the longer residence time and the much greater microbial numbers in the colon (Sender et al., 2016) mean that there is greater potential for microbial fermentation of starch. A number of factors can influence how much starch makes it to the colon, including the total amount consumed, other foods consumed along with the starch that may decrease enzyme accessibility, genetic factors that determine the amount of digestive enzymes produced and various factors that influence the rate of intestinal motility (Silvester et al., 1995; Carpenter et al., 2015). Additionally, the rising rate of diabetes across the developed world has meant an increasing number of drugs that aim to slow starch digestion and the end production of glucose to avoid blood sugar spikes (Tahrani et al., 2016). Drugs in this class, such as acarbose, typically act as inhibitors of the starch digestive enzymes, particularly maltase–glucoamylase (Joshi et al., 2015), but also to a lesser extent pancreatic amylase (Akkarachiyasit et al., 2010). This slowdown in digestion means that there is more starch that survives to the colon and has been found to influence the microbial population there (Baxter et al., 2019a). While these dietary, physiological and medical factors all play an important role in the digestion of starch and its transit through the intestinal tract, the structure of the starch itself plays an even more important role (Silvester et al., 1995).

Starch is produced by plants as an energy storage medium and plays an important role in leaves, roots, and seeds. This storage compound takes the form of semicrystalline granules made up of the glucose polymers amylose and amylopectin. Amylose is mostly linear α1,4-linked glucose, while amylopectin consists of linear chains joined via α1,6-linked branch points. Typical amylopectin:amylose ratios are around 70:30, though the exact ratio varies from plant to plant. The amylopectin dominates the gross structure of the granules, with alternating amorphous and crystalline regions forming depending on the presence of amylopectin branch points. Clusters of branch points form the amorphous regions, while linear regions wind together in double helices, which then pack together into crystalline arrays (Damager et al., 2010). There have been two predominant crystalline structures found for the packing of the double helices, termed the A-type and the B-type (Imberty et al., 1991). From a dietary perspective, grains contain the A-type crystalline structure, while tubers such as potatoes contain the B-type. Legumes contain what is termed the C-type, which is simply a mixture of the A- and

B-types (Bogracheva et al., 1998). The location of amylose within the granules is still an area of study and debate, though it seems to be dispersed through the granule, perhaps interacting with the amylopectin, with differences between starch sources (Bertoft, 2017). Certain lines of grains such as wheat and corn have been developed that are termed high amylose. In these lines the amylopectin:amylose ratio is essentially flipped to 30:70, and although the gross morphology is substantially different then is what is seen in tubers, it adopts the B-type crystalline form, with amylose playing a more important role in the crystalline regions (Shi et al., 1998).

For most human starch consumption, the starch source is first cooked, which to varying degrees depending on the starch and the cooking method, eliminates the starch granular structure in a process termed gelatinization (Wang et al., 2018). This gelatinized starch is what human enzymes are very adept at degrading and extracting energy from. However, the human enzymes can also degrade the A-type starch granules, although at a somewhat slower rate and it is referred to as slowly digested starch (Englyst et al., 1992). This in turn reduces the rate of glucose release, but also allows a greater fraction to make it to the colon. However, human enzymes have very little ability to degrade B-type starch granules and these thus make it to the colon almost untouched (Faisant et al., 1995). This starch that makes it through to the colon relatively unscathed by human enzymes has been termed resistant starch (Englyst & Cummings, 1985). Starches with B-type and C-type crystallinity are one of the most prominent examples of resistant starch, but there are other types as well (Figure 7.1).

Type 1 resistant starch is found in whole grains or other instances where the starch is surrounded by some material that blocks access of human enzymes (Englyst et al., 1992). Type 2 resistant starches are the B-type and C-type crystallinity starch granules, including the high amylose varieties of grains. Type 3 resistant starch is retrograded starch that has been gelatinized but then able to recrystallize over time (Wang et al., 2015). Several factors affect the rate of retrogradation, such as temperature, chain length and other materials that are present with the starch. It is a process that occurs with most starchy foods and can be associated with staling such as in breads. Type 4 resistant starch is chemically modified starch. This can encompass several different modifications, with cross-linking reactions with phosphate being among the most popular in food starches (Dupuis et al., 2014). There has also been a proposed type 5 resistant starch, amylose with fatty acid chains inserted into the helix formed by the amylose chain (Ai et al., 2013). There is still some debate as to whether the reduction in human enzymatic activity against this substrate is sufficient to class it as resistant starch, or if it is better thought of as a slowly digested starch. This last point illustrates the fact that the definition of resistant starch is largely tied to the kinetic phenomenon of it avoiding digestion for

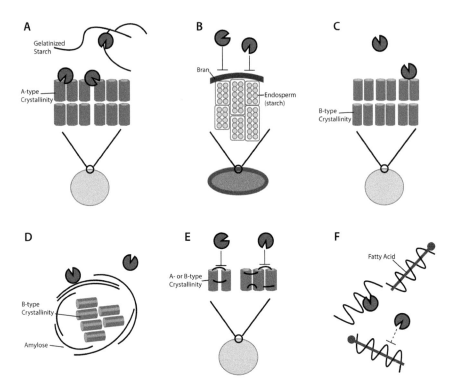

Figure 7.1 Mechanisms of starch digestion resistance in the different resistant starch types. (A) Either gelatinized or A-type crystallinity starch are readily attacked and degraded by human enzymes, though rates may vary. (B) Type 1 resistant starch. Nonstarch materials, such as the bran in whole grains, blocks access of enzymes to the starch. (C) Type 2 resistant starch. The presence of B-type or C-type crystallinity blocks binding of human enzymes or results in nonproductive binding leading to highly inefficient digestion. (D) Type 3 resistant starch. The B-type crystallinity formed during retrogradation of amylose and amylopectin prevents binding similar to what is seen in type 2 resistant starch. (E) Type 4 resistant starch. Chemical modification of starch (typically via cross-linking) closes gaps between starch crystallites and sterically blocks access to individual chains, preventing digestion. (F) Type 5 resistant starch (or slowly degraded starch). Helical amylose forms inclusion complexes with other molecules, typically lipids, impeding access by enzymes.

long enough to survive passage through the small intestine. Thus, the definition of what constitutes resistant starch may vary from person to person due to physiological factors or perhaps even meal to meal depending on total starch content and other constituents. Standardized methodologies for measuring the resistant starch content of foods have allowed agreed upon numbers to be put on the resistant starch content of a given food (McCleary et al., 2002). However, interindividual differences may play an important role in the outcome of resistant starch consumption.

7.2 MICROBIAL DIGESTION OF RESISTANT STARCH

The focus on the definition of resistant starch is whether it reaches the colon. Why is this important? The colon is home to one of the densest populations of microbes known, with a total number of microbial cells that equals or exceeds the number of human cells in the entire body (Sender et al., 2016). There are more microbes present in the colon than in the rest of the body combined and they collectively contain a metabolic repertoire that vastly exceeds the number of reactions that humans alone are capable of performing. Thus, it is unsurprising that among this repertoire are microbes that are capable of digesting what is to us resistant starch. Enzymes capable of digestion of starch or starch breakdown products are among the most ubiquitous throughout the gut microbiome (El Kaoutari et al., 2013). Despite this, the ability to grow on resistant starch as a sole carbohydrate source is a relatively rare trait. Certain *Bifidobacterium* species within the *B. adolescentis* group and the *B. pseudolongum* group (Centanni et al., 2018; Jung et al. 2020), along with the Firmicute *Ruminococcus bromii* (Ze et al., 2012) are the only organisms in the gut that have been confirmed to thoroughly digest and utilize resistant starch.

To better understand what sets the resistant starch degrading organisms apart, it is instructive to look at two other well-characterized starch degrading organisms in the gut: *Bacteroides thetaiotaomicron* and *Eubacterium rectale*. The starch utilization system (Sus) of *B. thetaiotaomicron* has been extensively studied as representative of a paradigm in polysaccharide digestion and utilization by members of the Bacteroidetes (Foley et al., 2016). *B. thetaiotaomicron* and many other members of this phylum are polysaccharide generalists and can degrade a wide variety of carbohydrates. Their machinery for accomplishing this are organized into operons termed polysaccharide utilization loci (PUL), which are expressed upon detection of the target polysaccharide (Terrapon et al., 2015). These PUL are characterized by the presence of a TonB-dependent outer membrane transporter (Glenwright et al., 2017; Pollet et al., 2021) specific to the breakdown products of the polysaccharide in question along with a series of binding proteins and enzymes localized to the cell surface along with accessory periplasmic enzymes. In the case of Sus there is the outer membrane transporter, three outer membrane starch binding proteins, two periplasmic enzymes and an amylase capable of hydrolyzing the α1,4 bonds of starch on the surface (Foley et al., 2016). This amylase, SusG, is a versatile enzyme able to efficiently degrade soluble starch and its breakdown products and derivatives (Koropatkin et al., 2010), with a flexible active site that can accommodate branch points (Arnal et al., 2018), though not hydrolyze them directly. Besides the binding proteins of Sus, SusG also contains a carbohydrate binding module (CBM) and a starch binding site on its catalytic domain (Koropatkin et al., 2010), outside of the active site, termed a surface binding site (Cockburn & Svensson, 2013). However, despite the binding versatility of its main starch degrading enzyme, *B. thetaiotaomicron* is incapable of growing on resistant starch sources (Ze et al., 2012). While some gram-positive organisms have similar PUL-like organization for some of their polysaccharide degrading components (Sheridan et al., 2016), many of the polysaccharide degradation genes are dispersed throughout the genomes of these organisms. This is the case with the starch utilizing components of *E. rectale*, which consists of three transporters and two enzymes that are upregulated when the organism is grown on starch (Cockburn et al., 2015). Both enzymes are α-amylases degrading α1,4 glucosidic bonds, with Amy13K localized to the cell wall and Amy13B localized to the membrane. Amy13K contains a suite of starch-specific CBMs, including two that seem to be found only among closely related gut microbes (Cockburn et al., 2018). These binding modules bind efficiently to corn starch, even high amylose varieties, but do not bind to potato starch or other resistant starches. Likewise, the enzyme's activity is driven by these binding specificities, as potato starch is a very poor substrate, while the corn starches are much better. Interestingly, removal of the binding domains reduces the enzyme activity towards corn starches down to the same level as for potato starch, indicating the importance of targeted binding for activity against these starches. Despite this ability to bind to the high-amylose corn starch and perform initial digestion, it is insufficient to support growth of *E. rectale* on high-amylose starches. The second enzyme Amy13B does not contain any binding domains and seems to play a role in degrading oligosaccharides to distribute them between the transporters (Cockburn et al., 2020).

While the starch degraders in the gut that do not grow on resistant starch are generally limited to one or two surface enzymes for starch digestion, resistant degraders display more complex systems. *Bifidobacterium adolescentis* has a suite of six enzymes that it expresses on its cell surface that may be involved in resistant starch digestion (Jung et al., 2020). These contain a variety of CBMs for starch binding including the intriguing CBM74. This family of CBMs seems to only be found in resistant-starch degrading organisms such as *B. adolescentis* and is the only characterized CBM identified so far that binds to potato starch better than to nonresistant starches (Valk et al., 2016). These enzymes include four α-amylases for α1,4 bond hydrolysis and a single polypeptide that has an α-amylase at the N-terminus and a pullulanase (debranching enzyme) at the C-terminus, with a series of CBMs in between. This is the only extracellular debranching enzyme in *B. adolescentis*; however, this is an extracellular activity that both *E. rectale* and *B. thetaiotaomicron* lack completely, suggesting it may be important for allowing more efficient digestion of resistant starch. In comparison *Ruminococcus bromii* takes the organization of its starch digestion system to another level of complexity by forming multienzyme complexes (Ze et al., 2015). Of the nine predicted extracellular starch degrading enzymes in *R. bromii*, five contain dockerin domains, which mediate protein–protein interactions with proteins containing complementary cohesin domains. This is akin to the cellulosome system found in some Clostridia for plant cell wall deconstruction, where many enzyme activities can be brought together to work in close proximity. In this case the system seems to be solely focused on starch digestion and has been termed the amylosome. The cohesin domains are located on what are known as scaffoldin proteins that can add additional functionalities and/or bring multiple enzymes together by having multiple cohesin domains. One of these contains a CBM74 allowing the addition of its unique binding properties to these enzymes. Additionally, one of the amylases of this system has both a dockerin and a cohesin domain, allowing other enzymes to partner with it directly. Of the five dockerin-containing enzymes, three are α-amylases and two are pullulanases with somewhat different substrate preferences (Mukhopadhya et al., 2018). In both cases these enzyme systems are considerably more complex than what is seen in non-resistant starch degrading organisms, so it is unclear what the minimum set of requirements are for resistant starch digestion or if it varies between starch sources. Based on these examples, however, it seems likely that pullulanase activity (along with the more common α-amylase activity) and sufficient starch binding capabilities are requirements, though it may be that other factors will be revealed upon further study.

7.3 RESISTANT STARCH MODULATION OF THE GUT MICROBIOME

The impact of resistant starch on members of the gut microbiome has been found to be variable both across starch types and even between different botanical sources of starch. This includes not only the resistant-starch degrading microbes, but also other members of the community that feed off the fermentation products of the primary degraders as well as released starch degradation products (Cockburn & Koropatkin, 2016). One general trend across different resistant starch sources is a decrease in diversity in the gut communities (Bendiks et al., 2020; Deehan et al., 2020). This may seem counterintuitive at first glance, as these are generally considered health-promoting foods and yet they have an effect that would appear to be unhealthy. However, the way most diversity metrics are calculated they take into account both how many different genera/species/strains are present and also how evenly they are distributed (Kim et al., 2017). Most supplementation trials with resistant starch use quite large doses to be certain of seeing an effect if one exists. This inevitably leads to a bloom in resistant-starch degraders as described earlier, which reduces the evenness of the community. It is likely that lower doses of resistant starch would not have this detrimental impact on the microbiome (Deehan et al., 2020).

7.4 RS1: WHOLE GRAIN STARCHES

The impact of whole grains on the gut microbiome has recently been reviewed (Koecher et al., 2019). A number of studies have found either no impact on the resistant-starch degraders (Vanegas et al., 2017; Ampatzoglou et al., 2015; Langkamp-Henken et al., 2012; Cooper et al., 2017; Bird et al., 2008; Vuholm et al., 2017; Vitaglione et al., 2015; Christensen et al., 2013; Lappi et al., 2013), or else have found an increase in *Bifidobacterium* species (Martínez et al., 2013; Costabile et al., 2008; Carvalho-Wells et al., 2010; Connolly et al., 2016) but not *Ruminococcus bromii*. This is interesting, as it has been found that while both *R. bromii* and *B. adolescentis* are dominant among species that adhere to starch particles, neither are enriched on bran (Leitch et al., 2007). The bran instead is dominated by Lachnospiraceae members related to *E. rectale* and *Roseburia* (Leitch et al., 2007). One confounding factor to consider is that many earlier studies based on fluorescent in situ hybridization (FISH) or qPCR did not include probes specific for the Ruminococcaceae, leading to their absence in these data sets. It is also possible that the mixture fibers present in these whole grains or even mixtures of whole grains, makes for a more general increase in the microbial population rather than enriching for more specific taxa. Other taxa that have been found in some studies to be enriched in people consuming whole grains include butyrate producers from the *Roseburia* genus and *Eubacterium rectale* (Martínez et al., 2013) or *Faecalibacterium prausnitzii* (Roager et al., 2019).

In the recent study by Roager et al. (2019), 16S sequencing and shotgun metagenomics were used to examine the impact of whole grains on the microbiome in overweight and obese individuals considered to be at risk of metabolic syndrome. Individuals consumed a diet rich in whole grains for 8 weeks and a diet rich in refined grains for 8 weeks in a random order and separated by a 6-week washout period. Subjects were given a choice of various whole grain or refined grain products in each period, which they consumed ad libitum, taking in 179 ± 50 g of whole grain or 13 ± 10 g of whole grain in the two diets. However, while the difference in whole grain consumption was substantial, this did not result in nearly as large of a change in resistant starch consumption. Based on reported diets, subjects consumed an average of 12.8 g of resistant starch per day in the whole grain diet and 9.4 g of resistant starch per day in the refined grain diet, excluding any potential nongrain sources. Measuring resistant starch in whole grains is relatively difficult as the resistance is based on the degree to which the grains are intact and preventing enzymatic access to the starch. In this case a protocol based on chewing the foods prior to in vitro measurement was utilized (Akerberg et al., 1998), as the chewing step is perhaps the most important point of digestion for releasing starch from the whole grains. No impact of diet was found on blood glucose, breath hydrogen, plasma short-chain fatty acids, intestinal integrity, or gut transit time, though there was a decrease in weight and inflammatory markers such as interleukin-6 and C-reactive protein. Changes to the microbiome were relatively minor and no significant changes to resistant-starch degrading organisms were noted, perhaps unsurprising given the relatively small change to resistant starch content of the diets. Compared to refined grain diets, butyrate levels were elevated by the end of the whole grain treatment and an increase in the butyrate producer *F. prausnitzii* was noted. However, it is possible these changes were more driven by other dietary fiber components such as arabinoxylan rather than resistant starch.

7.5 RS2: B-TYPE AND C-TYPE STARCHES

The impact of RS2 starches on the gut microbiome has recently been reviewed (Bendiks et al., 2020). Within RS2 there is a divergence in the impact of consuming these starches depending on the source. Potato starch strongly favors *B. adolescentis* and high-amylose maize starch strongly favors *R. bromii*. Despite both being B-type starches, there appears to be other important factors that influence their digestibility. Whether this is the amylose content per se, structural differences that it imparts or else factors unique to potato starch is currently unknown. In RS2 studies while the evidence for increases in resistant starch degraders is strong, the impacts on other organisms have been somewhat variable. Some studies have found an increase in the important butyrate producing organism *Faecalibacterium prausnitzii* (Maier et al., 2017; Laffin et al., 2019), or decreases among the *Bacteroides* (Hald et al., 2016; Zhang et al., 2019). The latter is interesting, as there are numerous starch degraders in this genus, such as *B. thetaiotaomicron*. These species may be disfavored by the pH drop that occurs with the increased production of short chain fatty acids during resistant starch fermentation, though the response at the phylum level of the Bacteroidetes and Firmicutes phyla overall is more variable between trials (Bendiks et al., 2020).

In the study by Baxter et al. (2019b) the differential impacts of potato starch and high-amylose maize starch were investigated in a healthy young adult population. The intervention length was relatively short at 2 weeks, however, significant changes were seen in several bacterial species. As with the studies noted earlier, potato starch was found to enrich for *B. adolescentis*, while high-amylose maize starch favored *R. bromii*. Additionally, an unknown member of the Ruminococcaceae and *Clostridium chartabidum* were found to be elevated in those consuming potato starch. They furthermore found a strong positive association between *Eubacterium rectale* and butyrate production across treatments. Looking at associations between primary degraders and butyrate producers illustrated significant positive associations between *R. bromii* and *E. rectale* as well as between the *Bifidobacterium* resistant-starch degraders and *Anaerostipes hadrus*. This suggests that different primary degraders need different butyrate producers to be present in order for resistant starch digestion to lead to butyrate production.

7.6 RS3: RETROGRADED STARCHES

In two human trials with RS3 consumption there has either been no significant change in resistant-starch degraders found (Klosterbuer et al., 2013) or an increase in *R. bromii* (Walker et al., 2011). It should be noted in the former case a restriction fragment length-based method was used that may have lower sensitivity to detect changes. During in vitro experiments it has been found that *Bifidobacterium* species such as *B. adolescentis* have been favored, though this may depend on the particular type of retrograded starch and its crystalline polymorph (Lesmes et al., 2008).

A recent study by Gu et al. (2020) investigated a series of different retrograded rice starches and their differing impacts on the gut microbiome during in vitro fermentations. The starches they used came from several rice varieties that differed in amylose content, molecular size and amylopectin chain length distributions. The produced retrograded starches then differed in their crystallinity type, degree of crystallinity and resistant starch content. Amylopectin chains in the degree of polymerization (DP) range of 36–100 seemed to have the greatest impact on the gut microbiome during in vitro fermentation, with two distinct populations of communities emerging based on this property. Furthermore, higher amounts of DP 36–100 chains, smaller total molecular size and C-type crystallinity favored *Blautia* species overall, while lower DP 36–100 chains, larger total molecular size and B-type crystallinity favored *Roseburia* species overall. However, despite these changes across butyrate-producing communities, there was no significant change in butyrate production between the starches, though all increased relative to controls for all three different fecal donors. This suggests some level of functional convergence in the microbiome, despite taxonomic divergence with these different RS3 sources. Another recent study by Giuberti and Gallo (2020) found similar short-chain fatty acid profiles for four different RS3 sources during in vitro fermentation, though they did find some small but significant differences in butyrate production for the RS3 sources. In this case microbiome changes were not measured and the RS3 sources did not overlap with those used in the study by Gu et al. (2020), but it does suggest that similar to the situation with RS2, differences in microbiome impacts exist between starch sources within this resistant starch type.

7.7 RS4: CHEMICALLY MODIFIED STARCHES

Among the resistant-starch degraders, it has been *B. adolescentis* that has mostly found to have been enriched on RS4 starches during human feeding trials with either Fibersym (a cross-linked wheat starch) (Martínez et al., 2010; Upadhyaya et al., 2016) or a tapioca-based resistant starch (Deehan et al., 2020). *R. bromii* (or at least a closely related *Ruminococcus* species) has been found to be enriched on a chemically cross-linked high-amylose maize starch (Deehan et al., 2020). Outside of these known resistant-starch degraders, *Parabacteroides distasonis* has been consistently identified as enriched in RS4 human trials (Deehan et al., 2020; Martínez et al., 2010; Upadhyaya et al., 2016), though it is not clear at this point to what extent it participates in resistant starch digestion. Overall, the RS4 starches are an interesting case, as in some cases the degree of cross-linking has the potential to render these starches almost inert to digestion causing them to act more like a fiber such as cellulose that is subject to only limited microbial fermentation during its transit through the human intestinal tract.

In the recent study by Deehan et al. (2020) three different RS4 sources were fed to people at four different doses in a clinical trial, and the impacts on the microbiome and fecal short-chain fatty acids were measured. The resistant starches were chemically cross-linked maize, potato or tapioca starches. Butyrate increased significantly only for the maize starch, while propionate increased significantly for the tapioca starch, both only at the higher end of the dose scale. The potato starch showed a general trend in decreasing amounts of short-chain fatty acids, with no species significantly enriched during the feeding trial, suggesting that it was highly resistant to fermentation. The maize starch showed significant increases in a *Ruminococcus* species, *Eubacterium rectale* and an *Oscillibacter* species, while the tapioca starch elicited an increase in *P. distasonis*, another *Parabacteroides* species, an *Eisenbergiella* species and *Faecalibacterium prausnitzii*. *B. adolescentis* showed an increasing trend on this starch, but it did not reach the level of statistical significance. Across the study, *P. distasonis* was positively associated with propionate levels, and this relationship was particularly strong in the maize and tapioca starch arms of the study. Similarly, *E. rectale* was positively associated with butyrate levels across the study. The authors also took the interesting step of investigating the ability of cultured representatives of certain key taxa to bind to and grow on the starches used in this study. They found that *B. adolescentis*, *R. bromii* and *P. distasonis* could bind to all three starches, while *E. rectale* could bind only to the maize starch. This result or *E. rectale* aligns well with data that showed that the carbohydrate binding modules from its cell-wall linked amylase could bind to and facilitate the utilization of regular and high-amylose maize starch, but not potato starch or a RS4 wheat starch (Cockburn et al., 2018). Each of the bacteria could use the maize starch for growth to some degree (Deehan et al., 2020), while the tapioca starch was used weakly by *R. bromii* and robustly by *B. adolescentis*. Only *R. bromii* showed significant growth increase on the potato starch after 24 hours, in line with the difficult-to-degrade nature of this particular RS4 source.

7.8 INTERINDIVIDUAL DIFFERENCES IN RESISTANT STARCH IMPACT ON THE MICROBIOME

An interesting aspect of the microbiome response to resistant starch is that there are interindividual differences to that response. It has been noted that while resistant starch increases fecal butyrate concentration at the population level, the individual level response is more variable with some individuals having no increase in butyrate or perhaps even decreases. Venkataraman et al. (2016) noted three categories of individuals in terms of butyrate response to potato starch consumption, those that were at low levels and increased to high levels, those that were at high levels and stayed at high levels, and those that were at low levels and did not increase. Out of 20 individuals in the study, 11 fell into the first group, 3 fell into the second group and 6 fell into the last group, which can be considered the nonresponders. It should be noted that all 20 individuals had at least one *R. bromii* and *B. adolescentis* present in their microbiomes. However, while these primary degraders increased significantly in both the responding categories, they did not increase in the nonresponders. In another study by the same research group (Baxter et al., 2019b), this time with study subjects examining potato starch or high-amylose maize starch, a similar phenomenon was observed, where increases in butyrate were inconsistent in the population. In this study there was a significant increase in butyrate for potato starch, but not for those consuming high-amylose maize starch. However, those consuming the potato starch were more likely to experience an increase in butyrate if their microbiome responded with an increase in *R. bromii* levels, rather than *B. adolescentis*. This suggests that the profile of resistant-starch degraders present in an individual may play an important role in determining the resulting butyrate production. In both studies (Baxter et al., 2019b; Venkataraman et al., 2016) the butyrate producer *E. rectale* was best correlated with butyrate production, suggesting that the profile of butyrate producers may be important as well.

Deehan et al. (2020) found similar interindividual differences in microbiome and butyrate response with RS4 starch sources. For instance, a pattern of co-exclusion was observed between two OTUs, one identified as *R. bromii* and the other as an unclassified *Ruminococcus* species, suggesting that they were in direct competition for the RS4 maize. The overall enhancement of butyrate for this condition was more consistent (8 out of 10 individuals) as well as the increase in *E. rectale* (10 out of 10 individuals), suggesting that these different *Ruminococcus* species behaved similarly. Additionally, both the maize RS4 and the tapioca RS4 displayed individuality in the optimum dose, with most having their short-chain fatty acid increases plateau at a dose of 35 g/day, with a few experiencing further increases at the 50 g/day dose.

A recent in vitro fermentation study by Teichmann and Cockburn (2021) also examined the phenomenon of interindividual differences in the butyrate response to resistant starch. This study examined 8 different resistant starch sources spanning RS2, RS3 and RS4 with 12 different starting microbiomes. From the perspective of butyrate responses, the results were highly individualized, with the microbiomes from no two individuals exhibiting the same pattern of significant butyrate increases across starch sources. However, all individuals had at least one starch source for which they experienced a significant butyrate increase, indicating that a lack of response to one resistant starch source does not mean a lack of response to all resistant starch sources. Conversely, only banana starch was a near universal butyrate enhancer (11 out of 12 microbiomes), with the others being sparser in response, suggesting that the microbiome does need to be considered in determining the optimum diet for an individual. Intriguingly, similar to the result found by Baxter et al. (2019b), *R. bromii* was more associated with increases in butyrate production than was *B. adolescentis*. Illustrating this point, the one individual that produced increases in butyrate from nearly all resistant starch sources was the one individual in the study whose microbiome lacked *B. adolescentis* and only contained *R. bromii* among resistant-starch degraders. The reasons for this are currently unclear; however, it may be related to pH-dependent shifts in fermentation profiles induced by high levels of lactate production by *B. adolescentis*.

7.9 HEALTH IMPACTS OF RESISTANT STARCH

Special attention has been paid to the impact on butyrate levels and abundances of butyrate-producing organisms in resistant starch clinical trials and in vitro experiments. This is due to the health benefits that are perceived to be tied to butyrate as well as the status of resistant starch as one of the best dietary interventions for increasing butyrate levels (Smith et al., 1998). Butyrate is tied to reduced levels of inflammation (Bach Knudsen et al., 2018), improved gut barrier function (Yan & Ajuwon, 2017; Zheng et al., 2017), inhibition of colon cancer (Li et al., 2018) and is the preferred energy source of the cells of the colon (Roediger, 1980). Resistant starch consumption itself is associated with improvements in glucose control (Maziarz et al., 2017; Stewart & Zimmer, 2017), decreased serum cholesterol (Yuan et al., 2018), improved markers for cardiovascular health (Peterson et al., 2018) and improved markers in chronic kidney disease (Esgalhado et al., 2018; Khosroshahi et al., 2019). However, the impact of resistant starch on health is somewhat complex. One study found that the positive impacts of resistant starch on insulin sensitivity were completely independent of processing and fermentation by the gut microbiome (Bindels et al., 2017). On the other hand, studies of the impact of resistant starch on chronic kidney disease have shown the benefits more firmly tied to modulation of the gut microbiome (Laffin et al., 2019; Snelson et al., 2019). Further complicating matters is that many of these studies put all resistant starches in the same category. However, as explored earlier, resistant starches of different types and plant sources can have significantly different impacts on the gut microbiome. Even the connection to butyrate production is somewhat complex (Figure 7.2), as none of the resistant-starch degrading bacteria directly produce butyrate themselves. *R. bromii* is mainly a producer of acetate and lower amounts of ethanol and formate (Crost et al., 2018), while *B. adolescentis* is primarily a lactate producer, with some acetate and ethanol production (Amaretti et al., 2007). Instead, butyrate production from resistant starch fermentation in the colon relies on cross-feeding networks where the primary resistant-starch degrading organisms produce substrates that can be utilized by the butyrate-producing organisms. Bacteria such as *E. rectale*, members of the *Roseburia* genus and *F. prausnitzii* can all produce butyrate from starch-derived oligosaccharides and acetate (Duncan et al., 2002). Other bacteria, such as *Anaerostipes hadrus* and *Anaerobutyricum hallii*, utilize a combination of starch-derived sugars, acetate and lactate to produce butyrate (Duncan et al., 2004). In each case it is apparent that the profile of both the sugars (mono-/di-/oligosaccharides) and fermentation products (acetate/lactate) will play an important role in determining which butyrate producers can thrive and how much butyrate can be produced. In future studies seeking to advance the understanding of the impact of resistant starch on the gut microbiome, it will be critical to elucidate the range of

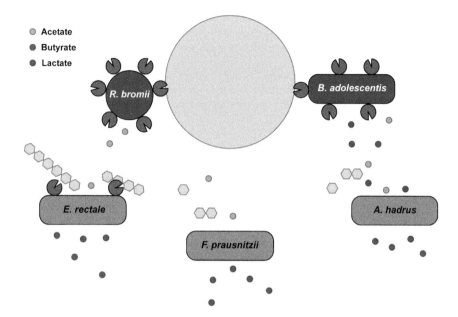

Figure 7.2   Cross-feeding between resistant-starch degraders and other members of the gut microbiome leads to butyrate production. None of the known resistant-starch degrading organisms directly produce butyrate; however, increasing fecal butyrate levels is a defining feature of resistant starch supplementation. The resistant-starch degraders can release soluble sugar chains or modify the starch surface to make it accessible to other organisms. Additionally, fermentation products, such as acetate and lactate, can serve as precursors of butyrate production in some organisms.

interactions between resistant-starch degraders and butyrate producers, and how these are impacted by species and even strain-level differences between individual microbiomes.

ACKNOWLEDGMENTS

This work was supported by the USDA National Institute of Food and Agriculture Federal Appropriations under Project PEN04650 and Accession number 1015962.

REFERENCES

Ai, Y., Hasjim, J., & Jane, J.-l. (2013). Effects of lipids on enzymatic hydrolysis and physical properties of starch. *Carbohydrate Polymers*, 92, 120–127.

Akerberg, A. K., Liljeberg, H. G., Granfeldt, Y. E., Drews, A. W., Bjorck, I. M. (1998). An in vitro method, based on chewing, to predict resistant starch content in foods allows parallel determination of potentially available starch and dietary fiber. *J Nutr*, 128, 651–660.

Akkarachiyasit, S., Charoenlertkul, P., Yibchok-Anun, S., & Adisakwattana, S. (2010). Inhibitory activities of cyanidin and its glycosides and synergistic effect with acarbose against intestinal α-glucosidase and pancreatic α-amylase. *Int J Mol Sci*, 11, 3387–3396.

Amaretti, A., Bernardi, T., Tamburini, E., Zanoni, S., Lomma, M., Matteuzzi, D., & Rossi, M. (2007). Kinetics and metabolism of Bifidobacterium adolescentis MB 239 growing on glucose, galactose, lactose, and galactooligosaccharides. *Appl Environ Microbiol*, 73, 3637–3644.

Ampatzoglou, A., Atwal, K. K., Maidens, C. M., Williams, C. L., Ross, A. B., Thielecke, F., Jonnalagadda, S. S., Kennedy, O. B., Yaqoob, P. (2015). Increased whole grain consumption does not affect blood biochemistry, body composition, or gut microbiology in healthy, low-habitual whole grain consumers. *J Nutr*, 145, 215–221.

Arnal, G., Cockburn, D. W., Brumer, H., & Koropatkin, N. M. (2018). Structural basis for the flexible recognition of alpha-glucan substrates by Bacteroides thetaiotaomicron SusG. *Protein Sci*, 27, 1093–1101.

Bach Knudsen, K. E., Lærke, H. N., Hedemann, M. S., Nielsen, T. S., Ingerslev, A. K., Gundelund Nielsen D. S., Theil, P. K., Purup, S., Hald, S., Schioldan, A. G., et al. (2018). Impact of diet-modulated butyrate production on intestinal barrier function and inflammation. *Nutrients*, 10.

Baxter, N. T., Lesniak, N. A., Sinani, H., Schloss, P. D., Koropatkin, N. M. (2019a). The glucoamylase inhibitor acarbose has a diet-dependent and reversible effect on the murine gut microbiome. *mSphere*, 4.

Baxter, N. T., Schmidt, A. W., Venkataraman, A., Kim, K. S., Waldron, C., & Schmidt, T. M. (2019b). Dynamics of human gut microbiota and short-chain fatty acids in response to dietary interventions with three fermentable fibers. *MBio*, 10.

Bendiks, Z. A., Knudsen, K. E. B., Keenan, M. J., & Marco, M. L. (2020). Conserved and variable responses of the gut microbiome to resistant starch type 2. *Nutr Res*, 77, 12–28.

Bertoft, E. (2017). Understanding starch structure: Recent progress. *Agronomy*, 7, 56.

Bindels, L. B., Segura Munoz, R. R., Gomes-Neto, J. C., Mutemberezi, V, Martínez, I, Salazar, N, Cody, E. A., Quintero-Villegas, M. I., Kittana, H, de Los Reyes-Gavilán C. G., et al. (2017). Resistant starch can improve insulin sensitivity independently of the gut microbiota. *Microbiome*, 5, 12.

Bird, A. R., Vuaran, M. S., King, R. A., Noakes, M, Keogh, J, Morell, M. K., & Topping, D. L. (2008). Wholegrain foods made from a novel high-amylose barley variety (Himalaya 292) improve indices of bowel health in human subjects. *Br J Nutr*, 99, 1032–1040.

Bogracheva, T. Y., Morris, V. J., Ring, S. G., & Hedley, C. L. (1998). The granular structure of C-type pea starch and its role in gelatinization. *Biopolymers*, 45, 323–332.

Brownlee, I. A., Gill, S., Wilcox, M. D., Pearson, J. P., & Chater, P. I. (2018). Starch digestion in the upper gastrointestinal tract of humans. *Starch/Stärke*, 70, 1700111.

Carpenter, D., Dhar, S., Mitchell, L. M., Fu, B., Tyson, J., Shwan, N. A., Yang, F., Thomas, M. G., & Armour, J. A. (2015). Obesity, starch digestion and amylase: association between copy number variants at human salivary (AMY1) and pancreatic (AMY2) amylase genes. *Hum Mol Genet*, 24, 3472–3480.

Carvalho-Wells, A. L., Helmolz, K., Nodet, C., Molzer, C., Leonard, C., McKevith, B., Thielecke, F., Jackson, K. G., & Tuohy, K. M. (2016). Determination of the in vivo prebiotic potential of a maize-based whole grain breakfast cereal: a human feeding study. *Br J Nutr*, 104, 1353–1356.

Centanni, M., Lawley, B., Butts, C. A., Roy, N. C., Lee, J., Kelly, W. J., & Tannock, G. W. (2018). Bifidobacterium pseudolongum in the ceca of rats fed hi-maize starch has characteristics of a keystone species in bifidobacterial blooms. *Appl Environ Microbiol*, 84.

Christensen, E. G., Licht, T. R., Kristensen, M., & Bahl, M. I. (2013). Bifidogenic effect of whole-grain wheat during a 12-week energy-restricted dietary intervention in postmenopausal women. *Eur J Clin Nutr*, 67, 1316–1321.

Cockburn, D., & Svensson, B. (2013). Surface binding sites in carbohydrate active enzymes: an emerging picture of structural and functional diversity. In T. K. Lindhorst, & A. P. Rauter (Eds.), *Carbohydrate Chemistry: Chemical and Biological Approaches: Volume 39* (pp. 204–221). Cambridge: Royal Society of Chemistry.

Cockburn, D. W., & Koropatkin, N. M. (2016). Polysaccharide degradation by the intestinal microbiota and its influence on human health and disease. *J Mol Biol*, 428, 3230–3252.

Cockburn, D. W., Orlovsky, N. I., Foley, M. H., Kwiatkowski, K. J., Bahr, C. M., Maynard, M., Demeler, B., & Koropatkin, N. M. (2015). Molecular details of a starch utilization pathway in the human gut symbiont Eubacterium rectale. *Molecular Microbiology*, 95, 209–230.

Cockburn, D. W., Suh, C., Medina, K. P., Duvall, R. M., Wawrzak, Z., Henrissat, B., & Koropatkin, N. M. (2018). Novel carbohydrate binding modules in the surface anchored alpha-amylase of Eubacterium rectale provide a molecular rationale for the range of starches used by this organism in the human gut. *Mol Microbiol*, 107, 249–264.

Cockburn, D. W., Cerqueira, F. M., Bahr, C., Koropatkin, N. M. (2020). The structures of the GH13_36 amylases from Eubacterium rectale and Ruminococcus bromii reveal subsite architectures that favor maltose production. *Amylase*, 4, 24–44.

Connolly, M. L., Tzounis, X., Tuohy, K. M., & Lovegrove, J. A. (2016). Hypocholesterolemic and prebiotic effects of a whole-grain oat-based granola breakfast cereal in a cardio-metabolic "at risk" population. *Front Microbiol*, 7, 1675.

Cooper, D. N., Kable, M. E., Marco, M. L., De Leon, A, Rust, B, Baker, J. E., Horn, W, Burnett, D, Keim, N. L. (2017). The effects of moderate whole grain consumption on fasting glucose and lipids, gastrointestinal symptoms, and microbiota. *Nutrients*, 9.

Costabile, A., Klinder, A., Fava, F., Napolitano, A., Fogliano, V., Leonard, C., Gibson, G. R., Tuohy, K. M. (2008). Whole-grain wheat breakfast cereal has a prebiotic effect on the human gut microbiota: A double-blind, placebo-controlled, crossover study. *Br J Nutr*, 99, 110–120.

Crost, E. H., Le Gall, G., Laverde-Gomez, J. A., Mukhopadhya, I., Flint, H. J., & Juge, N. (2018). Mechanistic insights into the cross-feeding of Ruminococcus gnavus and Ruminococcus bromii on host and dietary carbohydrates. *Frontiers in Microbiology*, 9.

Damager, I., Engelsen, S. B., Blennow, A., Møller, B. L., & Motawia, M. S. (2010). First principles insight into the alpha-glucan structures of starch: Their synthesis, conformation, and hydration. *Chemical Reviews*, 110, 2049–2080.

Deehan, E. C., Yang, C., Perez-Muñoz, M. E., Nguyen, N. K., Cheng, C. C., Triador, L., Zhang, Z., Bakal, J. A., & Walter, J. (2020). Precision microbiome modulation with discrete dietary fiber structures directs short-chain fatty acid production. *Cell Host Microbe*, 27, 389–404.e386.

DeMartino, P., & Cockburn, D. W. (2019). Resistant starch: Impact on the gut microbiome and health. *Curr Opin Biotechnol*, 61, 66–71.

Duncan, S. H., Barcenilla, A., Stewart, C. S., Pryde, S. E., & Flint, H. J. (2002). Acetate utilization and butyryl coenzyme A (CoA):acetate-CoA transferase in butyrate-producing bacteria from the human large intestine. *Appl Environ Microbiol*, 68, 5186–5190.

Duncan, S. H., Louis, P., & Flint, H. J. (2004). Lactate-utilizing bacteria, isolated from human feces, that produce butyrate as a major fermentation product. *Appl Environ Microbiol*, 70, 5810–5817.

Dupuis, J. H., Liu, Q., & Yada, R. Y. (2014). Methodologies for increasing the resistant starch content of food starches: A review. *Comprehensive Reviews in Food Science and Food Safety*, 13, 1219–1234.

El Kaoutari, A., Armougom, F., Gordon, J. I., Raoult, D., & Henrissat, B. (2013). The abundance and variety of carbohydrate-active enzymes in the human gut microbiota. *Nat Rev Microbiol*, 11, 497–504.

Englyst, H. N., & Cummings, J. H. (1985). Digestion of the polysaccharides of some cereal foods in the human small intestine. *Am J Clin Nutr*, 42, 778–787.

Englyst, H. N., Kingman, S. M., & Cummings, J. H. (1992). Classification and measurement of nutritionally important starch fractions. *European Journal of Clinical Nutrition*, 46(Suppl 2), S33–S50.

Esgalhado, M., Kemp, J. A., Azevedo, R., Paiva, B. R., Stockler-Pinto, M. B., Dolenga, C. J., Borges, N. A., Nakao, L. S., & Mafra, D. (2018). Could resistant starch supplementation improve inflammatory and oxidative stress biomarkers and uremic toxins levels in hemodialysis patients? A pilot randomized controlled trial. *Food & Function*, 9, 6508–6516.

Faisant, N., Buléon, A., Colonna, P., Molis, C., Lartigue, S., Galmiche, J. P., & Champ, M. (1995). Digestion of raw banana starch in the small intestine of healthy humans: Structural features of resistant starch. *Br J Nutr*, 73, 111–123.

Foley, M. H., Cockburn, D. W., & Koropatkin, N. M. (2016). The sus operon: A model system for starch uptake by the human gut Bacteroidetes. *Cellular and Molecular Life Sciences: CMLS*, 73, 2603–2617.

Giuberti, G., & Gallo, A. (2020). In vitro evaluation of fermentation characteristics of type 3 resistant starch. *Heliyon*, 6, e03145.

Glenwright, A. J., Pothula, K. R., Bhamidimarri, S. P., Chorev, D. S., Baslé, A., Firbank, S. J., Zheng, H., Robinson, C. V., Winterhalter, M., Kleinekathöfer, U., et al. (2017). Structural basis for nutrient acquisition by dominant members of the human gut microbiota. *Nature*, 541, 407–411.

Gu, F., Li, C., Hamaker, B. R., Gilbert, R. G., & Zhang, X. (2020). Fecal microbiota responses to rice RS3 are specific to amylose molecular structure. *Carbohydr Polym*, 243, 116475.

Hald, S., Schioldan, A. G., Moore, M. E., Dige, A., Laerke, H. N., Agnholt, J., Bach Knudsen, K. E., Hermansen, K., Marco, M. L., Gregersen, S., et al. (2016). Effects of arabinoxylan and resistant starch on intestinal microbiota and short-chain fatty acids in subjects with metabolic syndrome: A randomised crossover study. *PLoS One*, 11, e0159223.

Imberty, A., Buléon, A., Tran, V., & Péerez, S. (1991). Recent advances in knowledge of starch structure. *Starch/Stärke*, 43, 375–384.

Joshi, S. R., Standl, E., Tong, N., Shah, P., Kalra, S., Rathod, R. (2015). Therapeutic potential of α-glucosidase inhibitors in type 2 diabetes mellitus: an evidence-based review. *Expert Opin Pharmacother*, 16, 1959–1981.

Jung, D. H., Seo, D. H., Kim, Y. J., Chung, W. H., Nam, Y. D., Park, C. S. (2020). The presence of resistant starch-degrading amylases in Bifidobacterium adolescentis of the human gut. *Int J Biol Macromol*, 161, 389–397.

Khosroshahi, H. T., Abedi, B., Ghojazadeh, M., Samadi, A., Jouyban, A. (2019). Effects of fermentable high fiber diet supplementation on gut derived and conventional nitrogenous product in patients on maintenance hemodialysis: a randomized controlled trial. *Nutr Metab*, 16, 18.

Kim, B. R., Shin, J., Guevarra, R., Lee, J. H., Kim, D. W., Seol, K. H., Lee, J. H., Kim, H. B., & Isaacson, R. (2017). Deciphering diversity indices for a better understanding of microbial communities. *J Microbiol Biotechnol*, 27, 2089–2093.

Klosterbuer, A. S., Hullar, M. A., Li, F., Traylor, E., Lampe, J. W., Thomas, W., & Slavin, J. L. (2013). Gastrointestinal effects of resistant starch, soluble maize fibre and pullulan in healthy adults. *Br J Nutr*, 110:1068–1074.

Koecher, K. J., McKeown, N. M., Sawicki, C. M., Menon, R. S., & Slavin, J. L. (2019). Effect of whole-grain consumption on changes in fecal microbiota: a review of human intervention trials. *Nutr Rev*, 77, 487–497.

Koropatkin, N. M., & Smith, T. J. (2010). SusG: a unique cell-membrane-associated alpha-amylase from a prominent human gut symbiont targets complex starch molecules. *Structure*, 18, 200–215.

Laffin, M. R., Tayebi Khosroshahi, H., Park, H., Laffin, L. J., Madsen, K., Kafil, H. S., Abedi, B., Shiralizadeh, S., & Vaziri, N. D. (2019). Amylose resistant starch (HAM-RS2) supplementation increases the proportion of Faecalibacterium bacteria in endstage renal disease patients: Microbial analysis from a randomized placebo-controlled trial. *Hemodial Int*, 23, 343–347.

Langkamp-Henken, B, Nieves, C, Jr., Culpepper, T, Radford, A, Girard, S. A., Hughes, C, Christman, M. C., Mai, V, Dahl, W. J., Boileau, T, et al. (2012). Fecal lactic acid bacteria increased in adolescents randomized to whole-grain but not refined-grain foods, whereas inflammatory cytokine production decreased equally with both interventions. *J Nutr*, 142, 2025–2032.

Lappi, J., Salojärvi, J., Kolehmainen, M., Mykkänen, H., Poutanen, K., de Vos, W. M., & Salonen, A. (2013). Intake of whole-grain and fiber-rich rye bread versus refined wheat bread does not differentiate intestinal microbiota composition in Finnish adults with metabolic syndrome. *J Nutr*, 143, 648–655.

Leitch, E. C., Walker, A. W., Duncan, S. H., Holtrop, G., & Flint, H. J. (2007). Selective colonization of insoluble substrates by human faecal bacteria. *Environ Microbiol*, 9, 667–679.

Lesmes, U., Beards, E. J., Gibson, G. R., Tuohy, K. M., & Shimoni, E. (2008). Effects of resistant starch type III polymorphs on human colon microbiota and short chain fatty acids in human gut models. *J Agric Food Chem*, 56, 5415–5421.

Li, Q., Cao, L., Tian, Y., Zhang, P., Ding, C., Lu, W., Jia, C., Shao, C., Liu, W., Wang, D., et al. (2018). Butyrate suppresses the proliferation of colorectal cancer cells via targeting pyruvate kinase M2 and metabolic reprogramming. *Molecular & Cellular Proteomics*, 17, 1531–1545.

Mackie, A., & Rigby, N. (2015). InfoGest consensus method. In K. Verhoeckx, P. Cotter, I. López-Expósito, C. Kleiveland, Lea, T., Mackie, A., Requena, T., Swiatecka, D., Wichers, H. (Eds.), *The Impact of Food Bioactives on Health: In Vitro and Ex Vivo Models* (pp. 13–22) New York: Springer International Publishing.

Maier, T. V., Lucio, M., Lee, L. H., VerBerkmoes, N. C., Brislawn, C. J., Bernhardt, J., Lamendella, R., McDermott, J. E., Bergeron, N., Heinzmann, S. S., et al. (2017). Impact of dietary resistant starch on the human gut microbiome, metaproteome, and metabolome. *mBio*, 8.

Martínez, I., Kim, J., Duffy, P. R., Schlegel, V. L., Walter, J. (2010). Resistant starches types 2 and 4 have differential effects on the composition of the fecal microbiota in human subjects. *PloS one*, 5, e15046.

Martínez, I., Lattimer, J. M., Hubach, K. L., Case, J. A., Yang, J., Weber, C. G., Louk, J. A., Rose, D. J., Kyureghian, G., Peterson, D. A., et al. (2013). Gut microbiome composition is linked to whole grain-induced immunological improvements. *ISME J*, 7, 269–280.

Maziarz, M. P., Preisendanz, S., Juma, S., Imrhan, V., Prasad, C., & Vijayagopal, P. (2017). Resistant starch lowers postprandial glucose and leptin in overweight adults consuming a moderate-to-high-fat diet: A randomized-controlled trial. *Nutr J*, 16, 14.

McCleary, B. V., McNally, M., & Rossiter, P. (2002). Measurement of resistant starch by enzymatic digestion in starch and selected plant materials: Collaborative study. *J AOAC Int*, 85, 1103–1111.

Mukhopadhya, I., Morais, S., Laverde-Gomez, J., Sheridan, P. O., Walker, A. W., Kelly, W., Klieve, A. V., Ouwerkerk, D., Duncan, S. H., Louis, P., et al. (2018). Sporulation capability and amylosome conservation among diverse human colonic and rumen isolates of the keystone starch-degrader Ruminococcus bromii. *Environ Microbiol*, 20, 324–336.

Peterson, C. M., Beyl, R. A., Marlatt, K. L., Martin, C. K., Aryana, K. J., Marco, M. L., Martin, R. J., Keenan, M. J., & Ravussin, E. (2018). Effect of 12 wk of resistant starch supplementation on cardiometabolic risk factors in adults with prediabetes: A randomized controlled trial. *Am J Clin Nutr*.

Pollet, R. M., Martin, L. M., & Koropatkin, N. M. (2021). TonB-dependent transporters in the bacteroidetes: Unique domain structures and potential functions. *Mol Microbiol*.

Roager, H. M., Vogt, J. K., Kristensen, M., Hansen, L. B. S., Ibrügger, S, Mærkedahl, R. B., Bahl, M. I., Lind, M. V., Nielsen, R. L., Frøkiær, H., et al. (2019). Whole grain-rich diet reduces body weight and systemic low-grade inflammation without inducing major changes of the gut microbiome: A randomised cross-over trial. *Gut*, 68, 83–93.

Roediger, W. E. (1980). Role of anaerobic bacteria in the metabolic welfare of the colonic mucosa in man. *Gut*, 21, 793–798.

Sender, R., Fuchs, S., & Milo, R. (2016). Revised estimates for the number of human and bacteria cells in the body. *PLoS Biology*, 14, e1002533.

Sheridan, P. O., Martin, J. C., Lawley, T. D., Browne, H. P., Harris, H. M. B., Bernalier-Donadille, A., Duncan, S. H., O'Toole, P. W., Scott, K. P., & Flint, H. J. (2016). Polysaccharide utilization loci and nutritional specialization in a dominant group of butyrate-producing human colonic Firmicutes. *Microb Genom*, 2, e000043.

Shi, Y.-C., Capitani, T., Trzasko, P., & Jeffcoat, R. (1998). Molecular structure of a low-amylopectin starch and other high-amylose maize starches. *Journal of Cereal Science*, 27, 289–299.

Silvester, K. R, Englyst, H. N., & Cummings, J. H. (1995). Ileal recovery of starch from whole diets containing resistant starch measured in vitro and fermentation of ileal effluent. *Am J Clin Nutr*, 62, 403–411.

Smith, J. G., Yokoyama, W. H., German, J. B. (1998). Butyric acid from the diet: Actions at the level of gene expression. *Crit Rev Food Sci Nutr*, 38, 259–297.

Snelson, M., Kellow, N. J., & Coughlan, M. T. (2019). Modulation of the gut microbiota by resistant starch as a treatment of chronic kidney diseases: Evidence of efficacy and mechanistic insights. *Advances in Nutrition*, 10, 303–320.

Stewart, M. L., Zimmer, J. P. (2017). A high fiber cookie made with resistant starch type 4 reduces post-prandial glucose and insulin responses in healthy adults. *Nutrients*, 9.

Tahrani, A. A., Barnett, A. H., Bailey, C. J. (2016). Pharmacology and therapeutic implications of current drugs for type 2 diabetes mellitus. *Nat Rev Endocrinol*, 12, 566–592.

Teichmann, J., & Cockburn, D. W. (2021). In vitro fermentation reveals changes in butyrate production dependent on resistant starch source and microbiome composition. *Front Microbiol*, 12, 640253.

Terrapon, N., Lombard, V., Gilbert, H. J., Henrissat, B. (2015). Automatic prediction of polysaccharide utilization loci in Bacteroidetes species. *Bioinformatics*, 31, 647–655.

Upadhyaya, B., McCormack, L., Fardin-Kia, A. R., Juenemann, R., Nichenametla, S., Clapper, J., Specker, B., & Dey, M. (2016). Impact of dietary resistant starch type 4 on human gut microbiota and immunometabolic functions. *Scientific Reports*, 6, 28797.

Valk, V., Lammerts van Bueren, A., van der Kaaij, R. M., & Dijkhuizen, L. (2016). Carbohydrate-binding module 74 is a novel starch-binding domain associated with large and multidomain alpha-amylase enzymes. *Febs J*, 283, 2354–2368.

Vanegas, S. M., Meydani, M., Barnett, J. B., Goldin, B., Kane, A., Rasmussen, H., Brown, C., Vangay, P., Knights, D., Jonnalagadda, S., et al. (2017). Substituting whole grains for refined grains in a 6-wk randomized trial has a modest effect on gut microbiota and immune and inflammatory markers of healthy adults. *Am J Clin Nutr*, 105, 635–650.

Venkataraman, A., Sieber, J. R., Schmidt, A. W., Waldron, C., Theis, K. R., & Schmidt, T. M. (2016). Variable responses of human microbiomes to dietary supplementation with resistant starch. *Microbiome*, 4, 33.

Vitaglione, P., Mennella, I., Ferracane, R., Rivellese, A. A., Giacco, R., Ercolini, D., Gibbons, S. M., La Storia, A., Gilbert, J. A., Jonnalagadda, S., et al. (2015). Whole-grain wheat consumption reduces inflammation in a randomized controlled trial on overweight and obese subjects with unhealthy dietary and lifestyle behaviors: role of polyphenols bound to cereal dietary fiber. *Am J Clin Nutr*, 101, 251–261.

Vuholm, S., Nielsen, D. S., Iversen, K. N., Suhr, J., Westermann, P., Krych, L., Andersen, J. R., Kristensen, M. (2017). Whole-grain rye and wheat affect some markers of gut health without altering the fecal microbiota in healthy overweight adults: A 6-week randomized trial. *J Nutr*, 147, 2067–2075.

Walker, A. W., Ince, J., Duncan, S. H., Webster, L. M., Holtrop, G., Ze, X., Brown, D., Stares, M. D., Scott, P., Bergerat, A., et al. (2011). Dominant and diet-responsive groups of bacteria within the human colonic microbiota. *ISME J*, 5, 220–230.

Wang, S., Li, C., Copeland, L., Niu, Q., & Wang, S. (2015). Starch retrogradation: A comprehensive review. *Comprehensive Reviews in Food Science and Food Safety*, 14, 568–585.

Wang, S., Chao, C., Xiang, F., Zhang, X., Wang, S., & Copeland, L. (2018). New insights into gelatinization mechanisms of cereal endosperm starches. *Scientific Reports*, 8, 3011.

Yan, H., & Ajuwon, K. M. (2017). Butyrate modifies intestinal barrier function in IPEC-J2 cells through a selective upregulation of tight junction proteins and activation of the Akt signaling pathway. *PLoS One*, 12, e0179586.

Yuan, H. C., Meng, Y., Bai, H., Shen, D. Q., Wan, B. C., & Chen, L. Y. (2018). Meta-analysis indicates that resistant starch lowers serum total cholesterol and low-density cholesterol. *Nutrition Research*, 54, 1–11.

Ze, X., Duncan, S. H., Louis, P., Flint, H. J. (2012). Ruminococcus bromii is a keystone species for the degradation of resistant starch in the human colon. *ISME Journal*, 6, 1535–1543.

Ze, X., David, B., Laverde-Gomez, J. A., Dassa, B., Sheridan, P. O., Duncan, S. H., Louis, P., Henrissat, B., Juge, N., Koropatkin, N. M., et al. (2015). Unique organization of extracellular amylases into amylosomes in the resistant starch-utilizing human colonic firmicutes bacterium Ruminococcus bromii. *mBio*, 6, 1–11.

Zhang, L., Ouyang, Y., Li, H., Shen, L., Ni, Y., Fang, Q., Wu, G., Qian, L., Xiao, Y., Zhang, J., et al. (2019). Metabolic phenotypes and the gut microbiota in response to dietary resistant starch type 2 in normal-weight subjects: a randomized crossover trial. *Scientific Reports*, 9, 4736.

Zheng, L., Kelly, C. J., Battista, K. D., Schaefer, R., Lanis, J. M., Alexeev, E. E., Wang, R. X., Onyiah, J. C., Kominsky, D. J., & Colgan, S. P. (2017). Microbial-derived butyrate promotes epithelial barrier function through IL-10 receptor–dependent repression of claudin-2. *Journal of Immunology*, 199, 2976–2984.

Chapter 8

Starchy Flour Modification to Alter Its Digestibility

Edith Agama-Acevedo, Madai Lopez-Silva
and Andres Aguirre-Cruz

CONTENTS

8.1 INTRODUCTION

Starchy flours of different botanical sources have been widely used for food production. Starchy flours can be classified into two groups: conventional sources, such as maize, rice, wheat and potato; and nonconventional sources, such as amaranth, banana, pea, sorghum and chickpea. The unit operations used to elaborate foods with starchy flour raw materials as the main ingredient include mixing, sheeting and thermal treatment (autoclaving, extrusion) that can modify the starch structure or interactions between the different macromolecules (e.g., starch, proteins, lipids, nonstarch polysaccharides, poly-phenols). Native or unmodified starchy flours have been used in foods such as pasta, nutrition bars, bread, nuggets and meat products. Starchy flours are used to reduce fat content, and increase the dietary fiber, protein, poly-phenol content, water retention and viscosity of the product. However, some drawbacks were found, such as decreased hardness, increased cooking loss, and dark color, which should be improved to increase the acceptability of the product. Currently, the development of functional foods is a growing sector of the food industry, where consumers prefer foods with high dietary and poly-phenol contents. Recently, the role of the food matrix type and food processing on the chemical interactions between the different compounds present in food and how they can influence the bioaccessibility of dietary phytochemicals has been a topic of interest (Shahidi & Pan, 2021). The use of technology for food processing is suggested to increase the bioaccessibility of polyphenols present in the food matrix.

Starch gelatinization and reorganization (retrogradation), as well as the interactions between different macromolecules, play an important role in the digestibility of this polysaccharide. Starchy foods are traditionally considered to have a high available starch content that is rapidly digested after ingestion with an increase in glucose in the blood. The peak glucose after ingestion of starchy food produced using traditional processes (e.g., baking, boiling, fry-ing) is related to some health problems, such as overweight, obesity, and dia-betes, which are related to metabolic syndrome and cardiovascular diseases. To reduce starch digestion and its impact on human health, diverse strategies have been proposed to develop foods with low-fat, low-carb, and high dietary fiber contents, which reduce their sensory and texture characteristics. These strategies only partially produce a positive effect in the prevention of afore-mentioned diseases. Blending starchy flours from diverse botanical origins has been a strategy to develop new foods where the available carbohydrates and fat content are reduced, and gluten is absent.

8.2 USE OF UNMODIFIED STARCHY FLOURS

Many studies have reported the use of starchy flours to elaborate foods with diverse objectives, such as improving functional and nutritional characteris-tics. Among those foods, gluten-free products, foods with high dietary fiber content, high antioxidant capacity, and high protein content are preferred (Figure 8.1).

The main challenge in the development of "new" foods is that the texture and sensory characteristics should be similar to those of the traditional or con-ventional foods. Table 8.1 summarizes studies using starchy flours to produce

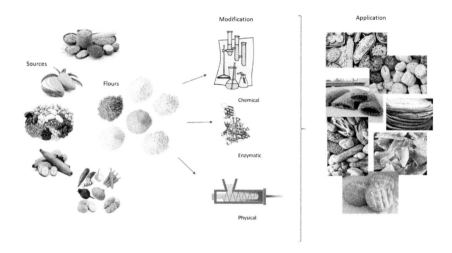

Figure 8.1 Starchy flours from diverse botanical sources, their modifications and end uses.

foods without previous treatment, such as autoclaving, extrusion, chemical and enzymatic treatments.

Green banana flour and soybean hulls were used to increase the dietary fiber content and functionality when the blend was used as a frying batter for chicken nuggets; the texture of the nuggets was poor, which produced low acceptability by consumers (Kumar et al., 2011). Other end uses of unconventional flours (green banana) were in beef burgers and mortadella sausages to reduce the fat content (Bastos et al., 2014; Alves et al., 2016).

Unripe plantain flour was blended with various starchy flours to produce gluten and gluten-free products such as pasta, snacks and bakery products. In general, the use of unripe plantain flour increased the undigestible carbohydrate content (dietary fiber) and reduced the starch hydrolysis rate with a reduction in the glycemic index. However, some drawbacks were found with the use of unripe banana and plantain flours. For example, spaghetti supplemented with different levels of unripe plantain flours (15, 30 and 45 g/100 g flour) showed an increase in cooking loss when plantain flour rose in the formulation and an increase in undigestible carbohydrate content and antioxidant capacity (Ovando-Martinez et al., 2009).

Unripe plantain flour obtained from the pulp was partially used in cereal bars and spaghetti. Bars were elaborated with the blend of four flours, three flours (wheat, barley and oat) were always used in the formulations, and the fourth flour was maize (white or blue) or unripe plantain flour (Utrilla-Coello et al., 2011). The use of wheat, barley and oat from a functional perspective was

TABLE 8.1 USE OF STARCHY FLOURS TO MODIFY THE FUNCTIONAL AND NUTRITIONAL CHARACTERISTICS OF FOODS

Flour	Product	Objective	Reference
Whole amaranth	Bread	Amaranth flour added up to 40% in the mix increased the DF and RS content.	Sanz-Penella et al. (2013)
Beans	Spaghetti	White-seeded low phytic acid and lectin-free bean cultivar were used to elaborate pasta. Increased DF and RS, but decreased SDS and RDS (50% approximately). The HI and pGI decreased at half.	Giuberti et al. (2015)
Plantain	Spaghetti	Increase DF with a blend of gluten-free flours such as chickpea and maize.	Agama-Acevedo et al. (2019)
Whole plantain and Hi-Maize	Spaghetti	The incorporation of both ingredients as DF in semolina-based pasta reduces starch digestibility more than wheat semolina (control).	Garcia-Valle et al. (2020)
Whole and pulp plantain	Spaghetti	It was more evident that the presence of fiber in the pasta matrix elaborated with whole plantain flour reduce swelling starch granules, decreasing its digestibility.	Patiño-Rodriguez et al. (2019)
Plantain and Hi-Maize	Cakes	Use as RS nonthermostable and thermostable sources in blends of wheat or rice in batter's formulation.	Magallanes-Cruz et al. (2020)
Potato	Steamed bread	Enhance the nutritional and functional quality.	Lui et al. (2016)
Potato	Steamed bread	The 50% substitution with potato flour duplicates the RS, with the more evident reduction of RDS than SDS. The estimated glycemic index decreased 87% in control to 73%.	Cao et al. (2019)
Rice	Bread	Gluten-free bread elaborated with pigmented flour showed lower starch hydrolysis and GI than nonpigmented flour.	Thiranusornkij et al. (2019)

(Continued)

TABLE 8.1   (CONTINUED) USE OF STARCHY FLOURS TO MODIFY THE FUNCTIONAL AND NUTRITIONAL CHARACTERISTICS OF FOODS

Flour	Product	Objective	Reference
Rice	Cookies	Physical and enzymatic methods to produce RS. Blends of rice flour and RS (50:50). RS increased in all cases, but SDS decreased. The GI was high in all cookies.	Giuberti et al. (2017)
Rice, plantain and Hi-Maize	Cookies	Gluten-free; increased DF content.	Garcia-Solis et al. (2018)
Whole buckwheat	Cookies	Malt whole buckwheat blended with rice flour (30:70). The buckwheat flour decreases cookies' total starch hydrolysis (40%) and the GI to 100 in rice flour cookies to 58 in malt whole buckwheat flour.	Molinari et al. (2018)
Sorghum	Flatbread	The replaced refined wheat flour with whole grain white and red flour did not reduce bread acceptability, incorporate antioxidants and decreased RDS.	Yousif et al. (2012)
Blue maize	Spaghetti	Incorporate bioactive compounds (polyphenols) with antioxidant capacity and increased SDS and RS of gluten-free pasta.	Camelo-Méndez et al. (2017)
Whole grain flour	Muffins	The digestible starch of muffins prepared with red barley, oat, yellow corn showed fewer RDS than muffins of rice and wheat.	Soong et al. (2014)
Chickpea	Bread	The replacement of rice flour with chickpea duplicated the RS content, reduced GI, and increased satiety.	Santo et al. (2021)
Whole barley	Bread	Increased DF and β-glucans.	Pejcz et al. (2017)
Purple yam	Bread	The purple yam flour decreased RDS and SDS; this was due to its starch granules being more resistant to thermal processing than wheat.	Liu et al. (2019)
Whole faba bean	Pasta	Decreased the starch digestibility, increased protein, and DF.	Gangola et al. (2021)

DF, dietary fiber; RS, resistant starch; SDS, slowly digestible starch; RDS, rapidly digestible starch; HI, hydrolysis index; pGI, predicted glycemic index; GI, glycemic index.

not clear; perhaps the role of wheat can be clearer for the gluten network, but that of barley and oat has not been completely elucidated. Bars with unripe plantain flour showed the lowest dietary fiber (DF) content, and a higher DF value was determined in bars with white and blue maize (11.6 and 10.2 g/100 g bar, respectively), with a high total undigestible fraction of 36.9–40.1 g/100 g bar, as determined by Saura-Calixto's method. A preference study showed global acceptability between 5.5 and 6.1, which was lower than the commercial bar score (7.6). More studies can be proposed to identify bars with higher acceptability by consumers to develop new products with the use of conventional and nonconventional starchy flours (Utrilla-Coello et al., 2011).

Unripe plantain flour (UPF) to prepare gluten-free spaghetti blended with chickpea and white and blue maize flours was prepared in twin-screw (Flores-Silva et al., 2015a) and single-screw extruders (Camelo-Mendez et al., 2017b). Flores-Silva et al. (2015a) prepared eight spaghettis with the blend UPF, chickpea and corn at different levels, and semolina spaghetti was used as a control. Chickpea flour was added at the highest level in the blend (between 60 and 70 g/100 g of the blend). The L^* value in the uncooked gluten-free spaghetti showed a slight decrease (78.5–80.1) compared with the control sample (83.8), and a higher decrease was observed in the cooked spaghetti (70.0–74.5) compared with the control spaghetti (81.7). No appreciable change in the diameter of cooked spaghetti was found between the gluten-free formulations and the control sample. However, gluten-free spaghettis showed greater hardness and chewiness and lower overall acceptability by consumers (4.8–5.8) than the control spaghetti (7.2). In this case, a twin-screw extruder was used to produce gluten-free spaghetti with characteristics comparable to semolina spaghetti.

Camelo-Mendez et al. (2017a) identified the phenolics and antioxidant capacity present in six samples. Fifteen phenolic compounds (one flavonol, one hydroxyphenylacetic acid, three hydroxycinnamic acids, five anthocyanins and five hydroxybenzoic acids) were found in spaghetti made with blue maize, and ten phenolics were identified in spaghetti prepared with white maize. The antioxidant capacity determined by the three methods (DPPH, ABTS and FRAP) increased when maize flour rose in the formulation. Spaghetti with blue maize had a higher antioxidant capacity than white maize. Ferulic acid was the main phenolic compound that was related to the antioxidant capacity. Gluten-free spaghetti with the formulation represents an alternative to produce functional foods with increased antioxidant characteristics. The optimum cooking time and cooking loss of gluten-free spaghetti ranged between 6.7 and 7.7 min and 9.2% and 11.1%, respectively. The cooking loss was lower than the 12% limit suggested for gluten-free spaghetti of good quality (Camelo-Mendez et al., 2018a). The DF content of gluten-free spaghetti ranged between 11.2 and 13.6 g/100 g, which is considered a food with high DF content (Camelo-Mendez et al., 2018b).

Traditionally, starch has been isolated from diverse botanical sources and modified to reduce its digestion rate. Studies on the use of starch from diverse botanical sources, and chemical, physical, enzymatic or a combination of these methods to modify the starch digestion rate have been widely reported. The use of isolated starch increases the final product cost that is used as an ingredient with reduced digestibility (slowly digested or resistant). In this sense, it was suggested that starchy flours be used to produce ingredients with reduced digestibility. Strategies such as the use of unconventional starch flours (e.g., unripe fruits, tubers, roots) and processing methods (chemical, physical and enzymatic methods) could be suitable, but they need to be evaluated.

In the treatments of starchy flours to modify the digestion starch rate, it is important to keep in mind the presence of different macromolecules (proteins, nonstarch polysaccharides, polyphenols) where the interactions with the starch decrease its hydrolysis by digestive enzymes (Bordenave et al., 2014; Ferruzzi et al., 2020).

The use of commercial starchy ingredients such as Hi-Maize 260®, which is suggested to increase DF content, can be considered an unmodified source because the raw material is a corn where starch has high amylose content, it is a maize variety developed with this characteristic. This type of starch has granules that exhibit low swelling during heating with water content, which are used in food development and produce partial gelatinization and a low hydrolysis rate. In this sense, there is interest in finding starchy sources that can give nutritional characteristics similar to those of Hi-Maize 260 and perhaps additional functional attributes such as antioxidant capacity. However, a basic study of the interactions between the macromolecules present in food formulations is mandatory to prepare acceptable products by consumers.

8.3 METHODS FOR MODIFICATION

The modification of starchy flours is a practice that has increased in recent years. The modification of starchy flours began when extruded starchy flours were produced as ingredients in bakery products and gluten-free products to decrease the fat content and texture characteristics of gluten-free products due to the absence of gluten (Roman et al., 2015a).

The objective of modifying starchy flours is to produce changes in the physicochemical and functional characteristics for the development of new or improved products. It is important to keep in mind that the modification of starchy flours can produce changes in major components, such as proteins, lipids and nonstarch polysaccharides, which can be beneficial in some functional and sensory characteristics of the products made with those modified starchy flours. The modification of starchy flours, mainly alternative sources such as amaranth, pulses and unripe fruits (mango, plantain), can be important in producing raw materials with modulated starch digestibility and gluten-free characteristics. Food products such as pasta, snacks and bakery products can be produced with gluten-free flours (Mandala & Kapsokefalou, 2011).

Methods to modify the starch structure have been also used to modify starchy flours (Figure 8.1). Single, dual and triple modifications may be used to produce starchy flour with specific functionalities with the objective of producing new or improved foods.

In terms of physical methods, emerging technologies considered nonthermal, such as ultrasound, high pressure, pulsed electric field, ozone and cold plasma, were reported to modify starch. The pros and cons of these technologies and the combination of two or more of them open a new possibility to modify starchy flours (Castanha et al., 2019). The study of starchy flours is incipient, and it is necessary to explore these technologies at the pilot plant and semi-industrial levels to increase the availability of modified starchy flours. Emerging technologies have been increasing in the last decade because they are considered "green" processes for starch modification as isolated polysaccharides or as the main component of starchy flours. However, in starchy protein-rich flour, the modification of this macromolecule can be important in the functionality of starchy flour (Bonto et al., 2021; Maniglia et al., 2020, 2021; Mhaske et al., 2021).

Environmentally friendly methods include oxidation with UV irradiation and thermal treatment. This oxidation method is comparable to chemical modification with NaClO. UV irradiation was an effective oxidizing agent when combined with thermal treatment that generated depolymerization of starch chains with the production of more lineal chains for oxidation. The modification of starch without chemical reagents alters its physicochemical properties as the pasting profile. However, when the modified starch was compared with the traditional oxidation with NaClO, the changes were more evident with the traditional method (Kurdziel et al., 2019). The "green" method is promising for modifying starch structure. More studies with different conditions, both UV irradiation and thermal treatment, can produce other functionalities in starch.

The organocatalytic esterification of corn starch (Imre & Vilaplana, 2020) and other polysaccharides (pectin, gums, cellulose) emerged in the last decade to modify the structure of those polysaccharides and, consequently, the physicochemical and functional characteristics (Ragavan et al., 2021). Organocatalytic esterification is classified as "green chemistry" due to the use of organic reagents. The organocatalytic esterification of polysaccharides as cellulose has the objective of increasing the prebiotic effect. Cellulose can be grafted by short-chain fatty acids (SCFAs) by organocatalytic esterification. Organocatalytic cellulose is not hydrolyzed in the small intestine and reaches the colon, where the microbiota metabolizes it with the liberation of those SCFAs that pass into the bloodstream, with health benefits such as a reduction in cholesterol levels (Gutierrez & Tovar, 2021). Starch is esterified with acetic, propionic and butyric acids as donor agents. The esterified starches show a structure that is not recognized by digestive enzymes and increase the dietary fiber content, which is fermented by bacteria classified in the Firmicutes group (Hamaker & Tuncil, 2014).

Additionally, the use of this environmentally friendly method for modifying starchy flours can be an alternative for their end uses because physicochemical and functional properties are changing and can produce foods with improved or different characteristics.

8.4 APPLICATIONS OF MODIFIED FLOURS

Modified starch flours can have different end uses. The modification methods, especially those carried out by chemical modification due to restrictions imposed by agencies of drug and foods of the countries, botanical sources and physicochemical characteristics are important during the processing of foods. There is growing interest in the use of modified starchy flours because they supply "new" or different functionalities in foods and can be exploited for the other components present in flours, such as proteins, lipids and nonstarch polysaccharides.

8.4.1 Chemical Treatment

Chemical treatment was proposed to modify starchy flours as reported in unripe plantain flour. Acid treatment with HCl at different concentrations was applied to produce a fiber-rich powder. This treatment of starch was used to produce glucose syrup for the longest time. In the first step of the acid treatment, the acid hydrolyzed the amorphous areas of starch with an increase in the crystalline zones; these zones are ordered regions that are not hydrolyzed by amylolytic enzymes (resistant starch) and are included in the DF fraction. Acid concentrations between 0.6 and 2.2 M were used. The total dietary fiber (TDF) ranged between 19.8 and 60 g/100 g, and the highest concentration was obtained at 37.5°C, 10.5 days and 1.6 M. The fiber-rich powder showed a 50°C water solubility index of approximately 50% and a water absorption index of 1%, both of which are important during the end use of modified unripe plantain flour (Aguirre-Cruz et al., 2008).

Esterification with citric acid was performed by dissolving 40 g of citric acid in 100 mL of distilled water and adding 100 g of flour; the moistened flour was placed in stainless steel trays and maintained at room temperature for 16 h.

After the tray was placed in an oven at 40°C for 24 h and then in a forced-air oven at 140°C for 7 h. The unreacted citric acid was removed by washing with distilled water. The washed flour was dried at 40°C and milled. The control flour was produced with the same procedure without citric acid. The raw (uncooked) starchy flour has a high resistant starch (RS) content (85.7 g/100 g) that decreases after cooking (12.4 g/100 g). The esterification with citric acid of UPF showed a high RS content (93.9 g/100 g), which was maintained after cooking (93.7 g/100 g). The esterified UPF showed almost complete starch disorganization due to the low enthalpy value (0.97 J/g) determined by differential scanning calorimetry (DSC) and slight disorganization by the hydrothermal treatment (10.2 J/g) compared with the nontreated UPF (12.7 J/g) (Sanchez-Rivera et al., 2017). The combination of esterification with citric acid and thermal treatment produced a synergistic effect that modified the starch structure that was not hydrolyzed by digestive enzymes. The authors reported that esterified UPFs can be used in bakeries, snacks and pastas, and can maintain a high RS content after cooking. Then, the use of esterified cookies to increase the indigestible carbohydrate content to reduce the glucose supply after consumption was reported. Four formulation cookies were created: control sample of wheat flour, control with nonmodified UPF, esterified UPF, and esterified UPF (40 g)–wheat flour (60 g). The esterified–UPF wheat flour cookie showed the highest total indigestible fraction (35.7 g/100 g). The hardness of the cookie decreased in the sample of nonmodified UPF, and the lowest value was found in the esterified and blended (esterified–wheat flour) cookies. Sensory characteristics such as color, texture and taste decreased with the addition of UPF compared with wheat control cookies. However, the blend esterified-wheat flour improved the three sensory characteristics, an issue that deserves more investigation to produce a new cookie with higher acceptability by consumers (Sanchez-Rivera et al., 2019).

Unripe plantain fruits with a green maturity of 1 (Von Loesecke, 1950) were used to produce flour, which was modified by diverse chemical methods, such as acetylation, carboxymethylation, methylation, oxidation and phosphation. The total dietary fiber determined by the AACC method where nonstarch polysaccharides are included ranged between 6.29 and 10.18 (g/100 g). The amylose content, which is important in the functional and digestibility characteristics, was determined by the DSC method with pure amylose as a reference. The amylose content for the native flour (unmodified) was 25.8%, and for those chemically modified between 26.6% and 51%; the latter was obtained for methylated flour with a high standard deviation of about 15. It is important to note that no appreciable difference in amylose content was calculated in the different modified plantain flours. Plantain flour modified by acetylation, carboxymethylation, oxidation and phosphation presented a higher whiteness index than the unmodified counterpart; the average gelatinization temperature of the modified flour did not change in relation to unmodified flour, except methylated flour. The characteristics mentioned of the modified plantain flour are important in the end uses of those flours. The modified plantain was utilized as an ingredient in gluten-free cookies and bread. The RS content increased in the foods supplemented with chemically modified plantain flour compared with the unmodified flour, and the highest value was determined with foods made with phosphate flour; no difference in RS content was found between cookie and bread for the same modified flour (Gutierrez, 2018). Other end uses of chemically modified unripe plantain flour deserve to be studied.

8.4.2 Physical Treatments

Physical treatments that modify starch and other components in flour (proteins, lipids, nonstarch polysaccharides) are widely used. Physical methods are preferred over chemical modifications because there are no restrictions imposed by government agencies (e.g., US Food and Drug Administration) and because they are environmentally friendly.

Various studies of starchy flours modified by extrusion, hydrothermal treatments, radiation, high pressure, combinations of two or more of those treatments, and pretreatments are found in the scientific literature.

Pregelatinized flours mainly produced by drum drying at the commercial level have been used in the food industry as fat substitutes (Lee et al., 2013: Roman et al., 2015a) and gluten-free products to improve the texture (Martinez et al., 2014). Additionally, extrusion has been used to prepare pregelatinized flours of different botanical sources, such as maize, wheat and rice. Although the extruded flours can be named pregelatinized, those commercially available do not show thermal transition in the DSC analysis, which indicates that starch was completely disorganized. Complete starch gelatinization produces a gel at room temperature as a function of the flour concentration. The food formulation, which means the diverse raw materials (ingredients) incorporated in the preparation, determines the food's functionality, sensory characteristics and digestibility. Commercial pregelatinized flour of maize (Zoc 90) was used to prepare mayonnaise with reduced fat content (30%, 50% and 70%) of the original fat content. Although an important reduction in the fat content is produced, the substitution of fat by maize extruded flour impacts the microstructure, freeze–thaw stability and rheological characteristics of the mayonnaise. However, a positive impact was obtained in the emulsion stability of fat-reduced mayonnaise (Roman et al., 2015).

The reduction of fat with extruded flours has been applied with cake. Traditionally, different carbohydrates are used, such as inulin, β-glucan, modified starches, fiber, maltodextrins, crystalline cellulose and polydextrose (Luo et al. 2019). Commercial extruded wheat flour (Roman et al., 2015b) and extruded amaranth flour (unpublished data) were used to replace fat in cakes. The amaranth cakes were classified as low-fat and gluten-free. The wheat flour cakes replaced with the extruded wheat flour did not show similar characteristics to the full-fat cake because an emulsifier was added. The addition of emulsifiers improved the quality of the cakes, maintaining the volume and reducing the hardness; these characteristics increased the acceptability of the cakes by consumers. The amaranth cakes were replaced with extruded amaranth cakes. The dietary fiber content in the three formulations ranged between 6.2 and 8.2 g/100 g, and gluten-free cakes can be classified as a high dietary food. Consumers showed good overall acceptability for gluten-free cakes (unpublished data).

Other applications of extruded wheat flour with different severities of extrusion treatment (1, mild; 2, intermediate; 3, high) were used to prepare batter for chicken nuggets. Chicken nuggets coated with 15% of the extruded flours showed similar appearance, texture, taste and overall acceptability to the control nuggets (Roman et al., 2018).

The advantage of extruded (pregelatinized) flours is that they produce a gel under cold conditions. Cold preparation of products such as creams and sauces are necessary. To improve the sensory characteristics of this kind of food, particle size is important to maintain the texture and mouthfeel of the product. Commercial wheat (Toc 190) and rice (Ooc 190) extruded flours with different particle sizes (fine, intermediate and coarse) were used. The flour (wheat and rice) botanical source played an important role in the resistance to the freeze–thaw process because those sauces prepared with extruded rice flour showed a less compact structure (Roman et al., 2018b).

Nonconventional flour, similar to that of unripe plantain, has been extruded. Unripe plantain flours, whole (pulp + peel) and pulp, were produced and extruded with a single-screw machine. The degree of gelatinization of the whole and pulp extruded flours was 63% and 68%, respectively, which means that the starch present in the flour was not completely disorganized and can present specific functionality and digestibility. Application in nonheated food, such as yogurt, to increase dietary fiber consumption showed lower overall acceptability than control yogurt. More investigation is necessary to improve the sensory characteristics of yogurt supplemented with extruded unripe plantain flour (Garcia-Valle et al., 2019).

Thermal treatments to modify starch and starchy flours are preferred because they can be cheaper and more environmentally friendly than chemical treatments. Hydrothermal and heat-moisture treatments (HMT) have been reported to modify starchy flours of different botanical origins. Unripe plantain flour (pulp) was prepared after boiling the whole fruit in a water bath for 10 min and compared to a control flour (no cooking). Both flours were conditioned at 30% moisture content for 24 h after incubation at 120°C for 24 h in a convection oven. The modified flour showed crystallinity peaks that indicate no complete disorganization of the starch structure. Although the starchy flour was submitted to heat treatment, it was cooked before heat treatment to analyze the starch digestion fractions. RS content in the cooked flours ranged between 8.1 and 18.4 g/100 g, and for uncooked flours, it ranged between 21.7 and 65.6 g/100 g (Rodriguez-Damian et al., 2013). Annealing (ANN) of the unripe plantain flour (70% moisture, 65°C for 24 h in a convection oven) showed some crystallinity peaks that were smaller than those obtained in the HMT flours; cooked samples presented RS contents ranging between 6.8 and 11.3 g/100 g, and those uncooked between 12.1 and 68.8 g/100 g (De la Rosa-Millan et al., 2014). HMT produced unripe plantain flours with a higher RS content than ANN samples. No end uses in foods were reported in both studies and can be an opportunity to use a food crop that is underutilized in some tropical countries.

Hydrothermal treatment under different conditions (moisture, temperature and time) was applied to sorghum, cassava and rice flours with the objective of modifying the functionality in the production of cake and bread. Cake added with modified sorghum did not produce changes in the cake's crust with a uniform size; the pore size increased, and consumers accepted the cake. The addition of modified cassava to wheat flour produced higher dough stability; the bread elaborated showed crumb porosity and lower crumb hardness. Wheat bread supplemented with modified cassava flour is an alternative to produce bakery products with acceptable physical characteristics (Dudu et al., 2020).

Microwaves were applied to rice flour to prepare gluten-free bread. The absence of gluten decreases the textural characteristics of bread. Rice flour was treated by microwave heating before bread making. The bread quality was not appreciably modified by the microwave heating of the rice flour because a slightly higher loaf-specific volume was determined compared with the bread with untreated rice flour. The results of this study can contribute to improving the textural characteristics of gluten-free bread (Perez-Quirce et al., 2018). The physical treatment of starchy flours modifies the starch structure, and consequently, interactions with other macromolecules can be produced during the processing of foods. These changes and interactions can alter starch digestibility because more studies on starch digestion are necessary.

The physical modification of flours by HMT and extrusion and sourdough fermentation can be proposed to alter the starch structure and produce slow digestion characteristics (Giuberti & Gallo, 2018). The challenge is to find conditions (e.g., temperature, moisture content, time, screw speed in the extrusion) that produce slow digestion characteristics. The starchy flours contain other macromolecules, such as proteins, nonstarch polysaccharides (pectin, gums, hemicellulose) and polyphenols. During physical treatments such as HMT and extrusion, interactions of protein–starch, starch–nonstarch polysaccharides and polyphenol–starch (noncovalent interactions) can be produced to develop a matrix. The interactions modify the starch structure that is not recognized or partially recognized by digestive enzymes and can decrease starch hydrolysis with a concomitant decrease in the glucose supply in the human body.

8.4.3 Enzymatic Modification

Other groups of modifications include enzymatic treatments with transglutaminase. Treatment with this proteolytic enzyme modifies the proteins present in gluten-free flours, such as rice, to produce protein aggregates that mimic gluten functionality. The replacement of wheat flour is a challenge due to the functionality of gluten in bakery products. The change in the protein conformation after enzymatic treatment with transglutaminase has been combined with the use of pregelatinized starches, hydrocolloids and other novel ingredients to improve the texture characteristics of bread. The interactions between the different macromolecules produce a food matrix that can show different digestion starch rates, an issue that has not been completely studied.

The use of enzymes to modify the starch structure present in flours has been rarely studied; enzymes to modify starch structure has been reported in isolated starch, as shown in Chapter 6. Enzymes such as β-amylase, transglucosidase, amylosucrase, pullulanase and isoamylase have been reported to modify starch structure and modify starch digestion rate. Application of enzymes is an alternative to producing SDS and RS in isolated starch; however, the addition of enzymes in starchy flours with diverse macromolecules and polyphenols can produce an important effect in the modification of the starch structure.

8.5 STARCH DIGESTIBILITY

The use of unmodified starchy flours has been studied (Section 8.2). The main goals of the use of diverse starchy flours, unlike those conventionally used as wheat, rice, potato, maize, are to utilize underused regional food crops with low commercial value, decrease the waste of those crops and increase the functional characteristics of traditional or regional foods such as bakery products, pasta, snacks, and porridges. Several studies evaluated the use of nonconventional food crops (see Chapter 9), emphasizing the functional and sensory characteristics. One of the objectives of the use of nonconventional starchy flours is to decrease the undigestible carbohydrate content (increasing the dietary fiber content) (Table 8.2).

TABLE 8.2	IMPACT OF FLOUR MODIFICATION ON STARCH DIGESTIBILITY		
Flour	**Modification**	**Effect on digestibility**	
Plantain pulp vs. whole	Extrusion	Extruded flour showed 12% RS content.	Garcia-Valle et al. (2019)
Plantain pulp precooked	HMT/storage	Precooked sample after HMT and HMT storage increases SDS notably, reducing RDS. When the samples were hydrothermally treated and storage was gelatinized, they preserve SDS and RS fraction compared with untreated samples.	Rodriguez-Damian et al. (2013)
Plantain pulp precooked	Annealing/storage	Precooked sample after annealing increased the RS, which did not change during storage. The effect of annealing and storage on RS was nullified when the samples were analyzed gelatinized.	De la Rosa-Millan et al. (2014)
Plantain	Chemical modification: Acetylation Carboxymethylation Methylation Oxidation Phosphation	The methylated and phosphate flours showed the lowest in vitro digestibility due to the highest RS content. Phosphated flour showed promising alternative to produce bread and cookies with high RS content.	Gutierrez, (2018)
Plantain	Citric acid esterification	Cookies with esterified plantain flour showed a few digestible starches compared with wheat and raw plantain flour.	Sanchez-Rivera et al. (2019)
Wheat	Extrusion	Enzymatic susceptibility depends on extrusion parameters.	Martínez et al. (2014)

HMT, high-moisture treatment; RS, resistant starch; SDS, slowly digestible starch; RDS, rapidly digestible starch.

It is important to keep in mind the chemical composition of unmodified starchy flours and the changes in those components after the modifications. Legumes, pulses cereal (oat, maize) flours contain oligosaccharides (raffinose, stachyose), oligofructose, inulin, β-glucans, and arabinoxylans, which are associated with an increase in viscosity in the intestine and slowing intestinal transit that delays gastric emptying with concomitant glucose and sterol absorption in the small intestine. These events decrease serum cholesterol, postprandial blood glucose and insulin levels (Korczak & Slavin, 2020).

Flour of cooked beans was blended with semolina at different levels to produce pasta, and the effect on chemical composition, especially protein content, and cooking quality (i.e., cooking time, cooking loss, firmness) were analyzed. However, starch digestibility analysis was not achieved (Gallegos-Infante et al., 2010). Other studies on the use of nonconventional starchy flours, such as sorghum in bread (Yousif et al., 2012) and noodles (Liu et al., 2012), evaluated the quality characteristics of the foods, such as texture, viscosity, protein content, and dietary fiber. White and red sorghum were blended with wheat flour at 30%, 40% and 50% to prepare flatbread. Sorghum flours showed lower protein content and higher ash content than wheat flour, without appreciable difference in DF content (3.5–4.0 g/100 g). However, the sorghum flatbreads showed higher DF levels (5.7 and 6.5 g/100 g) than wheat flatbread (4.2 g/100 g), but the authors provided no explanation. Sorghum flatbreads presented a lower starch digestion rate than wheat flatbread (control), which was attributed to a higher RS content (36.4 and 38.9 g/100 g dry starch) than wheat bread (30.1 g/100 g dry starch). It is important to note that the acceptability of new bread formulations is an important issue; white sorghum bread at 30% and red sorghum at 40% and 50% showed a higher overall acceptability than control bread (Yousif et al., 2012).

UPF and commercial wheat flour were used to prepare bread. Unripe plantain bread was added with commercial gluten to improve the characteristics of the network. This type of bread presented 5.1 g/100 g of DF, which is considered a high DF food, with 6.7 g/100 g of resistant starch. Both values were higher than those for wheat bread and produced a decrease in the starch hydrolysis rate (Juarez-Garcia et al., 2006). UPF was blended with semolina to produce spaghetti in a single-screw extruder (Ovando-Martinez et al., 2009). The spaghetti with the highest substitution level of UPF (45%) showed 12.4 g/100 g of spaghetti, dry basis, of RS, with 26.1 g/100 g of insoluble indigestible fraction and 3.3 g/100 g of soluble indigestible fraction, where resistant protein, RS, and other nonstarch polysaccharides were included (Saura-Calixto et al., 2000). The high RS content in the composite pasta produced the lowest starch hydrolysis percentage. Unmodified starchy flours (maize, chickpea and UPF) were blended to prepare spaghetti. The composite spaghetti showed RS contents between 3.6 and 5.0 g/100 g of dry spaghetti and slowly digestible starch contents between 12.8 and 27.3 g/100 g of dry spaghetti. Both fractions can be considered high, and their consumption is associated with health benefits (see Chapter 6). The maximum starch hydrolysis percentage for composite spaghetti ranged between 24% and 32%, which produced a low predicted glycemic index (60–64). The blend of different unmodified starchy flours produces a food matrix that can be associated with low starch digestion. Additionally, the use of nonconventional starchy flours that present high DF content, protein content (pulses) and polyphenols can produce interactions that restrict the starch hydrolysis rate (Camelo-Mendez et al., 2018).

Unconventional flours such as chickpea and UPF were blended with maize flour to produce deep-fried gluten-free snacks. Snacks with the highest chickpea flour (60 g/100 g) and with similar flour content (33 g/100 g of each flour) showed the highest DF level that included nonstarch polysaccharides, and both showed the highest RS content (3.2 and 2.8 g/100 g, respectively). However, the snacks with the highest UPF (50 g/100 g) and with similar flour level presented the highest slowly digestible starch content (13.4 and 11.6 g/100 g, respectively). The snacks showed a low glycemic index (28–35) that reflects resistant and slow starch digestion and the DF content (Flores-Silva et al., 2015b).

Semolina was blended with chickpea flour to increase the indigestible carbohydrate content in extruded pasta. Semolina pasta and chickpea pasta were used as references, and blends of semolina:chickpea (80:20, 60:40, 40:60 and 20:80) were elaborated. After the extrusion process, pasta was analyzed by DSC to evaluate the gelatinization degree, where semolina pasta showed the highest value and chickpea pasta (Garcia-Valle et al., 2021a) showed the lowest value; the gelatinization degree decreased with increasing chickpea flour in the composite. The DF content in the cooked composite pasta where RS was included ranged between 32.9 and 51.6 g/100 g, with the highest value in those composites for the semolina:chickpea blend of 20:80, which agrees with the lowest starch hydrolysis rate. Confocal laser microscopy showed the presence of nonstarch polysaccharides and a protein network that can be responsible for the slow starch digestion characteristics. The composite pasta showed texture characteristics that deviated further from the semolina pasta when increasing the amount of chickpea flour (Garcia-Valle et al., 2021b). Acceptability studies with consumers may grant the promotion of composite pasta as a good fiber source.

The food matrix (snack) can play an important role in the starch digestion characteristics of starchy flour and is an issue that deserves more investigation.

UPFs with peel (WUPF) and without peel (UPF) were extruded to produce partial starch gelatinization (63% and 68%, respectively) to modify the functional and digestibility characteristics. Dietary fiber content with a method that included RS was 18.5 g/100 g and 12.4 g/100 g for whole and pulp extruded flour, respectively. The undigestible carbohydrate content produced slow starch digestion (Garcia-Valle et al., 2019).

RS powders were commercially available for use in foods some years ago (Raigond et al., 2015) and are recommended as a dietary fiber source. The manufacturer reported that those RS powders are classified as type 2, type 3, and type 4, and studies to produce RS type 5 are underway. Commercially available RS powders are suggested in bakery products, snacks, pasta and batters, among others. Hi-Maize 260 is produced by Ingredion with a high-amylose starch; the manufacturer reports that this powder contains approximately 53% RS (determined as DF) and 40% digestible starch. Gluten-free cookies were prepared with unripe plantain flour (85 g/100 g) or rice flour (50 g/100 g), with 15 and 50 g/100 g of Hi-Maize 260. Although lower Hi-Maize 260 was added to UPF, it showed lower rapidly digestible starch (14.1 g/100 g) and higher RS (35.7 g/100 g) than rice cookies (28.8 and 13.2 g/100 g, respectively). Total dietary fiber was determined with an integral method that is used in products "as eaten" and included RS content because no additional cooking is necessary. Cookies were hydrolyzed for 4 and 16 h; UPF cookies showed 62.2 g/100 g DF at 4 and 43.7 g/100 g at 16 h of hydrolysis, values that were higher than those determined in rice cookies (21.6 and 19.1 g/100 g, respectively). The higher value of DF at 4 h can be related to slowly digestible starch that requires more time for its hydrolysis. Also, it is important that the difference in the DF and RS between the UPF and rice cookies is associated with the nonstarch polysaccharides present in the UPF (Garcia-Solis et al., 2018).

Cakes were elaborated with two sources of indigestible carbohydrates (UPF and Hi-Maize 260), and these ingredients were blended with wheat and rice flours. Hi-Maize 260 is a commercial source of undigestible carbohydrate (RS) that can be used in cakes to increase the DF level. The highest DF content (including RS content) was found in the cakes with UPF and Hi-Maize 260, and rice flour and Hi-Maize 260, 18.4 and 18.1 g/100 g, respectively. Also, the cake with the blend of wheat flour and Hi-Maize 260 presented an important DF content (11.6 g/110 g) (Magallanes-Cruz et al., 2020). Semolina pasta prepared by extrusion was replaced with two sources of resistant starch – WUPF and Hi-Maize 260 – at different levels (0.15, 0.25, and 0.35 w/w). Uncooked pasta with Hi-Maize 260 showed an increase in DF content between 15.7 and 30.2 g/100 g, which was higher than pasta with WUPF (between 5.5 and 6.7 g/100 g), which agrees with water absorption because of nonstarch polysaccharides (components of DF) present with hydrophilic characteristics. No appreciable difference in the texture characteristics (hardness and elasticity) was found between pasta replaced with both ingredients at the different levels of substitution. The starch hydrolysis rate was faster for semolina pasta and decreased for semolina–WUPF > semolina–Hi-Maize 260 (Garcia-Valle et al., 2020).

The addition of Hi-Maize 260 to increase DF content depends on the food matrix. More studies can be conducted in this sense.

Unconventional flour was extruded to produce disorganization and rearrangement of the main components of amaranth flour (starch and proteins). The extruded amaranth flour did not show thermal transition when analyzed by differential scanning calorimetry. The aim of that study was to prepare a gluten-free cake with amaranth flour blends and extruded amaranth flour at three levels (1/3, 2/3 and 3/3) to reduce the fat content. The control cake had 14 g/100 g of oil and was reduced to 9.3, 4.6 and 0.0 g oil/100 g of formula. The total reduction in oil produced a decrease in the starch hydrolysis rate, and the network produced by the extruded amaranth flour can be associated with this pattern, where a slight increase in the RS content was determined. The amaranth cakes showed an important amount of the two starch digestion fractions (SDS and RS) that are associated with health benefits (see Chapter 6). The acceptability by consumers and the texture characteristics of gluten-free and low-fat amaranth cakes is a challenge for this kind of product that should be considered. The use of some ingredients, such as hydrocolloids or proteins, to strengthen the food matrix can be an alternative.

8.6 FUTURE TRENDS

There is interest in the development of new or different functional foods. Starchy flours are widely used in different food matrices, such as pasta, bakery products and snacks. The functionality and nutritional characteristics can be improved after modifying the main component (starch), but the other macromolecules present in the formulation can interact and modify the starch digestibility and bioaccessibility of dietary phytochemicals. The use of nontraditional starchy flours and even more modification by different treatments with an emphasis in ecological processes is an opportunity to obtain raw material with improved functionality and nutritional features. The applications of alternative food processing methods to produce a specific food matrix and starchy flours, with actual poor end uses, are topics that deserve more investigation.

8.7 CONCLUSIONS

Interest in producing functional foods with starchy flours is growing in the food industry. Modification of starchy flours by different treatments or combinations can produce a raw material that can be considered for the production of different food matrices. These matrices prepared with modified starchy flours can give place to structures that modulate the bioaccessibility of dietary polyphenols and starch digestibility. The current use of environmental methods to modify starchy flours is preferred and can expand the future end uses due to changes in physicochemical, functional and nutritional properties. The future is now arriving.

ABBREVIATIONS

ANN:	annealing
DF:	dietary fiber
DSC:	differential scanning calorimetry
HMT:	heat-moisture treatment
RS:	resistant starch
SCFAs:	short-chain fatty acids
UPF:	unripe plantain flour

REFERENCES

Agama-Acevedo, E., Bello-Pérez, L. A., Pacheco-Vargas, G., Tovar J., & Sayago-Ayerdi, S. (2019). Unripe plantain flour as a dietary fiber source in gluten-free spaghetti with moderate glycemic index. *Journal of Food Processing and Preservation*, 43(8), e14012. https://doi.org/10.1111/jfpp.14012.

Aguirre-Cruz, A. Álvarez-Castillo, A., Yee-Madeira, H., & Bello-Pérez, L. A. (2008). Production of fiber-rich powder by the acid treatment of unripe banana flour. *Journal of Applied Polymer Science*, 109, 383–387.

Alves, L. A. A. S., Lorenzo, J. M., Goncalves, A. A. A., Santos, B. A., Heck, R. T., Cichoski, A. J., & Campagnol, P. C. B. (2016). Production of healthier bologna type sausages using pork skin and green banana flour as a fat replacer. *Meat Science*, 121, 73–78.

Bastos, S. C., Pimenta, M. E., Pimenta, C. J., Reis, T. A., Nunes, C. A., Pinheiro, A. C. M., Fabrício, L. F. F., & Leal, R. S. (2014). Alternative fat substitutes for beef burger: Technological and sensory characteristics. *Journal of Food Science and Technology*, 51(9), 2046–2053.

Bonto, A., Tiozon, R. N., Sreenivasulu, N., & Camacho, D. H. (2021). Impact of ultrasonic treatment on rice starch and grain functional properties: A review. *Ultrasonics Sonochemistry*, 71, 1005383.

Bordenave, N., Hamaker, B. R., & Ferruzzi, M. G. (2014). Nature and consequences of non-covalent interactions between flavonoids and macronutrients in foods. *Food & Function*, 5, 18–34.

Camelo-Méndez, G. A., Flores-Silva, P. C., Agama-Acevedo, E., & Bello-Pérez, L. A. (2017a). Multivariable analysis of gluten-free pasta elaborated with non-conventional flours based on the phenolic profile, antioxidant capacity and color. *Plant Foods for Human Nutrition*, 72, 411–417. https://doi.org/10.1007/S11130-017-0639-9.

Camelo-Méndez, G. A., Agama-Acevedo, E., Tovar, J., & Bello-Pérez, L. A. (2017b). Functional study or raw and cooked blue maize flour: Starch digestibility, total phenolic content and antioxidant capacity. *Journal of Cereal Science*, 76, 179–185.

Camelo-Méndez, G. A., Tovar, J., & Bello-Pérez, L. A. (2018a). Influence of blue maize flour on gluten free pasta quality and antioxidant retention characteristics. *Journal of Food Science and Technology*, 55(7), 2739–2748.

Camelo-Méndez, G. A., Flores-Silva, P. C., Agama-Acevedo, E., Tovar, J., & Bello-Pérez, L. A. (2018b). Incorporation of whole blue maize flour increases antioxidant capacity and reduces *In vitro* starch digestibility of gluten-free pasta. *Starch-Starke*, 70, 1–8.

Camelo-Méndez, G. A., Agama-Acevedo, E. Rosell, M. C., Perea-Flores, M. J., & Bello-Pérez, L. A. (2018c). Starch and antioxidant compound release during *In vitro* gastrointestinal digestion of gluten-free pasta. *Food Chemistry*, 263, 201–207 https://doi.org/10.1002/Star.201700126.

Cao, Y., Zhang, F., Guo, P., Dong, S., & Li, H. (2019). Effect of wheat flour substitution with potato pulp on dough rheology, the quality of steamed bread and in vitro starch digestibility. *LWT: Food Science and Technology*, 111, 527–533.

Castanha, N., Santos, D. N., Cunha, R. L., & Agusto, P. E. D. (2019). Properties and possible applications of ozone-modified potato starch. *Food Research International* 116, 1192–1201.

De La Rosa-Millan, E., Agama-Acevedo, E., Osorio-Diaz, P., & Bello-Pérez, L. A. (2014). Effect of cooking, annealing and storage on starch digestibility and physicochemical characteristics of unripe banana flour. *Revista Mexicana de Ingeniería Química*, 13, 151–163.

Dudu, O. E., Ma, Y., Adelekan, A., Oyedeji, A. B., Oyeyinka, S. A., & Ogungbemi, J. W. (2020). Bread-making potential of heat-moisture treated cassava flour-additive complexes. *LWT-Food Science and Technology*, 130, 109477.

Ferruzzi, M. G. Hamaker, B. R., & Bordenave N. (2020). Phenolic compounds are les degraded in presence of starch than in presence of proteins through processing in model porridges. *Food Chemistry*, 309, 125769.

Flores-Silva, P., Berrios J., Pan, J., Agama-Acevedo, E., Monsalve- González, A., & Bello-Pérez, L. (2015a). Gluten-free spaghetti with unripe plantain, chickpea and maize: Physicochemical, texture and sensory properties. *CyTA–Journal of Food*, 13, 159–166.

Flores-Silva, P. C., Rodríguez-Ambriz, S. L., & Bello-Pérez, L. A. (2015b). Gluten-free snacks using plantain, chickpea and maize blend: Chemical composition, starch digestibility, and predicted glycemic index. *Journal of Food Science*, 80(5), C961–C966.

Gallegos-Infante, J. A., Rocha-Guzmán, N. E., González-Laredo, R. F., Ochoa-Martínez, L. A., Corzo, N., Bello-Pérez, L. A., Medina-Torres, L., & Peralta-Álvarez, L. E. (2010). Quality of spaghetti pasta containing Mexican common bean flour (*Phaseolus vulgaris* L.). *Food Chemistry*, 119(4), 1544–1549.

Gangola, M. P., Ramadoss, R. B., Jaiswal, S. Chan, C., Mollard, R., Fabek, H., Tulbek, M., Jones, P., Sanchez-Hernandez, D., Anderson, G. H., & Chibbar, N. R. (2021). Faba bean meal, starch or protein fortification of durum wheat pasta differentially influences noodle composition, starch structure and in vitro digestibility. *Food Chemistry*, 349, 129167.

García-Valle, D., Bello-Pérez, L. A., Flore-Silva, P. C., Agama-Acevedo, E., & Tovar, J. (2019). Extruded unripe plantain flour as an indigestible carbohydrate-rich ingredient. *Frontiers in Nutrition*, 6, 2. https://doi.org/10.3389/fnut.2019.00002.

García-Valle, D. E., Agama-Acevedo, E., Álvarez-Ramírez, J., & Bello-Pérez, L. A. (2020). Semolina pasta replaced with whole unripe plantain flour: Chemical, cooking quality, texture, and starch digestibility. *Starch/Starke*, 72, 1900097. https://doi .org/10.1002/star.201900097.

García-Valle, D., Bello-Pérez, L. A., Agama-Acevedo, E., & Álvarez-Ramírez, J. (2021a). Structural characteristics an in vitro starch digestibility of pasta made with durum wheat semolina and chickpea flour. *LWT Food Science and Technology*, 145, 111347. https://doi.org/10.1016/j.lwt.2021.111347.

García-Valle, D., Bello-Pérez, L. A., Flore-Silva, P. C., Agama-Acevedo, E., & Tovar, J. (2021b). Addition of chickpea markedly increases the indigestible carbohydrate content in semolina pasta as eaten. *Journal of The Science of Food and Agriculture*, 101, 2869–2876.

García-Solís, S. E., Bello-Pérez, L. A., Agama-Acevedo, E., & Flores-Silva, P. (2018). Plantain flour: A potential nutraceutical ingredient to increase fiber and reduce starch digestibility of gluten free cookies. *Starch-Starke*, 70, 1–5.

Giuberti, G., & Gallo, A. (2018). Reducing the glycaemic index and increasing the slowly digestible starch content in gluten-free cereal-based foods: A review. *International Journal of Food and Technology*, 53, 50–60.

Giuberti, G., Gallo, A., Cerioli, C., Fortunati, P., & Masoero, F. (2015). Cooking quality and starch digestibility of gluten free pasta using new bean flour. *Food Chemistry*, 175, 43–49.

Giuberti, G., Marti, A., Fortunati, P., & Gallo, A. (2017). Gluten free cookies with resistant starch ingredients from modified waxy rice starches: Nutritional aspects and textural characteristics. *Journal of Food Science*, 76, 157–164.

Gutiérrez, T. J. (2018). Plantain flours as potential raw material for the development of gluten-free functional foods. *Carbohydrate Polymers*, 202, 265–279.

Gutiérrez, T. J., & Tovar, J. (2021). Update of the concept of type 5 resistant starch (RS5): Self-assembled starch V-type complexes. *Trends in Food Science and Technology*, 109, 711–724. https://doi.org/10.1016/j.tifs.2021.01.078.

Hamaker, B. R., & Tuncil, Y. E. (2014). A perspective on the complexity of dietary fiber structures and their potential effect on the gut microbiota. *Journal of Molecular Biology*, 426, 3838–3850.

Imre, B., & Vilaplana, F. (2020). Organocatalytic esterification of corn starches towards enhanced thermal stability and moisture resistance. *Green Chemistry*, 22, 5017–5031.

Juárez-García, E., Agama-Acevedo, E., Sayago-Ayerdi, S., Rodríguez-Ambriz, S., & Bello-Pérez, L. (2006). Composition, digestibility and application in breadmaking of banana flour. *Plant Foods for Human Nutrition*, 61, 131–137.

Korczak. R., & Slavin, J. L. (2020). Definitions, regulations, and new frontiers for dietary fiber and whole grains. *Nutrition Reviews*, 78(S1), 6–12.

Kumar, V., Biswas, A. K., Sahoo, J., Chatli, M. K., & Sivakumar, S. (2011). Quality and storability of chicken nuggets formulated with green banana and soybean hulls flours. *Journal of Food Science and Technology*, 50(6), 1058–1068. https://doi.org/10 .1007/s13197-011-0442-9.

Kurdziel, M., Łabanowska, M., Pietrzyk, S., Sobolewska-Zielińska, J., & Michalec, M. (2019). Changes in the physicochemical properties of barley and oat starches upon the use of environmentally friendly oxidation methods. *Carbohydrate Polymers*, 210, 339–349.

Lee, I. Lee, S., Lee, N., & Ko, S. (2013). Reduced-fat mayonnaise formulated whit gelatinized starch rice and xantan gum. *Cereal Chemistry*, 90, 29–34.

Liu, L., Herald, T. J., Wang, D. M., Wilson, J. D., Bean, S. R. and Ramouni, F. M. (2012). Characterization of sorghum grain and evaluation of sorghum flour in a Chinese egg noodle system. *Journal of Cereal Science*, 55, 31–36.

Liu, X., Lu, K., Yu, J., Coupeland, L., Wang, S., & Wang, S. (2019). Effect of purple yam flour substitution for wheat flour on in vitro starch digestibility of wheat bread. *Food Chemistry*, 289, 118–124.

Lui, X.-L., Mu, T.-H., Sun, H.-N., Zhang, M., & Chen, J.-W. (2016). Influence of potato flour on dough rheological properties and quality of steamed bread. *Journal of Integrative Agriculture*, 15, 1666–2676.

Luo, X., Arcot, J., Gill, T., Louie, J. C. Y., & Rangan, A. (2019) A review of food reformulation of baked products to reduce added sugar intake. *Trends in Food Science and Technology*, 86, 412–425. https://doi.org/10.1016/j.tifs.2019.02.051.

Magallanes-Cruz, P. A., Bello-Pérez, L. A., Agama-Acevedo, E., Tovar, J., & Carmona-Rodríguez R. (2020). Effect of the addition of thermostable and non-thermostable type 2 resistant starch (RS2) in cake batters. *LWT-Food Science and Technology*, 119, 108834.

Mandala, I., & Kapsokefalou, M. (2011). Gluten-free bread: Sensory, physicochemical, and nutritional aspects. In Preedy, V. R., Watson, R. R., & Patel, V. B. (Eds.), *Flour and Bread and Their Fortification in Health and Disease Prevention* (pp. 235–245). London: Academic Press Elsevier.

Maniglia, B. C., Castana, N., Le-Bail, P. Le-Bail, A., & Augusto, P. E. D. (2020). Starch modification through environmentally friendly alternatives: A review. *Critical Reviews in Food Science and Nutrition*, 61(15), 2482–2505. https://doi.org/10.1080/10408398.2020.1778633.

Maniglia, B. C., Castanha, N., Rojas, M. L., & Augusto, P. E. D. (2021). Emerging technologies to enhance starch performance. *Current Opinion in Food Science*, 37, 26–36. https://doi.org/10.1016/j.cofs.2020.09.003.

Martinez, M. M., Oliete, B., Román, L., & Gómez, M. (2014). Influence of addition of extruded flours on rice bread quality. *Journal Food Quality*, 37, 83–94.

Mhaske, P., Wang, Z., Farahnaky, A., Kasapis, S., & Majzoobi, M. (2021). Green and clean modification of cassava starch-effects on composition, structure, properties and digestibility. *Critical Reviews in Foods Science and Nutrition*. https://doi.org/10.1080/10408398.2021.1919050.

Molinari, R., Constantini, L. Timperio, A. M., Lelli, V., Bonafaccia, F., Bonafaccia, G., & Merendino, N. (2018). Tartary buckwheat malt as ingredient of gluten- free cookies. *Journal of Cereal Science*, 80, 37–43.

Ovando-Martínez, M., Sayago-Ayerdi, S., Agama-Acevedo, E., Goñi, I., & Bello-Pérez, L. (2009). Unripe banana flour as an ingredient to increase the undigestible carbohydrates of pasta. *Food Chemistry*, 113, 121–126.

Patiño-Rodríguez, O., Agama-Acevedo, E. Pacheco-Vargas, G., Alvarez-Ramirez, J., & Bello-Pérez, L. A. (2019). Physicochemical, microstructural and digestibility analysis of gluten free spaghetti of whole unripe plantain flour. *Food Chemistry*, 298, 125085.

Pejcz, E., Czaj, A., Wojciechowics-Budzisz, A., Gil, Z., & Spychaj, R. (2017). The potential of naked barley sourdough to improve the quality and dietary fibre content of barley enriched wheat bread. *Journal of Cereal Science*, 77, 97–101.

Perez-Quirce, S., Caballero, P. A., Vela, A. J., Villanueva, M., & Ronda, F. (2018). Impact of yeast and fungi (1®3)(1®6)-b-glucan concentrates on viscoelastic behavior and bread making performance of gluten-free rice-based doughs. *Food Hydrocolloids*, 79, 382–390.

Ragavan, K. V., Hernandez-Hernandez, O., Martinez, M. M., & Gutiérrez, T. J. (2021). Organocatalytic esterification of polysaccharides for food applications: A review. *Trends in Food Science & Technology*, 119, 45–56.

Raigond, P., Ezekiel, R., & Raigond, B. (2015). Resistant starch in food: A review. *Journal of the Science of Food and Agriculture*, 95, 1968–1978.

Rodríguez-Damián, A. R., De La Rosa-Millán, J., Agama-Acevedo, E., Osorio-Diaz, P., & Bello-Pérez, L. A. (2013). Effect of different thermal treatments and storage on starch digestibility and physicochemical characteristics of unripe banana flour. *Journal of Food Processing and Preservation*, 37, 987–998.

Román, L. Santos, I., Martínez, M. M., & Gómez, M. (2015a). Effect of extruded wheat flour as a fat replacer on batter characteristics and cake quality. *Journal of Food Science and Technology*, 52, 8188–8195.

Román, L. Martinez, M. M., & Gómez, M. (2015b). Assessing of the potential of extruded flour paste as fat replacer in O/W emulsion: A rheological and microstructural study. *Food Research International*, 74, 72–79.

Román, L., Pico, J., Antolín, B., Martínez, M. M., & Gómez, M. (2018a). Extruded flour improves batter pick-up, coating crispness and aroma profiles. *Food Chemistry*, 260, 106–114.

Román, L., Reguilón, M. P., & Gómez, M. (2018b). Physicochemical characteristics of sauce model systems: Influence of particle size and extruded flour source. *Journal of Food Engineering*, 219, 93–100.

Sánchez-Rivera, M.M., Nuñez-Santiago, M.C., Bello-Pérez, L.A., Agama-Acevedo, E., & Alvarez-Ramírez, J. (2017). Citric acid esterification of unripe plantain flour: Physicochemical properties and starch digestibility. *Starch/Starke*, 69, 1700019.

Sánchez-Rivera, M. M., Bello-Pérez, L. A., Tovar, J., Martínez, M., & Agama-Acevedo, E. (2019). Esterified plantain flour for the production of cookies rich in digestible carbohydrates. *Food Chemistry*, 292, 1–5.

Santos, F. G., Aguilar, E. V., Rosell, C. M., & Capriles, V.D. (2021). Potential of chickpea and psyllium in gluten-free breadmaking: Assessing bread's quality, sensory acceptability, and glycemic and satiety indexes. *Food Hydrocolloids*, 113, 106487.

Sanz-Penella, J. M., Wronkowska, M., Soral-Smietana, M., & Haros, M. (2013). Effect of whole amaranth flour on bread properties and nutritive value. *LWT. Food Science and Technology*, 50, 679–685.

Saura-Calixto, F., García-Alonso, A., Goñi, I., & Bravo, L. (2000). In vitro determination of the indigestible fraction in foods: An alternative to dietary fiber analysis. *Journal of Agricultural and Food Chemistry*, 48, 3342–3347.

Shahidi, F., & Pan, Y. (2021). Influence of food matrix and food processing on the chemical interaction and bioaccessibility of dietary phytochemicals: A review. *Critical*

Reviews in Food Science and Nutrition. https://doi.org/10.1080/10408398.2021
.1901650.

Soong, Y. Y., Tan, S. P. Leong, L. P., & Henry, J. K. (2014). Total antioxidant capacity and starch digestibility of muffins baked with rice, wheat, oat, corn and barley flour. *Food Chemistry,* 164, 462–469.

Thiranusornkij, L., Thamnarathip, P., Chandrachai, A., Kuakpetoon, D., & Adisakwattana, S. (2019). Comparative studies on physicochemical properties, starch hydrolysis, predicts glycemic index of Hom Mali rice and Riceberry rice flour and their applications in bread. *Food Chemistry,* 283, 224–231.

Utrilla-Coello, R. G., Agama-Acevedo, E., Osorio-Diaz, P., Tovar, J., & Bello-Pérez, L. A. (2011). Composition and starch digestibility of whole grain bars containing maize or unripe banana flours. *Starch/Starke,* 63(7), 416–423.

Von Loesecke, H. V. (1950). *Bananas.* New York: Interscience Publisher Inc.

Yousif, A., Nhepera, D., & Johnson, S. (2012). Influence of sorghum flour addition on flat bread in vitro starch digestibility, antioxidant capacity and consumer acceptability. *Food Chemistry,* 134, 880–887.

Chapter 9

Use of Modified Starchy Flours

Pablo Martín Palavecino, Pedro Losano Richard,
Mariela Cecilia Bustos, Alberto Edel León
and Pablo Daniel Ribotta

CONTENTS

DOI: 10.1201/9781003088929-9

9.1 INTRODUCTION

Research has shown that major cereals such as wheat, rice and maize are the most conventional sources to produce starchy flour and its products. In this sense, diversity of food supply is one of the factors involved in food security and hardly accomplished, as most food calories came from a few sources. In addition, major cereals integrate fewer satiating diets than those based on less conventional crops; they are also not suitable for unfavorable environments. Some alternative crops with particularly high nutritional attributes form part of healthier food production. Despite their advantages, these crops pose processing problems leading to poor-quality products and low consumer acceptance (Taylor & Belton, 2002; Choquechambi et al., 2019; Rosell, Cortez, & Repo-Carrasco, 2009). Therefore, in this chapter, we will center not on the well-known "big three" cereals, but rather on flour from pulses, minor cereals, Andean grains, tubers, and roots, and their modification to be integrated into starchy food products. The widespread use of starchy flour in food is well known both at home and at an industrial level. Yet, this raw material needs to be modified in order to improve its properties and increase possible applications. Accordingly, in this chapter we will focus on starchy food products that require these modified raw materials, like bread, pasta, batter-type products, and extruded snacks.

Grain processing includes primary treatments, such as dehulling and milling, to obtain flour or grits, and a secondary process to modify its functionality, such as fine grinding, extrusion and heat-moisture treatments. In addition to applying these physical processes, also performed are flour fermentation and grain malting, and some chemical modifications. The change in flours' physicochemical and functional properties (including water absorption, pasting properties and thickening power) altered their adequacy to the different elaborations (Shobana et al., 2013; Sun et al., 2018).

The use of flour instead of isolated starch relies on the differences found in nutrition and functionality. A starchy flour contains fibers, proteins, minerals and health-promoting compounds that were separated during starch isolation. This process often involves setting complex operational parameters and using large amounts of water that has to be removed to reach a low moisture product. In addition, those components present in starchy flour interact with the food matrix and determine its texture and sensory attributes.

Thus, and as introduced earlier, we discuss here the use of less conventional modified starchy flour. This chapter has been divided into four parts, one for each group of raw material. These are, in turn, subdivided into food products, starting from the most usual applications: starchy food like baked goods and pasta. We then describe the novel uses found in the different types of food. Finally, we address health issues and identify areas for further research. Then, we describe the current uses of unconventional starchy flour, encouraging the reader to develop new ones.

9.2 UNCONVENTIONAL CEREALS

In recent years, there has been an increasing interest in raw materials with high content of health-promoting compounds. Here, we selected sorghum, buck-wheat and millet due to their ability to take full advantage of nutrients in poor cultivation systems. They also account for underused food resources in most developed countries, showing considerable potential to be easily adopted by consumers and thus play an increasingly prominent role in new food products.

Sorghum (*Sorghum bicolor* L. Moench) is the main energy source in several regions of Africa and India, and mostly consumed daily as porridge, flatbread, grit or simply boiled. Its starch is slowly digested compared with other cereals, resulting in lower postprandial blood glucose in humans. One major drawback of this cereal is that cooking turns its proteins less digestible (Cisse et al., 2018; Galán et al., 2020). Sorghum and millets account for less than 5% of world grain production, but they are vital in low-humidity zones. Millet constitutes a staple food for millions of low-income people, being key for their food security. Among millet species, pearl millet (*Pennisetum typhoides*) is notable in production and consumption. Both sorghum and millet flours are relevant to the human diet due to their high fiber and antioxidant content (mostly polyphenols). Procedures to obtain sorghum flour include dry milling preceded by several cleaning and dehulling steps, resulting in flour with different degrees of extraction (separation of outer layers from endosperm). Because of their small size, finger millet (*Eleusine coracana*) and other similar species are milled without previous dehulling, contrary to the production of pearl millet flour where decortication is a common practice (Obilana & Manyasa, 2002).

Buckwheat is grain rich in protein with a well-balanced amino acid composition; high levels of dietary fiber; vitamins; antioxidants; and relatively high copper, zinc and manganese content. The most extensively cultivated buckwheat varieties include common buckwheat (*Fagopyrum esculentum*) and tartary buckwheat (*Fagopyrum tataricum*) (Sun et al., 2018). Flour could be achieved by dry milling followed by sieving to separate the hull, but often the grains are hydrothermally processed to enhance dehulling and milling. In both cases, grain shape and structure lead to incomplete hull removal resulting in darker flour compared with wheat flour. Hydrothermal treatment promotes hull removal and consists of roasting and/or steaming the grains previously tempered to 20% humidity and inducing starch gelatinization and protein denaturation (Biacs et al., 2002).

9.2.1 Bread

Minor cereals alone lack the functional proteins for bread making. Hence, strategies to successfully develop bread comprise the addition of wheat flour, vital wheat gluten, modified starch, modified flour or additives like gum. The incorporation of modified starchy flour into bread formulations is a possible solution to fulfill the functionalities of gluten, especially in gluten-free products, along with nutritional improvement. Partial wheat replacement for millet, sorghum or buckwheat flour increases dietary fiber and antioxidant content, thus reducing caloric intake.

Flour from fermented or malted grains improves not only its nutritional contribution but also the technological properties of the product. The shelf-life extension of baked goods can be achieved by using additives such as chemically modified carboxy methyl starch, and malting or fermentation processes that allow elaborating clean-label food products (Dahir et al., 2015; Wang et al., 2019).

Cooked and fermented pearl millet flour drastically reduces its antinutrients (phytic acid); up to 20% substitution in wheat bread slightly decreases specific loaf volume and almost doubles its hardness (Ranasalva & Visvanathan, 2014). On the other hand, malted finger millet flour, also replacing 20% of wheat, resulted in bread with improved sensory and texture attributes with intact bread loaf but lower volume than that of the control due to protein matrix disruption (Bhol & John Don Bosco, 2014).

Sourdough fermentation by lactic acid bacteria produces a high level of dextrans (α-glucans polysaccharides). In bread from pearl millet wheat flour, equal parts enhance its technological and nutritional properties; increase bread specific volume; and decrease crumb firmness, moisture loss and staling rate; while improving in vitro protein digestibility and decreasing estimated glycemic index (Wang et al., 2019).

Bread can also be produced only with millet, sorghum or buckwheat. Yet, due to its low volume, most of these products are flatbreads with dark and nonelastic bread crumb (Anglani, 1998; Collar, 2007).

Tortilla is a nonfermented flatbread initially consumed in Mexico but has now spread all over the world. It is made of corn grains cooked in calcium hydroxide (nixtamalization). Sorghum is also used alone or mixed with maize, showing similar dough rheological properties and acceptable quality to those of tortillas: color, flavor and texture (Ratnavathi & Patil, 2014; Anglani, 1998). The nixtamalization process modifies starch properties, increases mineral content and reduces insoluble fiber while increasing soluble fiber. However, it notably decreases (around 50%) the digestibility of proteins. On the other hand, the hot-press method used with untreated sorghum flour generally results in lower quality products, as in wheat tortilla with addition of sorghum flour and sorghum gluten-free tortilla (Winger et al., 2014; Torres et al., 1993).

The enzymatic modification of flour could improve bread characteristics. The transglutaminase treatment of buckwheat flour improves its bread-making potential through the polymerization of protein into a high-molecular-weight complex. It enhances bread specific volume thanks to higher amounts of crumb pores, which are also better structured and distributed (Renzetti et al., 2008; Sadowska & Diowksz, 2020).

Sorghum bread could be made with up to 10% of high pressure–treated sorghum batter. Extreme conditions (>300 MPa) promote starch gelatinization with resulting increased viscosity and low replacement (2%), yielding good quality bread with high specific volume and delayed staling (Vallons et al., 2010). Wheat replacement with sorghum and millet high pressure–treated batter (300 MPa 10 min) increases bread antiradical activity but increases staling kinetics (Angioloni & Collar, 2012).

9.2.2 Cookies

Cookies are one of the most commonly consumed snacks worldwide, generally prepared from soft wheat flour. Their high levels of sugar and fat prevents gluten development and allows wheat replacement without significantly altering their texture. Despite this, partial substitution (25%–50%) with non-wheat cereals like millet, sorghum and buckwheat is the most common formulation approach in cookies. This does not affect product quality and process, reducing costs and improving nutritional profiles. The resulting cookies are generally darker, hard, gritty and mealy, with a flavor different to that of wheat counterparts. However, grittiness could be reduced through an increase in the pH of the cookie dough (Anglani, 1998; Ranasalva & Visvanathan, 2014; Biacs et al., 2002).

Milling and classification also play an important role in cookies, as they determine particle size and starch damage, affecting, in turn, water absorption and dough rheology (Gómez & Martínez, 2016). Flour fractions with larger particles showed higher oil absorption and lower water absorption, probably due to greater protein and less damaged starch content. Consequently, biscuits made from these coarser fractions were less hard and more accepted by consumers (Dayakar Rao et al., 2016).

9.2.3 Batters

The preparation of cakes and muffins is similar in all types of grain flour, as its structure is mainly determined by egg proteins and gluten development should be avoided. On the other hand, starch plays a major role in batter preparation since it governs the mixture rheology during preparation and helps to set the structure throughout the cooking process.

Treatments like ozonation (oxidation) enhance the starch swelling capacity of sorghum and increase batter viscosity, which prevents cake collapse during cooling, yielding products with more cells and low crumb firmness (Marston, Khouryieh, & Aramouni, 2015).

Proso millet flour treated with heat-moisture produces low specific gravity and high consistency batter, resulting in high volume and low hardness cakes (Fathi et al., 2016). Thermomechanically treated flour could also help to reduce the shortening content, as in the case of steam jet–cooked buckwheat flour showing similar specific gravity and textural properties to those of the control (Min et al. 2010).

In summary, flour modification that considerably damages starch enhances both batter and final product properties. Likewise, the use of non-wheat flour like sorghum to produce muffins has shown lower glucose and insulin responses after consumption, highlighting its functional properties (Poquette, Gu, & Lee, 2014).

Fine particle size flour is required to obtain good quality batter products with low hardness, high specific volume and pleasant mouthfeel. Then, coarsest fractions should be removed and/or reground (Gómez & Martínez, 2016; Ratnavathi & Patil, 2014).

9.2.4 Pasta and Noodles

Pasta and noodles make up the staple food of many cultures and countries, due to their long shelf-life, low cost, easy preparation and versatility. Both products are usually made from wheat and their characteristic textures come from the gluten matrix. However, there is non-wheat alternatives like starch noodles and gluten-free pasta where gluten functionality was successfully fulfilled by starch, gums, egg proteins and pulse flour (Thilagavathi et al., 2020; Yadav et al., 2012). Indeed, buckwheat noodles are popular in Italy, Japan and China (Li & Zhang, 2001).

Good quality pasta and noodles can be obtained from buckwheat, sorghum and millet using composite wheat flour. This approach brought about cheaper and healthier products such as dietary fibers, antioxidants, minerals and increased vitamins (Ratnavathi & Patil, 2014; Yadav et al., 2012). Particularly, an increase in fiber intake decreases the glycemic index of pasta by dropping blood sugar levels (Ilamaran et al., 2017).

The major part of wheat dry pasta and noodles is produced through extrusion process at low temperatures, generally below 45°C, to prevent protein (gluten) denaturation. Millet–wheat mix was cold-extruded resulting in good quality pasta (Jalgaonkar et al., 2019). When the structural role of the products is played by starch, more drastic extrusion conditions are required (extrusion cooking) in order to gelatinize starch. This is a common process to produce starch noodles, gluten-free pasta, and pregelatinized starch and flour.

The treated starchy material could serve as an ingredient for non-wheat pasta, yielding high quality products (Palavecino et al., 2017).

Fermentation and sprouting were tested as alternative treatments for sorghum flour to enhance pasta properties. Marengo et al. (2015) demonstrated that it was not possible to produce rice-based pasta with sprouted flour, although fermentation improved its cooking properties. Microwave application on dough produced partial gelatinization of starch, which led to good quality 100% sorghum noodles (Suhendro et al., 2000), and composite sorghum, millet and wheat pasta (Kamble et al., 2019). Depigmentation of pearl millet grains (acid treatment) not only improves color sensory attributes, it also increases in vitro starch and protein digestibility, although it decreases dietary fiber content (Rathi, Kawatra, & Sehgal, 2004).

Dehulled buckwheat was successfully used to replace up to 60% of wheat in fresh laminated pasta and to produce gluten-free pasta mixed with precooked rice. The drying process needs to be controlled as it could produce browning of the products due to Maillard reaction (high lysine and reducing sugar content) (Alamprese, Casiraghi, & Pagani, 2007).

Buckwheat is the only cereal that contains rutin, a polyphenolic compound with health benefits such as antioxidant and anti-inflammatory activity. However, rutin in flour is relatively low and rapidly hydrolyzed, therefore, it must be concentrated to meet the criteria for the claims associated with its intake. Rutin was successfully extracted from tartary buckwheat milling fractions using ethanol and ultrasonic methods, and added to the formulation of fried wheat noodles to reach its recommended daily dose (Cho & Lee, 2015). Flour hydrothermal treatments (steaming and autoclaving) were useful to preserve rutin and improve its technological properties for wheat noodle application, where rutin content remained high after cooking (Yoo et al., 2012). Alternatively, autoclaving the grains cause the migration of rutin from the bran to the endosperm, reaching a high level of this compound into dehulled flour (Oh et al., 2019). Once hydrothermal treatment was applied to buckwheat, rutin was embedded in the flour matrix and not transformed into quercetin (Lukšič et al., 2016).

Pregelatinized buckwheat flour, produced from roasting, steaming, extrusion, boiling or microwaving, showed potential for elaborating high quality and functional noodles with improved textural properties (Sun et al. 2018). The use of ozone in semidried buckwheat noodles promoted protein cross-linking by disulfide bond formation, thus enhancing cooking and textural attributes while reducing microbial charge (Bai & Zhou, 2021).

9.2.5 Extruded Products

The extrusion technique involves different processes such as variation in temperature and pressure, enabling the formulation of a wide range of food products. Particularly, extrusion-cooking processes bring about major physical and chemical changes into raw materials, promoting starch gelatinization, protein denaturation, enzyme inactivation and microorganism destruction (Collar, 2007). These complex transformations occur almost simultaneously in one (apparently) simple device: the extruder, which is capable of producing food ranging from crispy snacks to tender vegan meat. Most commercially available extruded products derive from major cereals. Sorghum, millet and buckwheat exhibit similar composition, thus analogue food stuffs could be developed (Ratnavathi & Patil, 2014).

Red and white sorghum flour from extruded whole grain has shown to provide fiber and phenolic compounds with colonic antioxidant effects. However, extrusion of refined sorghum flour kept most phenolic compounds, having potential beneficial health effects (Galán et al., 2020; Llopart et al., 2017). Extruded products from whole-grain flour are darker than its refined flour counterparts; they also exhibit a low expansion ratio and high enzymatic activity and lipid oxidation, nevertheless these products contain health-promoting ingredients (Kharat et al., 2019). Expanded sorghum crisps produced from white and red sorghum flour required similar extrusion conditions, leading to promising products that allow introducing sorghum into this market (Pezzali et al., 2020).

Pearl millet grits were fortified and extruded with up to 7.5% whey protein concentrate, yielding acceptable texture snacks from addition of the dairy by-product (Yadav et al., 2014). To improve the expansion ratio and reduce product harness, kodo millet was mixed with chickpea flour (70:30) (Geetha, Mishra, & Srivastav, 2014). Comparing the three major millets (finger, foxtail and pearl) to produce direct expanded extruded products, foxtail showed the highest expansion with lower specific mechanical energy input (Kharat et al., 2019).

Delicious and crispy extruded products could be obtained from buckwheat (Biacs et al., 2002). Extrusion of buckwheat dehulled groats increased phenolic acid content, but decreased, among others, total polyphenols and tocopherols, resulting in a slight decrease of antioxidant activity (Zieliński et al., 2006).

9.3 ROOTS AND TUBERS

Several plants store high amounts of edible carbohydrates in underground organs like roots and tubers, but also contain proteins, vitamins and minerals. Therefore, they constitute a valuable resource that should be preserved and popularized due to their nutritional and functional properties. The major part of the world production of amylaceous roots and tubers is found in potato (*Solanum tuberosum*), sweet potato (*Ipomoea batatas*), cassava (*Manihot esculenta*), yam (*Dioscorea* spp.) and taro (*Colocasia esculenta* and *Xanthosoma sagittifolium*); yet there are thousands of other underexploited species. High water content shortens the shelf life of roots and tubers, hence, enhancing flour production, starch isolation and improving its functional properties are vital to these crops economy (Pérez, 2007). Flour is usually made from cleaning, slicing, drying and milling process.

Tubers are largely produced in tropical and subtropical regions, while roots commonly adapt to marginal conditions, consequently, they are geographically widespread. Yet, they both represent a staple food in numerous countries. Their starch granules are easily disrupted when cooked in water and form a clear paste with low resistance to heat and shear. However, their flour could be rich in minerals, β-carotene, antioxidants (anthocyanins, phenolic compounds and ascorbic acid) and dietary fiber (Wongsagonsup et al., 2014; Singla et al., 2020).

9.3.1 Bread

Potato bread is highly popular around the world, particularly in fast-food chains, mixing wheat flour with steamed and mashed root or dehydrated flakes. This practice of partially replacing wheat flour with pregelatinized starchy flour was replicated in similar products with dissimilar results. Sorghum-based bread made from native cassava starch showed better crumb properties than those containing pregelatinized starch due to the rheological properties provided in batters (Onyango et al., 2011). Sweet potato paste made from shredded, steamed and blended tubers was incorporated to wheat bread, showing a reduce in loaf volume and increased staling rate when tuber addition where higher than 20%, highlighting its great aptitude to toast making (Wu, Sung, & Yang, 2009). Batter bread made from sorghum, amaranth and heat-moisture treated cassava starch (40:10:50 ratio) showed a texture softer than those containing native starch (Onyango, 2013). Wheat bread with 20% of citric acid and replacement of heat-moisture treated sweet potato starch had lower specific volume and more hardness; however, it showed similar sensory attributes than those of the control bread. In addition, modified starch bread exhibited significantly high resistant starch (RS) (34.2%); its consumption brings low blood glucose response in mice and low in vitro glycemic index (Tien et al., 2018).

Gluten-free bread was made using taro flour. To develop a light bread crumb structure, partially pregelatinized taro flour, enzymes, hydrocolloids and potato blends were assessed. This formulation, including thermal-treated flour, did not significantly improve bread volume and texture, different to what was produced using endoproteases. Most of these products showed a better starch digestibility pattern (higher slowly digestible starch and RS) than those reported in gluten-free (GF) bread (Calle, Benavent-Gil, & Rosell, 2020). A mix of fermented cassava–sweet potato–sorghum flour was used with xanthan gum in GF bread production. Considering dough rheological behavior and bread textural properties, the optimal formulation contained 75% fermented cassava, 20% sweet potato flour and 5% sorghum (Monthe et al., 2019).

Chemically modified starch and its addition to bread have been extensively investigated as the characteristics of gelatinization and retrogradation are central for bread quality (Miyazaki et al., 2006). From several esterified cassava starches, only highly hydroxypropylated starch significantly retarded staling in frozen-dough wheat bread, but exhibited unpleasant mouthfeel (Miyazaki, Maeda, & Morita, 2008). The specific volume of wheat bread was improved using acetylated and oxidized cassava starch, compared to that of native and acid-thinned blended samples, although bread from native starch was the most acceptable one (Okereke, Iwe, & Agiriga, 2015).

9.3.2 Batters

Modified starch as a fat replacer to lower lipid content in muffins was achieved with cross-linked cassava starch (Rodriguez-Sandoval, Prasca-Sierra, & Hernandez, 2017). An 8% replacement in muffins showed no significant differences with the control in specific volume, color, crumb moisture and consumer acceptability; yet, it exhibited greater crumb firmness during storage. Taro flour modified by oxidation and irradiation improved cake quality, showing higher specific volume, lower hardness and chewiness, and sensory properties than those in unmodified flour (Ekafitri, Pranoto, & Herminiati, 2019). Cakes made with purple-fleshed sweet potato flour exhibited better antioxidant properties and organoleptic values with blanching pretreatment (Chuango, Julianti, & Ginting, 2019), and presented better color, aroma and volume when treated through dipping (in 0.5% citric acid) and steam blanching (Hutasoit, Julianti, & Lubis, 2018).

9.3.3 Pasta and Noodles

Particularly in GF pasta and starch noodles, starch plays an important role to guarantee product quality. Then, tuber or root starchy flour is a potential raw material. However, to obtain high quality products, they need to be treated before or during the elaboration process, and a protein source should be added (Palavecino et al., 2020; Obadi & Xu, 2021). For instance, physically pregelatinized cassava flour (70% starch and 30% bagasse) mixed with native cassava starch and amaranth flour yielded high protein and fiber GF pasta (Figure. 9.1) with a texture comparable to that of the wheat semolina sample (Fiorda et al., 2013). Chemically treated (sodium hydroxide) sweet potato flour was fortified with soy, although a decrease was found in the pasta cooking properties,

Figure 9.1 Vermicelli-type pastas elaborated with cassava pregelatinized flour, cassava starch and amaranth flour. (Reprinted with permission from Fiorda et al., 2013.)

firmness and β-carotene content (Limroongreungrat & Huang, 2007). When potato and sweet potato acetylated starch partially replaced wheat flour (up to 20%) in white salted noodle, cooking loss decreased and texture parameters improved (Chen, Schols, & Voragen, 2003). Applied to taro starch, the same modification in starch noodles increased swelling power, resistant starch, and amylose, and lowered oxalate content (Sisilia et al., 2018).

9.3.4 Dry Applications

The efforts made during World War II to extend vegetable life through dehydration led to the first mass-produced modified starch for human consumption: the instant mashed potato. Despite the process, this product consisted mainly of dry pregelatinized potato starch that could be rapidly rehydrated with hot water to produce potato puree. This is not the only product from dry modified starchy flour. Starch cross-linking allows weak granules to resist harsh processing conditions like high shear, intense heating or low pH without losing their thickening properties. The low cross-linking level minimizes granule disruption but provides similar peak viscosity to the nontreated sample and small breakdown, in other words, imparted constant viscosity. Thus, cross-linked tapioca starch could achieve high quality soup in terms of textural and sensory attributes (Wongsagonsup et al., 2014). Cleaned, grinded and cooked sweet potato (puree) could be dehydrated through a drum or spray-drying process to extend its preservation and reduce its size. Chemical and extrusion-cooking treatments were also applied to produce sweet potato powders with specific textural and antioxidant properties (Truong & Avula, 2010).

9.3.5 Cookies and Crackers

Crackers were produced from 70% wheat flour and 30% water yam flour modified through autoclaving–cooling cycles. The replacement of wheat flour increased resistant starch (up to 3.5%) and dietary fiber (up to 18%) and significantly decreased the glycemic index (to 34) (Rosida & Susiloningsih, 2020).

Modified cassava flour fermented with lactic acid bacteria was applied to GF cookies using different types of flour with up to 19-month shelf life in aluminum foil packages (Kurniadi et al., 2019) and with yellow pumpkin, showing similar sensory results but higher fiber content than those found in wheat counterparts (Mustika & Kartika, 2020).

9.4 PULSES

Pulses, also known worldwide as legumes, are a cheap source of proteins and fiber. Pulse plants belong to the Leguminosae family. They have root nodules formed by a symbiotic relationship between the plant and a soil-borne rhizobia bacteria that absorbs inert nitrogen from air, improving soil nutrition in dry areas, reducing the use of fuel-based chemical nitrogen and, consequently, mitigating global warming (Rawal & Navarro, 2019). Furthermore, due to low moisture and presence of seed tegument, legume grains are an excellent nonperishable protein source for areas with limited storage technology. This section will deal with the starchy flour of beans, chickpeas, lentils and peas, and how their modification influences certain food products, including GF food.

In recent years there has been a growing trend toward reducing the use of animal protein and additives in food formulations. Legumes are an excellent alternative source of highly digestible protein. Modified legume starchy flour could replace additives; thus, the industry will be able to simplify the labeling of its products. Most countries of South Asia consume legumes as a staple food, being the major consumers worldwide. However, countries without a consumption culture of legumes began to consume them as substitutes for cereals in many products (Monnet et al., 2019).

The starchy legume flour is rich in SDS, which generates low postprandial glucose and insulin response inducing prolonged satiety. This effect results from the high content of amylose and the interaction of starch with fibers. In legumes, starch granules are embedded in a protein matrix, surrounded by a cell membrane and a thick and mechanically strong outer cell wall. The unique structure of legume cotyledon cells provides effective encapsulation for starch against the extracellular environment. The same barrier restricts the activity of amylase, which takes longer to digest starch. In addition, the higher amylose content of pulses presents more amorphous regions, lowering the gelatinization temperature. The high fiber content has a positive impact on human health but a negative effect on food technology. Generally, legume-based products are precooked during production; this stage changes the starch structure and reduces antinutritional components. During gelatinization the strong cell wall acts as a barrier to impose restrictions on water availability, heat transfer and intracellular space, which are required for the swelling and gelatinization of the internal starch (Do et al., 2019).

The modification of starchy legume flour in the food industry is performed to improve its technological and nutritional quality. To expand technological suitability, heating methods such as steaming, microwaving, boiling, extrusion, infrared and roasting led to a complete or partial gelatinization of starch molecules, improving foaming, emulsifying, thickening and water retention capacity, and enhancing the quality of legume flour in food products. On the other hand, reduction of nutritional negative issues could be placed into two groups: (1) thermolabile, such as protease inhibitors, lectins and low protein digestibility, which could be reduced through thermal treatments; and (2) thermostable, like raffinose, phytic acid, saponins and tannins, which could be decreased by methods like dehulling, air classification, germination, extrusion and fermentation (Sun et al., 2020; De Pasquale et al., 2020).

9.4.1 Baked Goods

In general, higher amylose content in legume starch produces a higher pasting temperature, absence of peak viscosity and increased viscosity during the holding temperature period. In this sense, the development of low viscosity has been correlated with a softer crumb (Foschia et al., 2017). Legume protein has excellent water-binding, gel-forming and emulsifying capabilities, providing structure when gluten is replaced, whereas legume fiber can help reduce breakage and enhance dough yield in place of gums (Nachay, 2017). Compared to wheat flour, chickpea and pea flour change the color of cakes (brighter and yellowish), and lentil flour increases the specific volume of this product. Despite this, in many cases, raw legume flour increases the hardness and chewiness of baked goods (Gularte, Gómez, & Rosell, 2012); therefore, flour should be treated and additives should be used to improve the quality of baked goods.

Thermal treatments, like pregelatinization of legume grains before milling, affect the nutritional properties of flour, increasing protein digestibility and formation of resistant starch while decreasing trypsin inhibitors. Yet, this process negatively affects the antioxidant activity, while physical properties like water-binding capacity increase drastically (De Pasquale et al., 2020). This characteristic turns pregelatinized legume flour into a potential ingredient to substitute wheat in baked goods.

Pretreatments of legume flour, like those spontaneous or assisted with lactic acid bacteria fermentation, are nonthermal ones employed to reduce antinutritional factors that may limit the bioavailability of several nutrients. Contrary to thermal treatments, fermentation increased free amino acid content improving protein digestibility, degraded raffinose and condensed tannins by increasing the bioavailability of fibers and decreased the starch hydrolysis index (Sáez et al., 2018; De Pasquale et al., 2020; Rizzello et al., 2014).

9.4.2 Pasta

Legume pasta improves the nutrition of low-income people by increasing the amount of protein and fiber in their diet. Legume flour in pasta increased the resistant starch content reducing the in vitro glycemic index (Giuberti et al., 2015). Normally, legume pasta contains additives to replace gluten or modified flour.

Most treatments to improve the quality of legume flour for pasta involve starch gelatinization. Extrusion cooking during pasta production exhibited greater firmness, flavor and texture after cooking; yet, an excess of temperature during extrusion reduced the quality of the pasta. On the other hand, gluten-free pasta industries use pretreated flour where starch is gelatinized before pasta production. During these thermal modifications of legume flour, antinutritional factors were reduced (Marti & Pagani, 2013).

9.4.3 Extruded Products

Pulse extrudates have an excellent capacity to bind water and emulsify due to their high-protein content; they are chosen as animal meat analogues. Nutrient-dense, extruded multi-legume bars, mixed with whey protein concentrate, honey and palm oil, have been reported to mitigate malnutrition in developing countries (Pasqualone et al., 2020).

The fibrillary structure similar to that of animal meat is given at the end of the extrusion process on the fibrating die where the protein-based mixture exhibits a phase transition from liquid to gel, allowing the formation of melted fiber while it flows through the die (Bouvier & Campanella 2014).

Extruded legume-based snacks use the thermomechanical energy of extrusion to expand the volume of the melt and reduce the antinutritional factors of legumes, like trypsin inhibitors, lectins, phytic acid and tannins. There is a strong influence of fiber, protein and lipids on the expansion performance. Fiber surrounds air bubbles and reduces maximum expansion, while protein and lipids affect starch swelling during gelatinization (Pasqualone et al., 2020). Then, the dehulling process and milling conditions are used to improve the quality of expansion.

9.4.4 Other Products

Wet isolation of pulse starch is an excellent source to replace expensive additives, however, this process requires large quantities of water. The use of different dry mills and air classification reduces the water needed to isolate the starch of peas and lentils, even though it is not as efficient as wet isolation. Starch extraction is particularly difficult in pulses due to the presence of proteins and lipids, and to smaller starch granules compared to tuber and root starch (Kringel et al., 2020).

After harvesting, legumes are classified according to the suitability of the grain to different products. By-products of grain classification represent a cheap source of potentially functional ingredients. The production of hummus with discarded grains of chickpea proved an excellent alternative due to complex carbohydrate composition of grains (Losano Richard et al., 2020).

9.5 ANDEAN CROPS

Over the last decades, Andean crops like quinoa, amaranth and cañahua have grown in popularity due to their health benefits and crop adaptability to different agroecosystems (Hirich, Allah, & Ragab, 2020). Hence, they can be cultivated in poor soils and at high altitudes (Rosell, Cortez, & Repo-Carrasco, 2009). Their seeds are GF and have excellent nutrient profiles: generally higher in protein, fat, ash and fiber than cereals. In addition, they show a considerable quantity of phenolic compounds with antioxidant activity (caffeic acid, ferulic acid and vanillic acid), as well as a high proportion of vitamins (ascorbic acid, thiamin and α-tocoferol) and minerals (calcium, magnesium, iron and zinc) (Perez-Rea & Antezana-Gomez, 2018; Repo-Carrasco-Valencia & Serna, 2011).

Quinoa (*Chenopodium quinoa* Willd.) is a highly nutritious food product with an outstanding protein quality with small seeds (1.36–2.66 mm in diameter) that can be found in different colors, including yellow, orange, red, pink, purple, white and black. Quinoa seed needs to be washed before consumption due to the bitter-tasting saponins in the seed coat (Jacobsen, 2003). Quinoa is usually consumed in soup, prepared like rice using lightly roasted seed, or ground into flour to produce baked products (Perez-Rea & Antezana-Gomez, 2018). *Amaranthus* spp. (Amaranthaceae family), commonly known as amaranth, consists of 60 species, a plant whose leaves are eaten as vegetables and seeds are consumed as cereal (Coelho et al., 2018). Amaranth is well known for its high nutritional quality, despite having some antinutrient compounds (e.g., phytic acid, oxalates and tannins) (Rosell, Cortez, & Repo-Carrasco 2009). Seeds are small and lenticular (1–1.5 mm in diameter) and can be cooked, popped, toasted or grinded to be consumed in preparations or in suspensions with water or milk (Rastogi & Shukla, 2013). Their most common use involves grinding their seed into flour for the development of flour-based products. Cañahua (*Chenopodium pallidicaule* Aellen) is probably the strongest Andean crop due to its resistance to frost, drought, salt and pests (Pérez, Steffolani, & León, 2016). It is also the one most poorly studied between the three selected in this section. Its seeds (0.7 mm diameter), half the size of quinoa grains, vary from yellow, orange and brown to black (Rodríguez et al., 2017). Cañahua seeds are consumed mainly as kañiwako, which are toasted and milled grains, an aromatic product eaten with sugar, milk and water. Cañahua can also be expanded and toasted and included in snacks, or simply milled into flour to be added in the elaboration of cereal-based products (Repo-Carrasco, Espinoza, & Jacobsen, 2003).

Several authors have pointed out that it is especially difficult to produce bread or pasta with an aerated crumb structure or al dente texture using only Andean crop flour, quinoa being the one showing better results (Bustos et al., 2019; Schoenlechner et al., 2010; Rosell, Cortez, & Repo-Carrasco, 2009). As a result, partial substitution or inclusion of modified Andean flour constitutes the two most viable options to improve the technological and nutritional value of bread and pasta. In this sense, soaking, malting, hydrothermal processing, sprouting, popping and lactic acid fermentation of Andean crop flour have been identified as upcoming trends in the market, mainly to improve mineral accessibility (Castro-Alba et al., 2019; Rollán, Gerez, & Leblanc, 2019; Aguilar et al., 2019).

9.5.1 Bread

The importance of exploiting Andean grains for the elaboration of healthy and nutritious food has forced food producers to develop novel and adequate strategies for their processing. The first challenge met has been the degradation of phytate content, key to decrease its negative effect on mineral bioavailability. To reduce its content, endogenous phytase activity of raw crops can be increased using processing techniques such as lactic acid fermentation. This type of fermentation is carried out by microorganisms present in raw grains or by starter culture (e.g., *Lactobacillus plantarum*) (Rollán, Gerez, & Leblanc, 2019). After 24 h of lactic acid fermentation, phytate degradation in quinoa, cañahua and amaranth depended on the type of crop, the physical state of the grain or flour, and the type of fermentation (spontaneous or with *Lactobacillus plantarum*). Phytate reduction was more effective in flour than in grains, quinoa and cañahua showing better results, followed by amaranth. After the fermentation of the three types of flour, mineral accessibility and estimated mineral bioavailability were significantly improved (Castro-Alba et al., 2019). In addition, the malting process has been used to decrease phytate and improve the nutritional quality of grains, i.e., increased antioxidant activity, phenolic compounds, flavonoids and reducing sugars; and decreased ash, lipids, protein content and saponins (Gobbetti et al., 2020; Aguilar et al., 2019).

One way to explore the effect of lactic acid fermentation of flour in baking goods involves sourdough fermentation. Rizzello et al. (2016) investigated the effect of sourdough fermentation of quinoa on white bread. Quinoa sourdough included in bread formulation increased the specific volume and cell-total area of the crumb, and the product showed a softer and more elastic texture compared with those made with raw quinoa or without quinoa flour. Recently, opposite results were found in GF bread with the addition of quinoa sourdough (Jagelaviciute & Cizeikiene, 2020). Considering nutritional aspects, quinoa sourdough increased total phenolics, antioxidant activity and amino acids, and improved phytase activity in final products. Amaranth freeze-dried sourdough added to GF bread also significantly changed the pH of the dough, and increased bread volume and decreased crumb hardness with larger pores, as compared to the control sample (Rózyło et al., 2015).

The incorporation of sprouted grains into cereal products has been extensively explored over the last years. Horstmann et al. (2019) reported the application of flour from commercial sprouts of quinoa and amaranth in GF bread formulation at a level of 5%. Dough formulations containing sprouted flour showed a reduced viscosity profile and breakdown in comparison to the sample without sprouted flour, quinoa sprouts showing less effect. This result is explained by the influence of enzymes by decreasing viscosity profiles and thus changing the molecular structure of starch through the breakdown of polymer chains (Aguilar et al., 2019). Those authors found that either amaranth or quinoa sprout flour increased specific volume, cell diameter and cell size, yielding softer textures and darker crumbs as compared to those of the control GF bread (Figure 9.2). Similar results were reported in GF bread (Suárez-Estrella et al., 2020).

Amaranth flour from germinated seeds was shown to increase water content and decrease molecular mobility in wheat dough when added at 25%; it

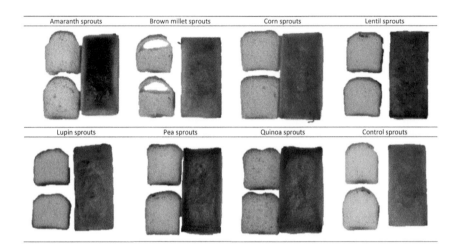

Figure 9.2 Cross-sections of baked breads formulated with the different sprouted flours. (Reprinted with permission from Horstmann et al., 2019.)

also maintained hardness with greater elasticity and viscosity compared to control dough. These results derive from modification of the amaranth protein during germination and the subsequent morphological changes in the gluten structure (Guardianelli, Salinas, & Puppo, 2019).

Some researchers have studied the enrichment of wheat bread with amaranth because its protein is one of the highest in nutritional quality. Particularly, amaranth is rich in lysine, an amino acid of which cereals are often deficient. In this sense, squalene (sterol precursor) plays a key role in decreasing low-density lipoprotein blood cholesterol, enhancing the immune system and even preventing cancer (Coelho et al., 2018). Popped amaranth flour was included up to 20% in wheat bread, contributing to a denser crumb structure, a more uniform porosity, and improving the crust color and flavor of final products, although negatively affecting crumb structure, elasticity and color. These authors showed that popping did not significantly reduce the content of squalene, which was around 8 to 12 times higher than the control bread made with wheat flour (Bodroža-Solarov et al., 2008).

Finally, extruded quinoa flour added to GF bread led to a product with a similar specific volume and lower firmness, with no significant changes in crumb, compared to the sample without extruded flour (Murgueytio & Santacruz, 2020).

9.5.2 Pasta

The addition of unmodified flour from Andean crops has been reported to decrease its cooking properties, also showing poor sensorial attributes

(Lorenzo, Sosa, & Califano, 2018). Thus, by applying extrusion cooking, flour components are modified to obtain good quality pasta. The incorporation of amaranth flour modified by extrusion cooking was used to make rice pasta. Despite the improvement in nutritional attributes observed for all amaranth-enriched pasta samples, pasta only with amaranth or extrusion-cooked rice flour showed low firmness or high cooking loss. On the other hand, pasta from an extrusion-cooked mixture of rice (75%) and amaranth (25%) flour showed the best textural and nutritional characteristics (Cabrera-Chávez et al., 2012).

The lactic acid fermentation of flour described in bread application has also been tested in pasta making. *Lactobacillus plantarum* and *Lactobacillus rossiae* were used as starters for quinoa flour fermentation, included in pasta formulation (20 g/100 g) (Lorusso et al., 2017). Results showed a pasta with improved protein digestibility and quality, low predicted glycemic index and high antioxidant potential, without significantly affecting cooking loss or firmness.

The inclusion of mixtures of raw and popped amaranth flour in semolina pasta showed a negative impact on their cooking properties due to protein matrix weakening. Accordingly, monoglycerides (1.2 g/100 g) and egg white powder (9 g/100 g) had to be included in pasta formulation to avoid pasta disintegration (Islas-Rubio et al., 2014). The firmness and sensorial properties of amaranth-enriched pasta were strongly affected, and the replacement of 25% of semolina with a mixture of 90:10 raw to popped amaranth grain flour allowed elaborating pasta with acceptable cooking quality and texture.

Germinated quinoa flour was used to make GF pasta with enhanced protein, total phenolics, antioxidant activity and mineral content, significantly decreasing phytic acid (Demir & Bilgiçli, 2020a). In this sense, up to 20% of a 50:50 mixture of raw:germinated quinoa flour showed a minimum effect on cooking loss and firmness compared to the control sample with no addition of quinoa (Demir & Bilgiçli, 2020b).

9.5.3 Extruded Products

The changes in amaranth flour after extrusion greatly increased water solubility index (from 11% to 61%), which is directly linked to the generation of water-soluble molecules during extrusion (Robin, Théoduloz, & Srichuwong, 2015). Researchers have suggested that this could be the result of the lower starch content in amaranth allowing greater amounts of free water to be available for starch transformation, resulting in low expansion volumes. The pasting profile of amaranth and quinoa extrudates was also studied in that research. As expected, the extrudates generated little viscosity, indicating that most starch crystallites and granular structures were disrupted in the extrusion process. Similar results were reported by Tobias-Espinoza et al. (2019) with the addition of amaranth to instant-extruded products.

Extrusion cooking was used to prepare expanded GF corn snacks with the addition of amaranth, quinoa and kañiwa flour (20% of solids) (Ramos Diaz et al., 2013, 2015, 2019). The greatest expansion was observed in extrudates containing amaranth, associated with the lowest dietary fiber content. It was observed that the addition of Andean crops produced the disruption of internal structures, reducing pores, crispiness and crunchiness linked to evenly distributed protein-associated structures (Ramos-Diaz, Rinnan, & Jouppila, 2019). Extrudates containing kañiwa showed higher gel-forming capacity associated with lower perception of hard particles and minor adhesiveness in extrudates containing 20% of kañiwa compared to an equal addition of amaranth or quinoa, which might be associated with more effective starch gelatinization during extrusion cooking (Ramos Diaz et al., 2015).

The effect of scarification and degermination of selected quinoa varieties on expansion properties was found to affect extrudate quality due to the removal of fat, protein and fiber. The magnitude of the effect also depends on grain hardness and its starch content, the main vehicle of expansion in the extrusion process (Aluwi et al., 2016). In general, authors indicate that degermed quinoa flour in which significant amounts of fiber, fat and protein are removed would likely perform better in the extrusion process, while scarified quinoa may present a low expansion ratio due to the increase in the protein content from the removal of the outer layer.

9.6 FUTURE RESEARCH

Alternative types of flour and modified processes improve the texture and cooking quality of gluten-free products, yet there are still some deficiencies in this field compared to those of wheat products, particularly pasta and bread. As a result, bioprocessing fermentation has recently drawn particular attention. Studies into the fermentation of different crops are still limited. Sensory analysis of products enriched with fermented flour should be carried out to find products with better market prospects.

Despite considerable efforts to improve the characteristics and applications of flour, further research is required to improve their quality features. Hence, the combination of raw materials, functional additives and processing technologies can synergistically fulfill this purpose. Additionally, by improving the nutritional quality of products with modified flour, consumers' health is also improved.

Some research has been conducted on several processes applied to Andean grains or flours to decrease antinutrient content. However, few studies have been done on the application and effects of such modified flour on products.

Other root and tuber crops like mashua (*Tropaeolum tuberosum* Ruíz & Pav.), arracacha (*Arracacia xanthorrhiza* Bancr.) and yacon (*Smallanthus sonchifolius* (Poepp.) H. Rob.) require further research. They also need to be introduced to common products, which take advantage of their nutritional values.

REFERENCES

Aguilar, J., Miano, A. C., Obregón, J., Soriano-Colchado, J., & Barraza-Jáuregui, G. (2019). Malting process as an alternative to obtain high nutritional quality quinoa flour. *Journal of Cereal Science*, 90(October). https://doi.org/10.1016/j.jcs.2019.102858.

Alamprese, C., Casiraghi, E., & Ambrogina Pagani, M.. (2007). Development of gluten-free fresh egg pasta analogues containing buckwheat. *European Food Research and Technology*, 225(2), 205–213. https://doi.org/10.1007/s00217-006-0405-y.

Aluwi, N. A., Gu, B. J., Dhumal, G. S., Medina-Meza, I. G., Murphy, K. M., & Ganjyal, G. M. (2016). Impacts of scarification and degermination on the expansion characteristics of select quinoa varieties during extrusion processing. *Journal of Food Science*, 81(12), E2939–E2949. https://doi.org/10.1111/1750-3841.13512.

Angioloni, A., & Collar, C. (2012). Effects of pressure treatment of hydrated oat, finger millet and sorghum flours on the quality and nutritional properties of composite wheat breads. *Journal of Cereal Science*, 56(3), 713–719. https://doi.org/10.1016/j.jcs.2012.08.001.

Anglani, C. (1998). Sorghum for human food: A review. *Plant Foods for Human Nutrition*, 52(1), 85–95. https://doi.org/10.1023/A:1008065519820.

Bai, Y. P., Zhou, H. M. (2021). Impact of aqueous ozone mixing on microbiological, quality and physicochemical characteristics of semi-dried buckwheat noodles. *Food Chemistry*, 336(January), 127709. https://doi.org/10.1016/j.foodchem.2020.127709.

Bhol, S., & Bosco, S. J. D. (2014). Influence of malted finger millet and red kidney bean flour on quality characteristics of developed bread. *LWT: Food Science and Technology*, 55(1), 294–300. https://doi.org/10.1016/j.lwt.2013.08.012.

Biacs, P., Aubrecht, E., Léder, I., & Lajos, J. (2002). Buckwheat. In Belton, P. S. & Taylor, J. R. N. (Eds.), *Pseudocereals and Less Common Cereals* (pp. 123–151). Berlin, Heidelberg: Springer Berlin Heidelberg. https://doi.org/10.1007/978-3-662-09544-7_4.

Bodroža-Solarov, M., Filipčev, B., Kevrešan, Ž., Mandić, A., & and Šimurina, O. (2008). Quality of bread supplemented with popped Amaranthus Cruentus grain. *Journal of Food Process Engineering*, 31(5), 602–618. https://doi.org/10.1111/j.1745-4530.2007.00177.x.

Bouvier, J.-M., & Campanella, O. H. (2014). The generic extrusion process III. In Bouvier, J.-M. & Campanella, O. H. (Eds.), *Extrusion Processing Technology: Food and Non-Food Biomaterials* (1st ed., pp. 243–309). Chichester, UK: John Wiley & Sons, Ltd. https://doi.org/10.1002/9781118541685.ch6.

Bustos, M. C., Ramos, M. I., Pérez, G. T., & León, A. E. (2019) Utilization of kañawa (Chenopodium Pallidicaule Aellen) flour in pasta making. *Journal of Chemistry*, 2019. https://doi.org/10.1155/2019/4385045.

Cabrera-Chávez, F., Calderón de la Barca, A. M., Islas-Rubio, A. R., Marti, A., Marengo, M., Ambrogina Pagani, M., Bonomi, F., & Iametti, S. (2012). Molecular rearrangements in extrusion processes for the production of amaranth-enriched, gluten-free rice pasta. *LWT: Food Science and Technology*, 47(2), 421–426. https://doi.org/10.1016/j.lwt.2012.01.040.

Calle, J., Benavent-Gil, Y., & Rosell, C. M. (2020). Development of gluten free breads from colocasia esculenta flour blended with hydrocolloids and enzymes. *Food Hydrocolloids*, 98(June 2019), 105243. https://doi.org/10.1016/j.foodhyd.2019.105243.

Castro-Alba, V., Lazarte, C. E., Perez-Rea, D., Carlsson, N. G., Almgren, A., Bergenståhl, B., & Granfeldt, Y. (2019) Fermentation of pseudocereals quinoa, canihua, and amaranth to improve mineral accessibility through degradation of phytate. *Journal of the Science of Food and Agriculture*, 99(11), 5239–5248. https://doi.org/10.1002/jsfa.9793.

Chen, Z., Schols, H. A., & Voragen, A. G. J. (2003). The use of potato and sweet potato starches affects white salted noodle quality. *Journal of Food Science*, 68(9), 2630–2637. https://doi.org/10.1111/j.1365-2621.2003.tb05781.x.

Cho, Y. J., & Lee, S. (2015). Extraction of rutin from tartary buckwheat milling fractions and evaluation of its thermal stability in an instant fried noodle system. *Food Chemistry*, 176, 40–44. https://doi.org/10.1016/j.foodchem.2014.12.020.

Choquechambi, L. A., Callisaya, I. R., Ramos, A., Bosque, H., Mújica, A., Jacobsen, S. E., Sørensen, M., & Leidi, E. O. (2019). Assessing the nutritional value of root and tuber crops from Bolivia and Peru. *Foods*, 8(11). https://doi.org/10.3390/foods8110526.

Chuango, K., Julianti, E., & Ginting, S. (2019). Effect of pre-treatment in the making of purple-fleshed sweet potato flour towards cake characteristics. *IOP Conference Series: Earth and Environmental Science*, 260(1). https://doi.org/10.1088/1755-1315/260/1/012090.

Cisse, F., Erickson, D., Hayes, A., Opekun, A., Nichols, B., & Hamaker, B. (2018). Traditional Malian solid foods made from sorghum and millet have markedly slower gastric emptying than rice, potato, or pasta. *Nutrients*, 10(2), 124. https://doi.org/10.3390/nu10020124.

Coelho, L.., Silva, P. M., Martins, J. T., Pinheiro, A. C., & Vicente, A. A. (2018). Emerging Opportunities in Exploring the Nutritional/Functional Value of Amaranth. *Food and Function*, 9 (11): 5499–5512. https://doi.org/10.1039/c8fo01422a.

Collar, C. (2007). Cereales Menores: Avena, Sorgo y Mijo. In León, A. E. & Rosell, C. M. (Eds.), *De Tales Harinas, Tales Panes. Granos, Harinas y Productos de Panificación En Iberoamérica* (1st ed., p. 480). Córdoba, Argentina: ISEKI-Food.

Dahir, M., Zhu, K.-X., Guo, X.-N., Aboshora, W., & Peng, W. (2015). Possibility to utilize sorghum flour in a modern bread making industry. *Journal of Academia and Industrial Research*, 4(4), 128-135.

Dayakar Rao, B., Anis, M., Kalpana, K., Sunooj, K. V., Patil, J. V., & Ganesh, T. (2016). Influence of milling methods and particle size on hydration properties of sorghum flour and quality of sorghum biscuits. *LWT: Food Science and Technology*, 67, 8–13. https://doi.org/10.1016/j.lwt.2015.11.033.

Demir, B., & Bilgiçli, N. (2020a). Changes in chemical and anti-nutritional properties of pasta enriched with raw and germinated Quinoa (Chenopodium Quinoa Willd.) flours. *Journal of Food Science and Technology*, 57(10), 3884–1892. https://doi.org/10.1007/s13197-020-04420-7.

Demir, B., Bilgiçli, N. (2020b). Utilization of quinoa flour (Chenopodium quinoa Willd.) in gluten-free pasta formulation: Effects on nutritional and sensory properties. *Food Science and Technology International*, 27(3), 242–250. https://doi.org 10.1177/1082013220940092.

Do, D. T., & Singh, J., Oey, I., & Singh, H. (2019). Modulating effect of cotyledon cell microstructure on in vitro digestion of starch in legumes. *Food Hydrocolloids*, 96(April), 112–122. https://doi.org/10.1016/j.foodhyd.2019.04.063.

Ekafitri, R., Pranoto, Y., & Herminiati, A. (2019). Baking quality, texture and sensory evaluation of gluten free cake made from modified taro flour. *AIP Conference Proceedings 2175*, 020001; https://doi.org/10.1063/1.5134565.

Fathi, B., Aalami, M., Kashaninejad, M., & Sadeghi Mahoonak, A. (2016). Utilization of heat-moisture treated proso millet flour in production of gluten-free pound cake. *Journal of Food Quality*, 39(6), 611–619. https://doi.org/10.1111/jfq.12249.

Fiorda, F. A., Soares, M. S., Flávio, A., Grosmann, M. V. E., & Souto, L. R. F. (2013). Microestructure, texture and colour of gluten-free pasta made with amaranth flour, cassava starch and cassava bagasse. *LWT: Food Science and Technology*, 54(1), 132–138. https://doi.org/10.1016/j.lwt.2013.04.020.

Foschia, M., Horstmann, S. W., Arendt, E. K., & Zannini, E. (2017). Legumes as functional ingredients in gluten-free bakery and pasta products. *Annual Review of Food Science and Technology*, 8(1), 75–96. https://doi.org/10.1146/annurev-food-030216-030045.

Galán, M. G., Cian, E., Albarracín, M., López-oliva Muñoz, M. E., Weisstaub, A., & Rosa, S. (2020). Refined sorghum flours precooked by extrusion enhance the integrity of the colonic mucosa barrier and promote a hepatic antioxidant environment in growing Wistar rats.. *Food & Function*, 11, 638–7650. https://doi.org/10.1039/d0fo01160f.

Geetha, R, Mishra H. N., & Srivastav, P. P. (2014). Twin screw extrusion of kodo millet-chickpea blend: Process parameter optimization, physico-chemical and functional properties. *Journal of Food Science and Technology*, 51(11), 3144–3153. https://doi.org/10.1007/s13197-012-0850-5.

Giuberti, G., Gallo, A., Cerioli, C., Fortunati, P., & Masoero, F. (2015). Cooking quality and starch digestibility of gluten free pasta using new bean flour. *Food Chemistry*, 175, 43–49. https://doi.org/10.1016/j.foodchem.2014.11.127.

Gobbetti, M., De Angelis, M., Di Cagno, R., Polo, A., & Rizzello, C. G. (2020). The sourdough fermentation is the powerful process to exploit the potential of legumes, pseudo-cereals and milling by-products in baking industry. *Critical Reviews in Food Science and Nutrition*, 60(13), 2158–2173. https://doi.org/10.1080/10408398.2019.1631753.

Gómez, M., & Martínez, M. M.. (2016). Changing flour functionality through physical treatments for the production of gluten-free baking goods. *Journal of Cereal Science*, 67, 68–74. https://doi.org/10.1016/j.jcs.2015.07.009.

Guardianelli, L. M., Salinas, M. V., & Puppo M. C. (2019). Hydration and rheological properties of amaranth-wheat flour dough: influence of germination of amaranth seeds. *Food Hydrocolloids*, 97(December 2018), 105242. https://doi.org/10.1016/j.foodhyd.2019.105242.

Gularte, M. A., Gómez, M, & Rosell, C. M. (2012). Impact of legume flours on quality and in vitro digestibility of starch and protein from gluten-free cakes. *Food and Bioprocess Technology*, 5(8), 3142–2150. https://doi.org/10.1007/s11947-011-0642-3.

Hirich, A., Choukr- Allah, R., & Ragab, R. (2020). *Emerging Research in Alternative Crops.* Berlin, Germany: Springer.

Horstmann, S. W., Atzler, J. J., Heitmann, M., Zannini, E., Lynch, K. M., & Arendt, E. K. (2019). A comparative study of gluten-free sprouts in the gluten-free bread-making process. *European Food Research and Technology,* 245(3), 617–629. https://doi.org/10.1007/s00217-018-3185-2.

Hutasoit, M. S., Julianti, E., & Lubis, Z. (2018). Effect of pretreatment on purple-fleshed sweet potato flour for cake making. *IOP Conference Series: Earth and Environmental Science,* 122(1). https://doi.org/10.1088/1755-1315/122/1/012086.

Ilamaran, M., Nousheen Noorul Iyn, I, Kanchana, S, Sivasankari, B., & Kendra, K. V. (2017). Anti-hyperglycemic effect of cereal and millet based modified starch substituted pasta foods. *Chemical Science Review and Letters,* 6(22), 704–709.

Islas-Rubio, A. R.., Calderón de la Barca, A. M., Cabrera-Chávez, F., Cota-Gastélum, A. G.., & Trust Beta. (2014). Effect of semolina replacement with a raw: popped amaranth flour blend on cooking quality and texture of pasta. *LWT - Food Science and Technology,* 57(1), 217–222. https://doi.org/10.1016/j.lwt.2014.01.014.

Jacobsen, S. E. (2003). The worldwide potential for quinoa (Chenopodium Quinoa Willd.). *Food Reviews International,* 19(1–2), 167–177. https://doi.org/10.1081/FRI-120018883.

Jagelaviciute, J., & Cizeikiene, D. (2020). The influence of non-traditional sourdough made with quinoa, hemp and chia flour on the characteristics of gluten-free maize/rice bread. *LWT,* 137, 110457. https://doi.org/10.1016/j.lwt.2020.110457.

Jalgaonkar, K., Jha, S. K., Mahawar, M. K., & Yadav, D. N. (2019). Pearl millet based pasta: Optimization of extrusion process through response surface methodology. *Journal of Food Science and Technology,* 56(3), 1134–1144. https://doi.org/10.1007/s13197-019-03574-3.

Kamble, D. B, Singh, R., Rani, S., Pal Kaur, B., Upadhyay, A., & Kumar, N. (2019). Optimization and characterization of antioxidant potential, in vitro protein digestion and structural attributes of microwave processed multigrain pasta. *Journal of Food Processing and Preservation,* 43(10), 1–11. https://doi.org/10.1111/jfpp.14125.

Kharat, S., Medina-Meza, I. G., Kowalski, R. J., Hosamani, A., Ramachandra, C. T., Hiregoudar, S., & Ganjyal, G. M. (2019). Extrusion processing characteristics of whole grain flours of select major millets (foxtail, finger, and pearl). *Food and Bioproducts Processing,* 114, 60–71. https://doi.org/10.1016/j.fbp.2018.07.002.

Kringel, D. H., El Halal, S. L. M., da Rosa Zavareze, E., & Guerra Dias, A. R. (2020). Methods for the extraction of roots, tubers, pulses, pseudocereals, and other unconventional starches sources: A review. *Starch/Stärke,* 72(11–12), 1900234. https://doi.org/10.1002/star.201900234.

Kurniadi, M., Khasanah, Y., Kusumaningrum A., Angwar, M., Rachmawanti, D., Parnanto, N. H. R., & Pratiwi, L. D. (2019). Formulation and shelf life prediction of cookies from modified cassava flour (mocaf) in flexible packaging. *IOP Conference Series: Earth and Environmental Science,* 251(1). https://doi.org/10.1088/1755-1315/251/1/012034.

Li, S.-Q., & Zhang, Q. H. (2001). Advances in the development of functional foods from buckwheat. *Critical Reviews in Food Science and Nutrition,* 41(6), 451–464. https://doi.org/10.1080/20014091091887.

Limroongreungrat, K., & Huang, Y. W. (2007). Pasta products made from sweetpotato fortified with soy protein. *LWT: Food Science and Technology*, 40(2), 200–206. https://doi.org/10.1016/j.lwt.2005.09.012.

Llopart, E. E., Cian, R. E., López-Oliva, M. M. E., Zuleta, Ángela, Weisstaub, A., & Drago, S. R. (2017). Colonic and systemic effects of extruded whole-grain sorghum consumption in growing Wistar rats. *British Journal of Nutrition*, 118(8), 589–597. https://doi.org/10.1017/S0007114517002513.

Lorenzo, G., Sosa, M., & Califano, A. (2018). Alternative proteins and pseudocereals in the development of gluten-free pasta. In *Food Control and Biosecurity* (pp. 433–458). Netherlands: Elsevier Inc. https://doi.org/10.1016/B978-0-12-811446-9/00015-0.

Lorusso, A., Verni, M., Montemurro, M., Coda, R., Gobbetti, M., & Rizzello, C. G. (2017). Use of fermented quinoa flour for pasta making and evaluation of the technological and nutritional features. *LWT: Food Science and Technology*, 78, 215–221. https://doi.org/10.1016/j.lwt.2016.12.046.

Losano Richard, P., Eugenia Steffolani, M., Allende, M. J., Carreras, J., & León, A. E.. (2020). By-products of the classification of chickpea as an alternative in the production of hummus. *International Journal of Food Science and Technology*, 56, 1759–1765. https://doi.org/10.1111/ijfs.14801.

Lukšič, L., Árvay, J., Vollmannová, A., Tóth, T., Škrabanja, V., Trček, J., Germ, M., & Kreft, I. (2016). Hydrothermal treatment of tartary buckwheat grain hinders the transformation of rutin to quercetin. *Journal of Cereal Science*, 72(November), 131–134. https://doi.org/10.1016/j.jcs.2016.10.009.

Marengo, M., Bonomi, F., Marti, A., Ambrogina Pagani, M., Elkhalifa, A. E. O., & Iametti, S. (2015). Molecular features of fermented and sprouted sorghum flours relate to their suitability as components of enriched gluten-free pasta. *LWT: Food Science and Technology*, 63(1), 1–8. https://doi.org/10.1016/j.lwt.2015.03.070.

Marston, K., Khouryieh, H., & Aramouni, F. (2015). Evaluation of sorghum flour functionality and quality characteristics of gluten-free bread and cake as influenced by ozone treatment. *Food Science and Technology International*, 21(8), 631–640. https://doi.org/10.1177/1082013214559311.

Marti, A., & Ambrogina Pagani, M. (2013). What can play the role of gluten in gluten free pasta? *Trends in Food Science & Technology*, 31(1), 63–71. https://doi.org/10.1016/j.tifs.2013.03.001.

Min, B., Lee, S. M., Yoo, S.-H., Inglett, G. E., & Lee, S. (2010). Functional characterization of steam jet-cooked buckwheat flour as a fat replacer in cake-baking. *Journal of the Science of Food and Agriculture*, 90(13), 2208–2213. https://doi.org/10.1002/jsfa.4072.

Miyazaki, M., Hung, P. V., Maeda, T., & Morita, N. (2006). Recent advances in application of modified starches for breadmaking. *Trends in Food Science & Technology*, 17(11), 591–599. https://doi.org/10.1016/j.tifs.2006.05.002.

Miyazaki, M., Maeda, T., & Morita, N. (2008). Bread quality of frozen dough substituted with modified tapioca starches. *European Food Research and Technology*, 227(2), 503–509. https://doi.org/10.1007/s00217-007-0747-0.

Monnet, A. F., Laleg, K., Michon, C., & Micard, V. (2019). Legume enriched cereal products: A generic approach derived from material science to predict their structuring by the process and their final properties. *Trends in Food Science and Technology*, 86, 131–143. https://doi.org/10.1016/j.tifs.2019.02.027.

Monthe, O. C., Grosmaire, L., Nguimbou, R. M., Dahdouh, L., Ricci, J., Tran, T., & Ndjouenkeu, R. (2019). Rheological and textural properties of gluten-free doughs and breads based on fermented cassava, sweet potato and sorghum mixed flours. *LWT*, 101(November 2018), 575–582. https://doi.org/10.1016/j.lwt.2018.11.051.

Murgueytio, E., & Santacruz, S. (2020). Volume, firmness and crumb characteristics of gluten-free bread based on extruded quinoa flour and lactic acid. *Brazilian Journal of Food Technology*, 23, 1–9. https://doi.org/10.1590/1981-6723.22019.

Mustika, A. R., & Kartika, W. D. (2020). Formulation of yellow pumpkin cookies with mocaf (modified cassava flour) flour addition as a snack for the obese community. *Food Research*, 4(S3), 109–113. https://doi.org/10.26656/fr.2017.4(S3).S02.

Nachay, K. (2017). The power of pulses. *Food Technology Magazine*, 2017.

Obadi, M., & Xu, B. (2021). Review on the physicochemical properties, modifications, and applications of starches and its common modified forms used in noodle products. *Food Hydrocolloids*, 112(August 2020), 106286. https://doi.org/10.1016/j.foodhyd.2020.106286.

Obilana, A. B., & Manyasa, E. (2002). Millets. In Belton, P. S. & Taylor, J. (Eds.), *Pseudocereals and Less Common Cereals* (pp. 177–217). Berlin, Heidelberg: Springer Berlin Heidelberg. https://doi.org/10.1007/978-3-662-09544-7_6.

Oh, M., Oh, I., Jeong, S., & Lee, S. (2019). Optical, rheological, thermal, and microstructural elucidation of rutin enrichment in tartary buckwheat flour by hydrothermal treatments. *Food Chemistry*, 300(December), 125193. https://doi.org/10.1016/j.foodchem.2019.125193.

Okereke, G. O., Iwe, M. O., & Agiriga, A. N. (2015). Production and evaluation of bread made from modified cassava starch and wheat flour blends. *Agrotechnology*, 4(1), 1–6. https://doi.org/10.4172/2168-9881.1000133.

Onyango, C. (2013). Effect of heat-moisture-treated cassava starch and amaranth malt on the quality of sorghum-cassava-amaranth bread. *African Journal of Food Science*, 7(5), 80–86. https://doi.org/10.5897/ajfs2012.0612.

Onyango, C., Mutungi, C., Unbehend, G., & Lindhauer, M. G. (2011). Rheological and textural properties of sorghum-based formulations modified with variable amounts of native or pregelatinised cassava starch. *LWT - Food Science and Technology*, 44(3), 687–693. https://doi.org/10.1016/j.lwt.2010.08.019.

Palavecino, P. M., Bustos, M. C., Alabí,, M. B. H., Nicolazzi, M. S., Penci, M. C., & Ribotta, P. D. (2017). Effect of ingredients on the quality of gluten-free sorghum pasta. *Journal of Food Science*, 82(9), 2085–2093. https://doi.org/10.1111/1750-3841.13821.

Palavecino, P. M., Curti, M. I., Bustos, M. C., Penci, M. C., & Ribotta, P. D. (2020). Sorghum pasta and noodles: Technological and nutritional aspects. *Plant Foods for Human Nutrition*, 75(3), 326–36. https://doi.org/10.1007/s11130-020-00829-9.

Pasquale, I. De, Pontonio, E., Gobbetti, M., & Rizzello, C. G.. (2020). Nutritional and functional effects of the lactic acid bacteria fermentation on gelatinized legume flours. *International Journal of Food Microbiology*, 316, 108426. https://doi.org/10.1016/j.ijfoodmicro.2019.108426.

Pasqualone, A., Costantini, M., Coldea, T. E., & Summo, C. (2020). Use of legumes in extrusion cooking: A review. *Foods*, 9(7), 1–17. https://doi.org/10.3390/foods9070958.

Pérez, E. (2007). Raices y Tuberculos. In Leon, A. E. & Rosell, C. M. (Eds.), *De Tales Harinas, Tales Panes. Granos, Harinas y Productos de Panificación En Iberoamérica* (p. 480). Córdoba, Argentina: ISEKI-Food.

Pérez, G. T., Eugenia Steffolani, M., & León, A. E. (2016). Cañahua: An ancient grain for new foods. In Kristbergsson, K. and Ötles, S.. (Eds.), *Functional Properties of Traditional Foods* (1st ed., pp. 119–130). New York: Springer Science+Business Media. https://doi.org/10.1007/978-1-4899-7662-8_9.

Perez-Rea, D., & Antezana-Gomez, R. (2018). *The Functionality of Pseudocereal Starches. Starch in Food: Structure, Function and Applications* (2nd ed., Vol. 570). New York: Elsevier Ltd. https://doi.org/10.1016/B978-0-08-100868-3.00012-3.

Pezzali, J. G., Suprabha-Raj, A., Siliveru, K., & Aldrich, C. G. (2020). Characterization of white and red sorghum flour and their potential use for production of extrudate crisps. *PLoS ONE*, 15(6 June). https://doi.org/10.1371/journal.pone.0234940.

Poquette, N. M, Gu, X., & Lee, S.-O. (2014). Grain sorghum muffin reduces glucose and insulin responses in men. *Food and Function*, 5(5), 894–899. https://doi.org/10.1039/C3FO60432B.

Ramos Diaz, J. M., Kirjoranta, S., Tenitz, S., Penttilä, P. A., Serimaa, R., Lampi, A. M., & Jouppila, K. (2013). Use of amaranth, quinoa and kañiwa in extruded corn-based snacks. *Journal of Cereal Science*, 58(1), 59–67. https://doi.org/10.1016/j.jcs.2013.04.003.

Ramos Diaz, J. M., Suuronen, J. P., Deegan, K. C., Serimaa, R., Tuorila, H., & Jouppila, K. (2015). Physical and sensory characteristics of corn-based extruded snacks containing amaranth, quinoa and kañiwa flour. *LWT: Food Science and Technology*, 64(2), 1047–1056. https://doi.org/10.1016/j.lwt.2015.07.011.

Ramos-Diaz, J. M., Rinnan, Å., & Jouppila, K. (2019). Application of NIR imaging to the study of expanded snacks containing amaranth, quinoa and kañiwa. *LWT*, 102(December 2018), 8–14. https://doi.org/10.1016/j.lwt.2018.12.029.

Ranasalva, N., & Visvanathan, R. (2014). Development of cookies and bread from cooked and fermented pearl millet flour. *African Journal of Food Science*, 8(6), 330–336. https://doi.org/10.5897/ajfs2013.1113.

Rastogi, A., & Shukla, S. (2013). Amaranth: A new millennium crop of nutraceutical values. *Critical Reviews in Food Science and Nutrition*, 53(2), 109–125. https://doi.org/10.1080/10408398.2010.517876.

Rathi, A., Kawatra, A., & Sehgal, S. (2004). Influence of depigmentation of pearl millet (Pennisetum Glaucum L.) on sensory attributes, nutrient composition, in vitro protein and starch digestibility of pasta. *Food Chemistry*, 85(2), 275–280. https://doi.org/10.1016/j.foodchem.2003.06.021.

Ratnavathi, C. V., & Patil, J. V. (2014). Sorghum utilization as food. *Journal of Nutrition & Food Sciences*, 4(1), 1–8. https://doi.org/10.4172/2155-9600.1000247.

Rawal, V., & Navarro, D. K. (2019). *The Global Economy of Pulses* (Rawal, V. & Navarro, D. K., Eds.). Rome: FAO. https://doi.org/10.4060/I7108EN.

Renzetti, S., Behr, J., Vogel, R. F., & Arendt, E. K. (2008). Transglutaminase polymerisation of buckwheat (Fagopyrum Esculentum Moench) proteins. *Journal of Cereal Science*, 48(3), 747–754. https://doi.org/10.1016/j.jcs.2008.04.005.

Repo-Carrasco, R., Espinoza, C., & Jacobsen, S. E. (2003). Nutritional value and use of the Andean crops quinoa (Chenopodium Quinoa) and kañiwa (Chenopodium Pallidicaule). *Food Reviews International*, 19(1–2), 179–189. https://doi.org/10.1081/FRI-120018884.

Repo-Carrasco-Valencia, R. A.-M., & Serna, L. A. (2011). Quinoa (Chenopodium Quinoa, Willd.) as a source of dietary fiber and other functional components.

Ciência e Tecnologia de Alimentos, 31(1), 225–30. https://doi.org/10.1590/s0101
-20612011000100035.

Rizzello, C. G., Calasso, M., Campanella, D., De Angelis, M., & Gobbetti, M. (2014). Use of
Sourdough fermentation and mixture of wheat, chickpea, lentil and bean flours
for enhancing the nutritional, texture and sensory characteristics of white bread.
International Journal of Food Microbiology, 180, 78–87. https://doi.org/10.1016/j
.ijfoodmicro.2014.04.005.

Rizzello, C. G., Lorusso, A., Montemurro, M., & Gobbetti, M. (2016). Use of sourdough
made with quinoa (Chenopodium Quinoa) flour and autochthonous selected
lactic acid bacteria for enhancing the nutritional, textural and sensory features
of white bread. *Food Microbiology*, 56, 1–13. https://doi.org/10.1016/j.fm.2015.11
.018.

Robin, F., Théoduloz, C., & Srichuwong, S. (2015). Properties of extruded whole grain
cereals and pseudocereals flours. *International Journal of Food Science and
Technology*, 50(10), 2152–2159. https://doi.org/10.1111/ijfs.12893.

Rodríguez, J. P., Aro, M., Coarite, M., Jacobsen, S. E., Ørting, B., Sørensen, M., &
Andreasen, C. (2017). Seed shattering of cañahua (Chenopodium Pallidicaule
Aellen). *Journal of Agronomy and Crop Science*, 203(3), 254–267. https://doi.org/10
.1111/jac.12192.

Rodriguez-Sandoval, E., Prasca-Sierra, I., & Hernandez, V. (2017). Effect of modified cassava starch as a fat replacer on the texture and quality characteristics of muffins.
Journal of Food Measurement and Characterization, 11(4), 1630–1639. https://doi
.org/10.1007/s11694-017-9543-0.

Rollán, G. C., Gerez, C. L., & Leblanc, J. G. (2019). Lactic fermentation as a strategy to
improve the nutritional and functional values of pseudocereals. *Frontiers in
Nutrition*, 6(September 2017). https://doi.org/10.3389/fnut.2019.00098.

Rosell, C. M., Cortez, G., & Repo-Carrasco, R. (2009). Breadmaking use of Andean crops
quinoa, kañiwa, kiwicha, and tarwi. *Cereal Chemistry*, 86(4), 386–392. https://doi
.org/10.1094/CCHEM-86-4-0386.

Rosida, R., & Susiloningsih, E. K. B. (2020). The characteristics of crackers made from
formulation of wheat and modified water yam flour. *FOODSCITECH*, 3(1), 18.
https://doi.org/10.25139/fst.v0i0.2681.

Rózyło, R., Rudy, S., Krzykowski, A., & Dziki, D. (2015). Novel application of freeze-dried
Amaranth Sourdough in gluten-free bread production. *Journal of Food Process
Engineering*, 38(2), 135–143. https://doi.org/10.1111/jfpe.12152.

Sadowska, A., & Diowksz, A. (2020). Effect of transglutaminase on specific volume and
crumb porosity of gluten-free buckwheat bread. *Zywnosc. Nauka. Technologia.
Jakosc/Food. Science Technology Quality*, 27(1), 74–84. https://doi.org/10.15193/
zntj/2020/122/323.

Sáez, G. D., Saavedra, L., Hebert, E. M., & Zárate, G. (2018). Identification and biotechno-
logical characterization of lactic acid bacteria isolated from chickpea sourdough
in Northwestern Argentina. *LWT*, 93, 249–256. https://doi.org/10.1016/j.lwt.2018
.03.040.

Schoenlechner, R., Drausinger, J., Ottenschlaeger, V., Jurackova, K., & Berghofer, E.
(2010). Functional properties of gluten-free pasta produced from amaranth, quinoa and buckwheat. *Plant Foods for Human Nutrition*, 65, 339–349. https://doi.org
/10.1007/s11130-010-0194-0.

Shobana, S., Krishnaswamy, K., Sudha, V., Malleshi, N. G., Anjana, R. M., Palaniappan, L., & Mohan, V. (2013). *Finger Millet (Ragi, Eleusine Coracana L.). A Review of Its Nutritional Properties, Processing, and Plausible Health Benefits. Advances in Food and Nutrition Research* (1st ed., Vol. 69). https://doi.org/10.1016/B978-0-12-410540 -9.00001-6.

Singla, D., Singh, A., Dhull, S. B., Kumar, P., Malik, T., & Kumar, P. (2020). Taro starch: Isolation, morphology, modification and novel applications concern: A review. *International Journal of Biological Macromolecules*, 163, 1283–1290. https://doi .org/10.1016/j.ijbiomac.2020.07.093.

Sisilia, F. Y., Herla, R., Elisa, J., Tampubolon, S. D. R., & Restuana, S. D. (2018). Physicochemical characteristics of tuber modified by acetylation method and its application in dry noodle product. *IOP Conference Series: Earth and Environmental Science*, 205(1). https://doi.org/10.1088/1755-1315/205/1/012044.

Suárez-Estrella, D., Cardone, G., Buratti, S., Ambrogina Pagani, M., & Marti, A. (2020). Sprouting as a pre-processing for producing quinoa-enriched bread. *Journal of Cereal Science*, 96, 103111. https://doi.org/10.1016/j.jcs.2020.103111.

Suhendro, E. L., Kunetz, C. F., McDonough, C. M., Rooney, L. W., & Waniska, R. D. (2000). Cooking characteristics and quality of noodles from food sorghum. *Cereal Chemistry Journal*, 77(2), 96–100. https://doi.org/10.1094/CCHEM.2000.77.2.96.

Sun, X., Li, W., Hu, Y., Zhou, X., Ji, M., Didi, Y., Fujita, K., Tatsumi, E., & Luan, G. (2018). Comparison of pregelatinization methods on physicochemical, functional and structural properties of tartary buckwheat flour and noodle quality. *Journal of Cereal Science*, 80(March), 63–71. https://doi.org/10.1016/j.jcs.2018.01.016.

Sun, X., Ohanenye, I. C., Ahmed, T., & Udenigwe, C. C. (2020). Microwave treatment increased protein digestibility of pigeon pea (Cajanus Cajan) flour: Elucidation of underlying mechanisms. *Food Chemistry*, 329(February), 127196. https://doi.org /10.1016/j.foodchem.2020.127196.

Taylor, J. R. N., & Belton, P. S. (2002). Sorghum. In Taylor, John R. N. & Belton, P. S. (Eds.), *Pseudocereals and Less Common Cereals* (pp. 25–91). Berlin, Heidelberg: Springer Berlin Heidelberg. https://doi.org/10.1007/978-3-662-09544-7_2.

Thilagavathi, S. Kanchana, P. B., & Ilamaran, M. (2020). Standardization of extruded products using modified millet flour and pulse flour. *International Journal of Food and Nutritional Sciences*, 4(4), 73. http://www.ijfans.org/article.asp?issn=2319 -1775;year=2015;volume=4;issue=4;spage=73;epage=79;aulast=Thilagavathi ;type=0.

Tien, N. N. T., Anh, T. N. Q., Phi, N. T. L., & Hung, P. V. (2018). In vitro and in vivo starch digestibility and quality of bread substituted with acid and heat-moisture treated sweet potato starch. *Starch/Staerke*, 70(9–10). https://doi.org/10.1002/star .201800069.

Tobias-Espinoza, J. L., Amaya-Guerra, C. A., Quintero-Ramos, A., Pérez-Carrillo, E., Núñez-González, M. A., Martínez-Bustos, F., Meléndez-Pizarro, C. O., Báez-González, J. G., & Ortega-Gutiérrez, J. A. (2019). Effects of the addition of flaxseed and Amaranth on the physicochemical and functional properties of instant-extruded products. *Foods*, 8(6), 1–14. https://doi.org/10.3390/foods8060183.

Torres, P. I., Ramírez-Wong, B., Serna-Saldivar, S. O., & Rooney, L. W. (1993). Effect of sorghum flour addition on the characteristics of wheat flour tortillas. *Cereal Chemistry*, 70(1), 8–13.

Truong, V. D., & Avula, R. Y. (2010). Sweet potato purees and dehydrated powders for functional food ingredients. In Ray, R. C. & Tomlin, K. I. (Eds.), *Sweet Potato: Post Harvest Aspects in Food, Feed and Industry* (pp. 117–161). Nova Science Publishers, Inc.

Vallons, K. J. R., Ryan, L. A. M., Koehler, P., & Arendt, E. K. (2010). High pressure–treated sorghum flour as a functional ingredient in the production of sorghum bread. *European Food Research and Technology*, 231(5), 711–717. https://doi.org/10.1007/s00217-010-1316-5.

Wang, Y., Compaoré-Sérémé, D., Sawadogo-Lingani, H., Coda, R., Katina, K., & Maina, N. H.. (2019). Influence of dextran synthesized in situ on the rheological, technological and nutritional properties of whole grain pearl millet bread. *Food Chemistry*, 285(October 2018), 221–230. https://doi.org/10.1016/j.foodchem.2019.01.126.

Winger, M., Khouryieh, H., Aramouni, F., & Herald, T. (2014). Sorghum flour characterization and evaluation in gluten-free flour tortilla. *Journal of Food Quality*, 37(2), 95–106. https://doi.org/10.1111/jfq.12080.

Wongsagonsup, R., Pujchakarn, T., Jitrakbumrung, S., Chaiwat, W., Fuongfuchat, A., Varavinit, S., Dangtip, S., & Suphantharika, M. (2014). Effect of cross-linking on physicochemical properties of tapioca starch and its application in soup product. *Carbohydrate Polymers*, 101, 656–665. https://doi.org/10.1016/j.carbpol.2013.09.100.

Wu, K. L., Sung, W. C., & Yang, C. H. (2009). Characteristics of dough and bread as affected by the incorporation of sweet potato paste in the formulation. *Journal of Marine Science and Technology*, 17(1), 13–22.

Yadav, D. N., Balasubramanian, S., Kaur, J., Anand, T., & Singh, A. K. (2012). Non-wheat pasta based on pearl millet flour containing barley and whey protein concentrate. *Journal of Food Science and Technology*, 2001, 1–8. https://doi.org/10.1007/s13197-012-0772-2.

Yadav, D. N., Tanupriya Anand, N., & Singh, A. K. (2014). Co-extrusion of pearl millet-whey protein concentrate for expanded snacks. *International Journal of Food Science and Technology*, 49(3), 840–846. https://doi.org/10.1111/ijfs.12373.

Yoo, J., Kim, Y., Yoo, S. H., Inglett, G. E., & Lee, S. (2012). Reduction of Rutin loss in buckwheat noodles and their physicochemical characterisation. *Food Chemistry*, 132, 2107–2111. Elsevier. https://doi.org/10.1016/j.foodchem.2011.12.065.

Zieliński, H., Michalska, A., Piskuła, M. K., & Kozłowska, H. (2006). Antioxidants in thermally treated buckwheat groats. *Molecular Nutrition & Food Research*, 50(9), 824–832. https://doi.org/10.1002/mnfr.200500258.

Chapter 10

Legume Starches, Their Use in Pasta Foods and Their Relationship with Human Health and Novelty Uses

José A. Gallegos-Infante, Claudia G. Félix-Villalobos,
Walfred Rosas-Flores, Nuria E. Rocha-Guzmán,
Martha R. Moreno-Jiménez and Rubén F. González-Laredo

CONTENTS

10.1 INTRODUCTION

Starch is an abundant polysaccharide in plants. It occurs in the form of granules in the chloroplast of green leaves and the amyloplast of seeds, legumes and tubers (Fuentes-Zaragoza et al., 2010). It is formed by glucose units accumulated in independent granules (Reddy et al., 2013; Kim et al., 2015).

DOI: 10.1201/9781003088929-10

Starch granules have a semicrystalline character, indicating a high degree of orientation of glucose molecules (Wani et al., 2010). Starch consists of amylose and amylopectin (Escarpa & Gonzalez, 1997). Amylose is a linear polymer of glucose, linked by α (1-4) bonds, and sometimes α (1-6) bonds may be present (McCready et al., 1950; Chel-Guerrero et al., 2011). Amylose is not soluble in water, but it can form hydrated micelles due to its ability to bind neighboring molecules by hydrogen bonds and generate a helical structure (Hernández-Medina et al., 2008). Amylopectin is a branched polymer of glucose units linked by α (1-4) (94%–96%) and α (1-6) bonds (4%–6%) (Manners & Matheson, 1981; Chaves Martins et al., 2018). These branches are located approximately every 15–25 glucose units. Therefore, the molecule is partially soluble in hot water (Hernández-Medina et al., 2008). Due to its abundance, starch can be easily and purely extracted by physical processes on a large industrial scale (French, 1984). Starch is a raw material with a wide field of application ranging from the imparting of texture and consistency in foods to the manufacture of biodegradable paper, adhesives and packaging. Starch is the main polysaccharide used at the industrial level, thanks to its properties modifying the texture, improving adhesion, enhancing moisture retention, increasing or decreasing viscosity, and increasing gel and film formation (Hernández-Medina et al., 2008).

10.2 STARCH SOURCES

The different vegetable sources such as cereals, tubers and legumes play an important role in the relationship of their properties and their applications in industry (Bajaj et al., 2018). Starch is organized into granules, where the diameters of the granules may vary between 1 and 200 nm (Tupa-Valencia, 2015). Forms of starch granules include the typical near-perfect spheres of small granules of wheat starch; discs, such as wheat and rye granules; polyhedral granules, as in rice and corn starch granules; irregular granules as seen in potato starch; cylindrical granules, often protruding on plant leaves; and filamentous granules in corn starch with a high degree of amylose. Granules of many plant species, such as oats, are composed of many starch particles that appear to have developed simultaneously within a single amyloplast (French, 1984).

The use of different sources of starch is related to different purposes. Tuber starch contains phosphorus in its esterified form, while cereal starches have large amounts of phospholipids. The starch from tubers is the most used due to its functional characteristics (swelling, high solubility, viscosity and gelling). Several reports indicate that legume starch showed better shear stability than wheat starches due to their higher maximum viscosity and lower degradation (Bajaj et al., 2018). Also, reports indicate that legume starches showed better retrogradation properties than cereal starches due to a higher amylose content, producing a more resistant starch (Bajaj et al., 2018).

10.3 RESISTANT STARCH

Resistant starch (RS) is recognized as a type of dietary fiber and is widely used as an ingredient in functional foods (Ashwar et al., 2016). RS has been defined as the part of the starch that is nondigestible in the small intestine of humans and may be fermented in the colon (Haralampu, 2000). Several reports have shown a relationship between the intake of fiber and a low risk of chronic degenerative diseases (Gemen et al., 2011; Fuller et al., 2016)

Five types of resistant starches have been reported (Birt et al., 2013). RS type 1 is physically entrapped in the cellular matrices, such as whole grains or partly milled cereal grains (Joye, 2019). RS type 2 is ungelatinized starch granules (Aribas et al., 2020). RS type 3 is a retrograded starch, commonly found in pasta (Snelson & Coughlan, 2019). RS type 4 is chemically modified starch typically through esterification, cross-linking or transglycosylation (Mah et al., 2018). RS type 5 is starch where the amylose component forms complexes with lipids (Ai et al., 2013).

Regardless of the type, resistant starch has been associated with several health benefits, such as slow glucose release; lower insulin response; inhibition or lower lipogenic enzymatic activity (i.e., all of them related to type 2 diabetes); and lower risk of obesity, coronary heart diseases and gastrointestinal disorders (Gelencsér et al., 2008; Ghodke & Ananthanarayan, 2008; Homayouni et al., 2014). Also, there are many reports about the use of RS as an ingredient to improve technological functionality in several food products (Buttriss & Stokes, 2008). Researchers have also found that the use of RS does not negatively affect mechanical and sensorial properties in comparison to traditional fibers.

10.4 STARCHY PASTA AND HEALTH

The use of legumes, specifically common beans, as ingredients of several food products, especially in pasta foods, has been reported in scientific literature for several years. The number of academic reports has increased over time, as can see in Table 10.1. Spaghetti, as a type of pasta, is considered to be a slowly digestible starch food, which is a feature ruled by the particular physical properties of the product. Several studies have been reported to increase the nutritional value of spaghetti by the addition of legume flours (Gallegos-Infante et

TABLE 10.1   PUBLISHED ARTICLES IN SCIENTIFIC JOURNALS ABOUT STARCH LEGUMES AND PASTA FROM 2010 TO 2021

Year	Articles published about legume starch	Articles published about pasta made with legume flour
2010	52	6
2011	49	8
2012	48	5
2013	61	5
2014	48	8
2015	52	4
2016	58	8
2017	71	9
2018	65	9
2019	96	16
2020	101	17
2021	81	10

al., 2010a). However, other reports also indicated the low quality of pasta when the amount of RS was increased (Gallegos-Infante et al., 2010b).

Manthey et al. (2004) reported that the use of buckwheat bran flour increases the cooking loss in comparison to traditional spaghetti made with semolina, similarly to a report by Gallegos-Infante et al. (2010b) who used common bean flour (30%) in spaghetti pasta. Likewise, Kim et al. (2006) reported that extruded pastry wheat flour can increase the resistant starch by modifying feed moisture and storage time.

10.4.1 Added Legume Flour to Pasta and Health

Gallegos-Infante et al. (2012) demonstrated that the incorporation of common bean flour to spaghetti at 30% (w/w) increased the presence of gallic acid, protocatechualdehyde, caffeic acid, vanillin, p-coumaric and coumaric acids, increasing the antioxidant activity of cooked pasta. Turco et al. (2019) reported that pasta made with 100% legume flours is a low glycemic food and can be considered as a suitable source of bioactive polyphenols and flavonoids in the human diet. Also, a report by Osorio-Díaz et al. (2008) indicated that in pasta made with durum wheat flour and chickpea flour, the dietary fiber content increased but the total starch decreased with the addition of chickpea flour content into the composite pasta. Greffeuille et al. (2015) reported in an in vivo study that the addition of 35% of legume flour to spaghetti pasta (65% semolina) showed low glycemic and insulin indexes. Similar results were reported in our laboratory by Garcia-Rivas (2011), who observed an important reduction of glucose after one month of standard diet based on spaghetti added with Bayo bean flour in lab animals (Wistar rats), as can be seen in Table 10.2.

Gallegos-Infante et al. (2012) reported that differences in antioxidant capacity and total phenolic content do not describe effects on glucose in the in vivo experiments; however, the previous experiment did not demonstrate the effect of resistant starch on the health benefits related to the reduced glucose level. About this fact, Ortega-Cordova (2013) did not observe any beneficial effect on glucose level, after diets with processed common beans. However, animals fed with processed beans showed a beneficial effect on triglycerides level in plasma and glomerular protection, though these effects could not be related to the polyphenol content of samples.

The main ingredient of traditional pasta is the durum wheat semolina. Unfortunately, several reports indicate that it lacks phenolic compounds because they are lost during the wheat grain milling (Ragaee et al., 2014). The usual strategy to produce functional pasta is to include whole wheat grain, and/or use alternative cereals and pseudocereals such as legumes in order to obtain a composite flour. Also, the main problem with this approach is the use of preprocessing technologies such as pressure cooking, drying, roasting, extrusion and infrared heating that might change starch properties and affect the polyphenolic profile and other bioactives present. Hence, efforts are still needed to evaluate the contribution of functional pasta consumption to maintain good health.

In general, reports about the impact on blood glucose of pasta starchy foods do not show this effect with the use of several types of sources (e.g., legumes, whole grain) (Jenkins et al., 1983). However, Kristensen et al. (2010) reported that the use of pasta made with whole wheat grain shows a higher increase of satiety in comparison with starchy foods made with refined wheat grain. In this sense, Eelderink et al. (2012) found that a slower intestinal uptake of

TABLE 10.2 GLUCOSE LEVEL IN THE BLOOD OF LAB ANIMALS (WISTAR RATS) FOLLOWING FEEDINGS OF SPAGHETTI WITH BAYO BEAN FLOUR ADDED

Sample	Glucose after 0 days	Glucose after 15 days	Glucose after 30 days
Spaghetti 0% Bayo bean flour	300±11 mg/dL	330 ± 9 mg/dL	360 ± 15 mg/dL
Spaghetti 30% Bayo bean flour	300 mg/dL	221 ± 13 mg/dL	192 ± 8 mg/dL
Data shown are mean ± standard deviation.			

glucose from a starchy food product can result in lower postprandial insulin and glucose-dependent insulinotropic polypeptide (GIP) concentrations, but not necessarily in a lower glycemic response because of a slower glucose clearance rate (GCR). Thus, even without being able to reduce postprandial glycemia, products with slowly digestible starch can have beneficial long-term effects. These authors indicate that starchy products with resistant starch cannot be identified by using the glycemic index and therefore another classification system may be necessary.

The study of nutrients in the real human diet is complex. In an interesting report by Pantophlet et al. (2017) it was indicated that consumption of pasta rich in fiber, followed by the consumption of whole wheat bread, showed that arabinoxylans (present in whole starchy foods) are degraded by the small intestinal microbiota. This action increases the solubility of arabinoxylans present in pasta, and then facilitates their degradation and the increase of small intestinal viscosity. These phenomena showed that wheat bran, depending on its processing, by means of increased viscosity, can interfere with macronutrient absorption or digestion in the small intestine, thereby influencing postprandial glucose and insulin responses. Thus, more research about the interaction of different foods and their processing could be done to understand the phenomena. In this sense, Robertson et al. (2021) showed a strong effect on the area under the curve (AUC) of freshly cooked starchy pasta food compared to several cycles of chilling and reheating. They indicated unclear results, claiming, although with no conclusive data, the possible formation of resistant starch type 5 after the oil and starch interactions.

An interesting proposal to the use of ingredients rich in soluble dietary fiber in blends with durum wheat semolina for the obtention of starchy foods such as pasta was made by Peressini et al. (2020). They induced large changes in dough viscoelastic properties and water absorption, which are key factors for the swelling of starch granules in cooked pasta. Also, they demonstrated that a swelling index obtained from rheometric data by a dynamic temperature sweep correlated with the in vitro glycemic response of pasta, suggesting that changes in food structure are important to modulate the glycemic response of pasta products.

10.5 NOVELTY USES OF LEGUME STARCH

Starch is the principal carbohydrate in legumes, and it is well known that this particular starch has unique structural features, making it attractive for different food and nonfood applications. Consequently, healthy habits in human nutrition have driven the food industries to search for new ingredients with functional characteristics. In this sense, legume starches have shown a poor digestibility property, which has resulted in a slow postprandial glucose promotion, affecting insulin and glycemic index responses. Complementary, another important factor is the current growing interest worldwide in finding new starch sources to satisfy market demand (Núñez-Bretón et al., 2019).

Traditionally, starch has been applied as a food additive controlling texture and consistency of different food products. The significance of starch lies in its important role as a thickener, water binder, emulsion stabilizer and gelling agent. Also in nonfood applications it has other key roles, such as in the paper industry as a surface coater; in the pharmaceutical sector as a drug and bioactive delivery system; in the texture industry, increasing the mechanical properties; and in the binder industry, taking advantage of its adhesive properties. Although there are many traditional starch applications, besides the commented ones, currently there is great interest to generate new starch applications to reach unique functional properties that could be adapted to different products and market niches. In the following some new developments in legume starch applications are described.

10.5.1 Starch Nanoparticles

Nowadays, starch nanoparticles have been gaining major attention due to the beneficial improvements they can bring into products of different sectors. Starch may be a key ingredient in composites, functional foods and cosmeceuticals; being the main area of development in the food packaging technology, increasing water vapor, and mechanical resistance of material matrixes. Basically, starch nanoparticles are formed from the breakdown of the semicrystalline structure of starch granules. There are several methods for starch nanoparticles preparation. One of the most utilized techniques is acid and enzymatic hydrolysis, allowing for nanocrystals with high crystallinity from the amorphous starch granule fraction. The principal disadvantage for the acid treatment is the need of long periods of time, leading to low yields. In this sense, new methodologies such as ultrasound, nanoprecipitation, reactive extrusion and other physicomechanical techniques are being experimented.

In recent years, starch nanocrystals have been used to reinforce biodegradable or nonbiodegradable matrix materials. As starch nanoparticles are smaller than 1000 nm, they exhibit many interesting properties. The main feature is their high surface-to-volume ratio, providing a major contact surface and promoting the transport phenomena.

Originally, the acid hydrolysis was by far the most used technique in starch nanocrystal formation, allowing the release of the crystalline sections from starch granules. At first, acid treatment was employed to understand the internal structure of starch granules, and then the methodology was adapted to obtain nanoparticles with an interesting grade of crystallinity. Acid hydrolysis can induce significant changes in the inner and outer structure of starch granules, modifying the physical and chemical properties of starch. This may become significant in making, for example, pastes with lower intrinsic viscosity values, reduced hot paste viscosity, increased gel strength, increased water solubility and better film-forming capabilities. Consequently, the use of a resistant starch ingredient is suitable in food industry applications as a gelling agent and a fat substitute (Wang & Copeland, 2015).

Legume starches have been poorly studied as sources of nanocrystals. These types of nanoparticles has been investigated using maize and potato starches as the main sources, which now represents a big opportunity to expand the knowledge about legume starch. Sharma et al. (2020) experimented with kidney bean (*Phaseolus vulgaris*) starch in the formation of nanoparticles or nanostarch. The formation of these particles was obtained through mixing the starch with sulfuric acid (3.16 M) for 7 days at 40°C. After that, it was centrifuged, and the precipitated washed and dried. The nanostarch obtained showed higher water solubility, swelling power, and water and oil absorption capacities.

Comparably, Xu et al. (2014) examined the morphological and thermal properties of nanocrystals of mung bean and chickpea starches. According to their results, the mung bean starch is a great candidate for the formation of starch nanocrystals. This starch apparently shows greater resistance to the hydrolysis process, maybe because the starch is less swollen in the process, making it less susceptible to degradation and reaching a yield of 45%. One of the changes registered in hydrolysis for these starches was the modification of the diffraction patterns. It is important to mention that comparatively to B-type starches, C-type starches are less susceptible to change. Subsequently, acid hydrolysis affects the morphology of the starch nanocrystals, but regardless of the botanical origin, round or oval shapes are always obtained.

In recent years, starch nanocrystals have been used to reinforce biodegradable or nonbiodegradable matrix materials. In this sense, Chen et al. (2008) studied pea starch nanocrystals in the quest to provide property-enhanced composites blending them with poly(vinyl alcohol), finding lower moisture, higher light transmittance, smoother fracture surfaces, higher tensile strength and elongations at the break. These changes may be due to the correct dispersion of the starch nanocrystals in the poly(vinyl alcohol) medium due to their smaller size (30–80 nm). Roy et al. (2020) studied the incorporation of starch nanoparticles obtained from mung beans through acid hydrolysis into a film formed with native starch, obtaining clear, transparent, thicker and flexible film with an improvement on its biodegradability properties. According to their results, reinforced films have considerable possibilities in the food industry as packaging material.

Structural studies have presented hydrolyzed starches with an interesting behavior. Ahmed and Auras (2011) found that the best way to describe starch gelatinization is through a second-order reaction model. In the same way, hydrolyzed starch showed lower activation energies than native starches, therefore, this change may be explained by the loss of one of the components along the hydrolysis process.

Another interesting application of legume starch nanocrystals is the stabilization of emulsions of different nature. Pickering emulsions are emulsions where the surfactants (stabilizers) are replaced by solid particles, usually in the micro- or nanoscale domain. Pickering emulsions have been gained a lot of attention in the industry, since the development of new materials could allow a wide diversity of applications, principally due to the increased instability and their low toxicity. Starch has been systematically used as an emulsion stabilizer in Pickering emulsions, due to their low cost, availability, biodegradability and nontoxicity. Starch nanocrystals maintain all their attributes and can improve them, since it is well known that the very low size is a key role in Pickering emulsions and therefore they can promote stable physical emulsions. In this sense, Gomes Daniel et al. (2020) studied starch nanocrystals at different concentrations from common bean (*Phaseolus vulgaris*) as stabilizers of Pickering emulsions, finding an increasing effect in the emulsion index (up to 21 days) when the nanocrystals were treated with sodium hypochlorite to produce an oxidized particle. It has been reported that starch nanocrystals tended to aggregate and form microstructures affecting their technological applications. Consequently, this effect may be reduced with a posterior treatment as oxidation.

Other novel applications of starch nanoparticles in drug delivery processes are based on many reasons such as small size, customized surface, multiple functionalities and improved solubility. In this sense, Farrag et al. (2018) used starch from peas in the formation of nanoparticles loaded with quercetin, a naturally occurring flavonoid with antioxidant activity, and anti-inflammatory and anticancer effects. They found a higher loading capacity of quercetin than in corn starch nanoparticles, and a similar released fraction of the flavonoid with respect to potato nanoparticles.

Starch nanoparticles have been investigated in novel nonfood applications, as for example in the formation of magnetic starch nanoparticles. Oladebeye (2020) synthesized these kind of nanoparticles using Jack bean (*Canavalia ensiformis*) through cross-linking with glutaraldehyde. He found that magnetic nanoparticles are affected by pH, having a rocklike morphology with a degree of crystallinity of 87.30% and mean size of 1.09 nm. Subsequently, these nanoparticles could be used in water treatment technology.

10.5.2 Legume Starch Films

The research to find new biopolymers capable of developing complex matrices that could be used as packaging materials is a growing and attractive field due to the sustainable quest of alternatives to petroleum-based derivates. Traditionally, starch has been used to develop materials with an important potential at low cost, forming transparent, flexible and biodegradables films. Normally, starch films are formed by blending starch suspensions with other components, principally plasticizers, by which they can modify the mechanical properties of the film. Starch films usually work as a physical barrier between the material of interest and the environment. Therefore, there is a growing interest in the development of starch films that besides protecting the product, as a mechanical barrier, are able to carry on flavors, nutraceuticals, antimicrobial and antioxidant agents, minerals, and vitamins; in other words, starches that are able to produce films with technofunctional properties.

Corrales et al. (2009) formed pea starch films with antimicrobial activity loading grape seed extracts into the matrix, finding a bacteriostatic effect against meat microbial surfaces at least for 4 days. They have prepared a remarkable system with antimicrobial capabilities in addition to the important fact that the films are totally biodegradable and edible, reducing any environmental constraints. Lee et al. (2020) enhanced water resistance of an edible film incorporating sunflower seed oil to the film matrix developed with mung bean starch. This effect may be attributed to the hydrophobicity caused by the sunflower seed oil, reducing the binding water sites into the filled matrix. Most edible films prepared with polysaccharides and proteins are relatively low water resistant, which is an undesirable feature in the food industry, since it accelerates the breakdown of edible films. In this sense, mung bean starch films enriched with sunflower seed oil arises as a great alternative, making a film with potential use as a new edible film in the food packaging industry.

Thermal properties of legume starch films is another important factor to be considered. Han et al. (2006) studied thermal properties of pea starch edible films containing beeswax emulsions, finding two endothermic peaks. One corresponded to the crystalline structure of the starch molecule and the second to the amylose–lipid complexes, resulting in a heterogeneous structure due to the physical reorganization of beeswax in matrices during storage time.

The thickness and mechanical properties of legume starch edible films are other key factors since films showing good elongation are desirable for the food packaging industry. Vanier et al. (2019) studied different genotypes of the common bean family (*Phaseolus vulgaris* L.). According to their results, starch with a high content of amylose exhibited higher tensile strength than other films since the tensile strength is favored by amylose crystallization.

Another important factor is the color and opacity of starch films since it influences consumer acceptance. Various factors affect the transparency of films, but principally the presence of insoluble components in the starch paste reduces the transparency. Also, legume starch usually requires longer heating times to ensure a complete gelatinization process and in consequence producing a transparent film (Zhang & Li, 2021).

10.6 FUTURE PROSPECTS IN NOVELTY APPLICATIONS OF LEGUME STARCH AND STARCHY FOODS

Legume starches have been interestingly studied from different points of view in the quest to satisfy the growing starch market. There are new starch applications, but they are still in development using common sources like corn, potato and cassava. What may be the most interesting application is the use of starch in 3D printing for different purposes, but scaffolds and food are still the principal ones, involving two very important sectors to humanity as health and nutrition.

More studies that emphasize nutrigenomics in order to understand the complex function of the legume starch in pasta foods and their relation to health beyond the glycemic index are needed. With no doubt, legume starches could play a key role in the future trends on functional foods and cosmeceuticals.

REFERENCES

Ahmed, J., & Auras, R. (2011). Effect of acid hydrolysis on rheological and thermal characteristics of lentil starch slurry. *LWT - Food Science and Technology*, 44(4), 976–983. https://doi.org/10.1016/j.lwt.2010.08.007.

Ai, Y., Hasjim, J., & Jane, J. L. (2013). Effects of lipids on enzymatic hydrolysis and physical properties of starch. *Carbohydrate Polymers*, 92(1), 120–127. https://doi.org/10.1016/j.carbpol.2012.08.092

Ashwar, B. A., Gani, A., Wani, I. A., Shah, A., Masoodi, F. A., & Saxena, D. C. (2016). Production of resistant starch from rice by dual autoclaving-retrogradation treatment: Invitro digestibility, thermal and structural characterization. *Food Hydrocolloids*, 56, 108–117. https://doi.org/10.1016/j.foodhyd.2015.12.004

Aribas, M., Kahraman, K., & Koksel, H. (2020). In vitro glycemic index, bile acid binding capacity and mineral bioavailability of spaghetti supplemented with resistant starch type 4 and wheat bran. *Journal of Functional Foods*, 65, 103778. https://doi.org/10.1016/j.jff.2020.103778

Bajaj, R., Singh, N., Kaur, A., & Inouchi, N. (2018). Structural, morphological, functional and digestibility properties of starches from cereals, tubers and legumes:

A comparative study. *Journal of Food Science and Technology*, 55(9), 3799–3808. https://doi.org/10.1007/s13197-018-3342-4

Birt, D. F., Boylston, T., Hendrich, S., Jane, J. L., Hollis, J., Li, L., McCleland, J., Moore S., Phillips, G. J., Rowling, M., Schalinske, K., Scott, M. P., & Whitley, E. M. (2013). Resistant starch: Promise for improving human health. *Advances in Nutrition*, 4(6), 587–601. https://doi.org/10.3945/an.113.004325

Buttriss, J. L., & Stokes, C. S. (2008). Dietary fibre and health: An overview. *Nutrition Bulletin*, 33(3), 186–200. https://doi.org/10.1111/j.1467-3010.2008.00705.x

Chaves-Martins, P., Gutkoski, L. C., & Martins, V. G. (2018). Impact of acid hydrolysis and esterification process in rice and potato starch properties. *International Journal of Biological Macromolecules*, 120(A), 959–965. https://doi.org/10.1016/j.ijbiomac.2018.08.170

Chel-Guerrero, L., Cruz-Cervera, G., Betancur-Ancona, D., & Solorza-Feria, J. (2011). Chemical composition, thermal and viscoelastic characterization of tuber starches growing in the Yucatan peninsula of Mexico. *Journal of Food Process Engineering*, 34(2), 363–382. https://doi.org/10.1111/j.1745-4530.2009.00362.x

Chen, Y., Xiaodong, C., Chang, P. R., & Huneault, M. A. (2008). Comparative study on the films of poly(vinyl alcohol)/pea starch nanocrystals and poly(vinyl alcohol)/native pea starch. *Carbohydrate Polymers*, 73(1), 8–17. https://doi.org/10.1016/j.carbpol.2007.10.015

Corrales, M., Han, J. H., & Tauscher, B. (2009). Antimicrobial properties of grape seed extracts and their effectiveness after incorporation into pea starch films. *International Journal of Food Science & Technology*, 44(2), 425–433. https://doi.org/10.1111/j.1365-2621.2008.01790.x

Eelderink, C., Schepers, M., Preston, T., Vonk, R. J., Oudhuis, L., & Priebe, M. G. (2012). Slowly and rapidly digestible starchy foods can elicit a similar glycemic response because of differential tissue glucose uptake in healthy men. *The American Journal of Clinical Nutrition*, 96(5), 1017–1024. https://doi.org/10.3945/ajcn.112.041947

Escarpa, A., & González, M. C. (1997). Tecnología del almidón resistente/Technology of resistant starch. *Food Science and Technology International*, 3(3), 149–161. https://doi.org/10.1177/108201329700300301

Farrag Y., Ide W., Montero B., Rico M., Rodriguez-Llamazares S., Barral L., & Bouza R. (2018). Preparation of starch nanoparticles loaded with quercetin using nano-precipitation technique. *International Journal of Biological Macromolecules*, 114, 426–433. https://doi.org/10.1016/j.ijbiomac.2018.03.134.

French D. (1984). Organization of starch granules. In Whistler R. L., Bemiller J. N., & Paschall E. (Eds.), *Starch Chemistry and Technology* (2nd ed). San Diego, CA, USA: Academic Press. https://doi.org/10.1016/B978-0-12-746270-7.50013-6

Fuentes-Zaragoza, E., Riquelme-Navarrete, M. J., Sánchez-Zapata, E., & Pérez-Álvarez, J. A. (2010). Resistant starch as functional ingredient: A review. *Food Research International*, 43(4), 931–942. https://doi.org/10.1016/j.foodres.2010.02.004

Fuller, S., Beck, E., Salman, H., & Tapsell, L. (2016). New horizons for the study of dietary fiber and health: A review. *Plant Foods for Human Nutrition*, 71(1), 1–12. https://doi.org/10.1007/s11130-016-0529-6

Gallegos-Infante, J. A., Bello-Perez, L. A., Rocha-Guzman, N. E., Gonzalez-Laredo, R. F., & Avila-Ontiveros, M. (2010a). Effect of the addition of common bean (*Phaseolus vulgaris* L.) flour on the in vitro digestibility of starch and undigestible carbohydrates in spaghetti. *Journal of Food Science*, 75(5), H151–H156. https://doi.org/10.1111/j.1750-3841.2010.01621.x

Gallegos-Infante, J. A., Rocha-Guzman, N. E., Gonzalez-Laredo, R. F., Ochoa-Martínez, L. A., Corzo, N., Bello-Perez, L. A., Medina-Torres, L. M., & Peralta-Alvarez, L. E. (2010b). Quality of spaghetti pasta containing Mexican common bean flour (*Phaseolus vulgaris* L.). *Food Chemistry*, 119(4), 1544–1549. https://doi.org/10.1016/j.foodchem.2009.09.040

Gallegos-Infante, J. A., García Rivas, M., Chang, S., Manthey, F., Yao, R. F., Reynoso-Camacho, R., Rocha-Guzman, N. E., & González-Laredo, R. F. (2012). Effect of the addition of common bean flour on the cooking quality and antioxidant characteristics of spaghetti. *Journal of Microbiology, Biotechnology and Food Sciences*, 2(2), 730–744.

Garcia-Rivas, M. (2011). *Efecto del consumo de pasta desarrollada a partir de trigo-frijol sobre los niveles hipoglucémicos e hipolipidémicos en ratas diabéticas* (Tesis de Maestría, Instituto Tecnológico de Durango, México).

Gelencsér, T., Gál, V., Hódsági, M., & Salgó, A. (2008). Evaluation of quality and digestibility characteristics of resistant starch-enriched pasta. *Food and Bioprocess Technology*, 1(2), 171–179. https://doi.org/10.1007/s11947-007-0040-z

Gemen, R., de Vries, J. F., & Slavin, J. L. (2011). Relationship between molecular structure of cereal dietary fiber and health effects: Focus on glucose/insulin response and gut health. *Nutrition reviews*, 69(1), 22–33. https://doi.org/10.1111/j.1753-4887.2010.00357.x

Ghodke, S. K., & Ananthanarayan, L. (2008). Health benefits of resistant starch. *Agro Food Industry Hi Tech*, 19(1), 26–28.

Gomes Daniel, T. H., Bedin, A. C., Souza, E. C. F., Lacerda, L. G., & Demiate, I. M. (2020). Pickering emulsions produced with starch nanocrystals from Cassava (*Manihot esculenta* Crantz), Beans (*Phaseolus vulgaris* L.), and Corn (*Zea mays*). Starch-Stärke, 72(11–12), 1900326. https://doi.org/10.1002/star.201900326

Greffeuille, V., Marsset-Baglieri, A., Molinari, N., Cassan, D., Sutra, T., Avignon, A., & Micard, V. (2015). Enrichment of pasta with faba bean does not impact glycemic or insulin response but can enhance satiety feeling and digestive comfort when dried at very high temperature. *Food & Function*, 6(9), 2996–3005. https://doi.org/10.1039/C5FO00382B

Han J. H., Seo G. H., Park I. M., Kim G. N., Lee D. S. (2006). Physical and mechanical properties of pea starch edible films containing beeswax emulsions. *Journal of Food Science*, 71(6):E290–E296. https://doi.org/10.1111/j.1750-3841.2006.00088.x.

Haralampu, S. G. (2000). Resistant starch: A review of the physical properties and biological impact of RS_3. *Carbohydrate Polymers*, 41(3), 285–292. https://doi.org/10.1016/S0144-8617(99)00147-2

Hernández-Medina, M., Torruco-Uco, J. G., Chel-Guerrero, L., Betancur-Ancona, D. (2008). Caracterización fisicoquímica de almidones de tubérculos cultivados en Yucatán, México. *Food Science and Technology*, 28(3), 718–726. https://doi.org/10.1590/S0101-20612008000300031

Homayouni, A., Amini, A., Keshtiban, A. K., Mortazavian, A. M., Esazadeh, K., Pourmoradian, S. (2014). Resistant starch in food industry: A changing outlook for consumer and producer. Starch-Stärke, 66(1–2), 102–114. https://doi.org/10.1002/star.201300110

Jenkins, D. J., Wolever, T. M., Jenkins, A. L., Lee, R., Wong, G. S., & Josse, R. (1983). Glycemic response to wheat products: Reduced response to pasta but no effect of fiber. *Diabetes Care*, 6(2), 155–159. https://doi.org/10.2337/diacare.6.2.155

Joye, I. J. (2019). Cereal biopolymers for nano-and microtechnology: A myriad of opportunities for novel (functional) food applications. *Trends in Food Science & Technology*, 83, 1–11. https://doi.org/10.1016/j.tifs.2018.10.009

Kim, H. Y., Park, S. S., Lim, S. T. (2015). Preparation, characterization and utilization of starch nanoparticles. *Colloids and Surfaces B: Biointerfaces*, 126(1), 607–620. https://doi.org/10.1016/j.colsurfb.2014.11.011

Kim, J. H., Tanhehco, E. J., Ng, P. K. W. (2006). Effect of extrusion conditions on resistant starch formation from pastry wheat flour. *Food Chemistry*, 99(4), 718–723. https://doi.org/10.1016/j.foodchem.2005.08.054

Kristensen, M., Jensen, M. G., Riboldi, G., Petronio, M., Bügel, S., Toubro, S., Tetens, I., Astrup, A. (2010). Wholegrain vs. refined wheat bread and pasta. Effect on post-prandial glycemia, appetite, and subsequent *ad libitum* energy intake in young healthy adults. *Appetite*, 54(1), 163–169. https://doi.org/10.1016/j.appet.2009.10.003

Lee J. S., Lee E. S., Han J. (2020). Enhancement of the water-resistance properties of an edible film prepared from mung bean starch via the incorporation of sunflower seed oil. *Science Reports*, 10(1), 13622. https://doi.org/10.1038/s41598-020-70651-5.

Mah, E., Garcia-Campayo, V., Liska, D. (2018). Substitution of corn starch with resis-tant starch type 4 in a breakfast bar decreases postprandial glucose and insulin responses: A randomized, controlled, crossover study. *Current Developments in Nutrition*, 2(10), nzy066. https://doi.org/10.1093/cdn/nzy066

Manners, D. J., & Matheson, N. K. (1981). The fine structure of amylopectin. *Carbohydrate Research*, 90(1), 99–110. https://doi.org/10.1016/S0008-6215(00)85616-5

Manthey, F. A., Yalla, S. R., Dick, T. J., Badaruddin, M. (2004). Extrusion properties and cooking quality of spaghetti containing buckwheat bran flour. *Cereal Chemistry*, 81(2), 232–236. https://doi.org/10.1094/CCHEM.2004.81.2.232

McCready, R. M., Guggolz, J., Silviera, V., & Owens, H. S. (1950). Determination of starch and amylose in vegetables. *Analytical chemistry*, 22(9), 1156–1158. https://doi.org/10.1021/ac60045a016

Núñez-Bretón L. C., Cruz-Rodríguez L. C., Tzompole-Colohua M. L., Jiménez-Guzmán J., Perea-Flores M. J., Rosas-Flores W., González-Jiménez F. E. (2019). Physicochemical, functional and structural characterization of Mexican *Oxalis tuberosa* starch modified by cross-linking. *Journal of Food Measurement and Characterization*, 13(4), 2862–2870. https://doi.org/10.1007/s11694-019-00207-3.

Oladebeye, A. O. (2020). Potentials of starch nanoparticles Jack Bean (*Canavalia ensifor-mis*) coprecipitated with iron (II, III) oxide. *International Research Journal of Pure and Applied Chemistry*, 17–24. https://doi.org/10.9734/irjpac/2020/v21i330157.

Ortega-Cordova, V. (2013). *Elaboración del enlatado de dos variedades de frijol común (Phaseolus vulgaris) y evaluación de su efecto antidiabético* (Tesis de Maestría, Universidad Autónoma de Querétaro (UAQ). Querétaro, México). http://ri.uaq.mx/handle/123456789/4571

Osorio-Diaz, P., Agama-Acevedo, E., Mendoza-Vinalay, M., Tovar, J., Bello-Perez, L. A. (2008). Pasta added with chickpea flour: Chemical composition, in vitro starch digestibility and predicted glycemic index. *Ciencia y Tecnología Alimentaria*, 6(1), 6–12. https://doi.org/10.1080/11358120809487621

Pantophlet, A. J., Wopereis, S., Eelderink, C., Vonk, R. J., Stroeve, J. H., Bijlsma, S., van Stee, L., Bobeldijk, I., Priebe, M. G. (2017). Metabolic profiling reveals differences in plasma concentrations of arabinose and xylose after consumption of fiber-rich pasta and wheat bread with differential rates of systemic appearance of exogenous glucose in healthy men. *The Journal of Nutrition*, 147(2), 152–160. https://doi.org/10.3945/jn.116.237404

Peressini, D., Cavarape, A., Brennan, M. A., Gao, J., Brennan, C. S. (2020). Viscoelastic properties of durum wheat doughs enriched with soluble dietary fibres in relation to pasta-making performance and glycaemic response of spaghetti. *Food Hydrocolloids*, 102, 105613. https://doi.org/10.1016/j.foodhyd.2019.105613

Ragaee, S., Seetharaman, K., Abdel-Aal, E-S. M. (2014). The impact of milling and thermal processing on phenolic compounds in cereal grains. *Critical reviews in food science and nutrition*, 54(7), 837–849. https://doi.org/10.1080/10408398.2011.610906

Reddy, C. K., Suriya, M., Haripriya, S. (2013). Physico-chemical and functional properties of Resistant starch prepared from red kidney beans (Phaseolus vulgaris. L) starch by enzymatic method. *Carbohydrate Polymers*, 95(1), 220–226. https://doi.org/10.1016/j.carbpol.2013.02.060

Robertson, T. M., Brown, J. E., Fielding, B. A., Robertson, M. D. (2021). The cumulative effects of chilling and reheating a carbohydrate-based pasta meal on the postprandial glycaemic response: A pilot study. *European Journal of Clinical Nutrition*, 75, 570–572. https://doi.org/10.1038/s41430-020-00736-x

Roy, K., Theory, R., Sinhmar, A., Pathera, A. K., & Nain, V. (2020). Development and characterization of nano starch-based composite films from mung bean (*Vigna radiata*). *International Journal of Biological Macromolecules*, 144, 242–251. https://doi.org/10.1016/j.ijbiomac.2019.12.113.

Sharma, I., Sinhmar, A., Theory, R., Sandhu, K. S., Kaur, M., Nain, V., Pathera, A. K., & Chavan, P. (2020). Synthesis and characterization of nano starch-based composite films from kidney bean (*Phaseolus vulgaris*). *Journal of Food Science and Technology*, 58, 21780–2185. https://doi.org/10.1007/s13197-020-04728-4.

Snelson, M., & Coughlan, M. T. (2019). Dietary advanced glycation end products: Digestion, metabolism and modulation of gut microbial ecology. *Nutrients*, 11(2), 215. https://doi.org/10.3390/nu11020215

Tupa Valencia, M. V. (2015). *Desarrollo de una metodología sostenible de síntesis de almidones acetilados* (Tesis de Maestría, Facultad de Ciencias Exactas y Naturales. Universidad de Buenos Aires).

Turco, I., Bacchetti, T., Morresi, C., Padalino, L., & Ferretti, G. (2019). Polyphenols and the glycaemic index of legume pasta. *Food & Function*, 10(9), 5931–5938. https://doi.org/10.1039/C9FO00696F

Vanier, N. L., de Oliveira, J. P., Bruni, G. P., El Halal, S. L.M, Villanova, F. A., Zavareze, E. D. R., Días, A. R. G., & Bassinello, P. Z. (2019). Characteristics of starch from different bean genotypes and its effect on biodegradable films. *Journal of the Science of Food and Agriculture*, 99 (3), 1207–1214. https://doi.org/10.1002/jsfa.9292.

Wang, S., & Copeland L. (2015). Effect of acid hydrolysis on starch structure and functionality: A review. *Critical Review on Food Science and Nutrition*, 55 (8), 1081–97. https://doi.org/10.1080/10408398.2012.684551.

Wani, I. A., Sogi, D. S., Wani, A. A., Gill, B. S., & Shivhare, U. S. (2010). Physico-chemical properties of starches from Indian kidney bean (*Phaseolus vulgaris*) cultivars. *International Journal of Food Science & Technology*, 45(10), 2176–2185. https://doi.org/10.1111/j.1365-2621.2010.02379.x

Xu, Y., Sismour, E. N., Grizzard, C., Thomas, M., Pestov, D., Huba, Z., Wang, T., & Bhardwaj, H. L. (2014). Morphological, structural, and thermal properties of starch nanocrystals affected by different botanic origins. *Cereal Chemistry*, 91(4), 383–388. https://doi.org/10.1094/CCHEM-10-13-0222-R.

Zhang, Y., & Li, Y. (2021). Comparison of physicochemical and mechanical properties of edible films made from navy bean and corn starches. *Journal of the Science and Food Agriculture*, 101(4), 1538–1545. https://doi.org/10.1002/jsfa.10772.

Chapter 11

Influence of the Food Matrix on the Digestibility and Metabolic Responses to Starchy Foods

Yolanda Elizabeth Pérez-Beltrán, Juscelino Tovar
and Sonia G. Sáyago-Ayerdi

CONTENTS

DOI: 10.1201/9781003088929-11

11.1 INTRODUCTION

The main function of food is to provide nutrients; however, a reductionist approach is to consider nutrients in an individual matter. To consider food as a sum of macro- and micronutrients is a limited way to understand their true effects on the metabolic response. The impact of carbohydrates, fats, proteins or even polyphenols can be different depending on the food in which they are present and on other products that are consumed at the same time. Therefore, it is essential to consider the interactions between nutrients and the food matrix when assessing and/or interpreting the metabolic response.

The term *food matrix* can be defined as the set of nutrient and non-nutrient components of foods and their molecular relationships. Capuano et al. (2018) described food matrix as the whole of the interactions between food components. These interactions occur at different scales and produce distinct microstructures and macrostructures (Aguilera, 2019). Digestion can be considered as a physiological interface between food and its effects on health. During digestion, the food matrix, which modulates the rate and degree to which nutrients and bioactive compounds are available for absorption, is broken down and the different nutrients and bioactive compounds can be absorbed through a synergy of mechanical, chemical and biochemical processes.

Therefore, it is important to understand the influence of the food matrix on the digestibility and metabolic responses to foods, especially those most consumed throughout the world. In this sense, starch is the most abundant food polysaccharide consumed worldwide and is thus an important macronutrient. Starch represents 45% to 65% of the caloric intake daily. It is also the main structure-building macroconstituent in many widely consumed foods, such as bread, pasta, tortilla, rice and pastries, and the main glucose provider in human diets.

Cereals and tubers are considered the main carbohydrate sources in different diets and are largely produced worldwide. They are not only rich in starch (~75%) and protein (8% to 14%), but also contain vitamins, minerals, phytoestrogens, and other phytochemicals (Martinez, 2021). Whole grains are characterized by the presence of all kernel constituents; they are rich in dietary fiber (DF), resistant starch (RS), antioxidants, and other important micronutrients such as folic acid and other vitamins. Altogether, these components of whole grains have relevant functional properties that can, at least in part, explain their health benefits (Della Pepa et al., 2018).

Legumes, on the other hand, are sources of complex carbohydrates, protein and DF, having significant amounts of vitamins and minerals, and high energy value (Verkempinck et al., 2020). Particularly, beans contain different types of phytochemicals (anthocyanins, flavonols, isoflavones) whose bioactivity has been proven in both in vitro and in vivo tests and supported by epidemiological data (Chávez-Mendoza & Sánchez, 2017; Monk et al., 2019). One of the mechanisms proposed for the health-promoting effect of legumes relates to its high DF content and relatively low glycemic index (GI). The latter is influenced by extrinsic and intrinsic factors of foods that alter gastrointestinal motility and influence the rate of starch digestion and absorption of the released glucose (Rovalino-Córdova et al., 2018). It is important to define the effects and influence of these factors in the broad scenario that encompasses the impact of the food matrix on starch digestion and thus the metabolic response derived from those processes.

11.2 STARCH DIGESTION

From the nutritional perspective, starch can be classified as glycemic or nonglycemic, depending on its bioavailability (Miao & Hamaker, 2021). Also, on the basis of the rate and extent of glucose liberation from starch during digestion, starch can be classified into different fractions using in vitro procedures, such as the one proposed by Englyst et al. (1992). According to that experimental protocol, three main nutritionally relevant starch fractions can be quantified:

- Rapidly digestible starch (RDS) – amount of glucose released after 20 min of in vitro digestion.
- Slowly digestible starch (SDS) – amount of glucose released between 20 and 120 min.
- Indigestible or resistant starch (RS) – the total starch content minus the amount of glucose released within 120 min of *in vitro* digestion. This fraction can be defined by the following equation: RS = Total starch – (RDS + SDS).

The indigestible starch fractions may arise from different reasons, such as physical inaccessibility (RS1), presence of incompletely gelatinized granules (RS2), retrograded starch (RS3) or the presence of unnatural chemical groups in modified starches (RS4). The formation of different types of complexes between starch and other food components such as lipids, proteins and phenolics, also confers indigestible character to starch molecules. Those complexes are referred to as type 5 resistant starch (RS5) (Gutiérrez & Tovar, 2021).

Starch digestion and utilization in the gastrointestinal tract are important physiological processes that together consist of three phases: the intraluminal phase, the brush border phase and the glucose absorption phase. Starch digestion starts in the mouth, where the polysaccharide is partly hydrolyzed by salivary α-amylase (E.C. 3.2.1.1). The process is largely completed in the duodenum by means of pancreatic α-amylase. Human salivary and pancreatic α-amylases are α-1,4 endoglucosidases, which have five subsites to bind the starch substrate and then cleave the α-1,4 glycosidic linkages using a multiple-attack mechanism (Sun & Miao, 2020). Despite the short duration of the oral phase and inactivation of salivary amylase at pH < 4, acidification in the stomach may still be sufficiently slow to allow salivary amylase to significantly contribute to starch digestion. Pancreatic α-amylase breaks the starch molecules into smaller oligosaccharides, maltose, maltotriose (G3), maltotetrose (G4) and other branched oligosaccharides called α-limit dextrins, which are further converted into glucose in the small intestine by the action of the brush border enzymes maltase–glucoamylase (MGAM) and sucrose–isomaltase (SI), which account for more than 85% of α-glucogenesis from starch. The action of α-amylase can thus be seen as an amplifier of mucosal starch digestion. MGAM has a higher α-glucogenic activity than SI, but it is inhibited by mealtime concentrations of luminal maltodextrins. In this way, MGAM regulates the total rate of starch α-glucogenesis. Compared to its N-terminal subunit, the C-terminal subunit of MGAM has greater catalytic efficiency due to its higher affinity for α-glucan substrates and greater number of binding configurations at the active site. Differences in the substrate concentration lead to changes in the formation of α-limit dextrins, which can influence the hydrolytic release of glucose monomers by the brush border enzymes before absorption (Lee et al., 2016; Lin et al., 2015).

The glucose released by starch digestion is continuously absorbed by enterocytes and transported into blood circulation by the action of glucose transporters, such as sodium-dependent glucose transporter 1 (SGLT1) and glucose transporter 2 (GLUT2) (Kellett & Brot-Laroche, 2005). Glucose absorption is closely associated with the activity and abundance of SGLT-1, which is hormonally regulated by incretin hormones in the digestive tract in response to distal glucose stimuli, such as glucose-dependent insulinotropic polypeptide (GIP) and glucagon-like peptide-1 (GLP-1) (Edwards et al., 2015a; Sun & Miao, 2020). These hormones, together with the vague nerve, are the main regulators of the proximal–distal feedback loop, called the gut–brain axis, which control the coordination of gastrointestinal tract activity and internal metabolism (Hasek et al., 2018). Released ileum glucose and short-chain fatty acids (SCFAs) from colonic fermentation have been shown to promote the so-called ileal brake and colonic brake, by stimulating the release of GLP-1 and peptide YY (PYY), which inhibit the upper gut motility, slow the gastric emptying rate and decrease food intake with associated health benefits (Pletsch & Hamaker, 2018).

As mentioned earlier, RDS causes sharp postprandial blood glucose rises, while SDS leads to a more gradual increase. RS is neither digested nor absorbed in the small intestine and does not lead to an increase in blood sugar levels, but it is fermented to a variable extent by the microbiota within large gut lumen (Gao et al., 2019). Thus, the location of digestion, and colonic fermentation of digestible starch not only mediates brake effects but also improves the physiological processes via the maintenance of glucose homeostasis and energy balance to achieve a wide range of health benefits. Indeed, different studies suggest that SDS may play a positive role in the prevention of diabetes and cardiovascular diseases, and provide beneficial effects on colorectal cancer, diabetes, and chronic kidney diseases (Hardy et al., 2020; Martinez, 2021).

In human diets, starch is almost uniquely consumed as gelatinized starch. Gelatinization is the irreversible physical change that occurs when starch is heated in the presence of enough water. During this process starch absorbs water and swells, disassembling and losing crystallinity and birefringence. Starch is much more susceptible to amylases when it is gelatinized compared to the native semicrystalline state (Capuano & Janssen, 2021). The main factors influencing the digestibility of starch are its morphology (granular size and shape), state (granular versus molecular) and structural organization (crystalline and amorphous) (Xu et al., 2019); and other factors, including starch modification (partial hydrolysis, heat-moisture treatment, annealing, chemical modification), as well as interactions between starch and other components of the food matrix (Toutounji et al., 2019; Chen et al., 2017).

11.3 INTERACTIONS OF STARCH WITH THE FOOD MATRIX

In a practical way, all parameters and processes involved in starch digestion can be classified as chemical or physical, or extrinsic and intrinsic factors (see Figure 11.1), which will be described next.

11.3.1 Extrinsic Factors Influencing Starch Digestion

11.3.1.1 Mechanical Processing

Food structure is key to starch digestion properties, and any disruption of the physical or botanical structure of cereal grains, legumes or roots can increase the rate of starch digestion and trigger larger metabolic responses. Hence, mechanical disruption of physical structures of foods is one of the diverse processing operations that impact food microstructure and therefore starch digestion and bioavailability. Mechanical processes to reduce particle size of foods, such mincing, crushing and milling, are effective ways of increasing the rate of starch digestion and postprandial glycemic response (Li et al., 2014a). Particle size reduction can occur at different stages of the processing chain, and this determines some specific changes in food. Milling, the first step in handcrafted grain and industry processing, often disrupts the physical barriers in the whole grain kernel by breaking the pericarp, bran layer, endosperm

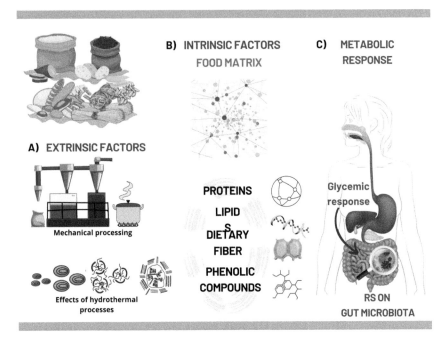

Figure 11.1   Factors influencing starch digestion. (A) Extrinsic factors: mechanical processing, effects of hydrothermal processes and storage on starch (native starch granule, granule swelling, gelatinization, retrogradation). (B) Intrinsic factors: Interaction of starch with the food matrix. Proteins (e.g., protein network embedding starch granules), lipids (e.g., starch–lipids complexes), dietary fiber and phenolic compounds. (C) Influence of the food matrix on the metabolic response (regulation of glycemic response, effect of resistant starch on human microbiota and metabolic functions).

tissue and embryo, and produces flours with wide variations in grain particle size, changes that consequently reduce the RS fraction, whereas the RDS fraction increases, with little change in the SDS fraction (Yu et al., 2015).

In vitro and in vivo studies have shown that particle size is inversely related to starch digestibility. Guo et al. (2018) investigated the physicochemical properties and in vitro starch digestibility of milled durum wheat grains with different particle sizes after heating with a wide range of flour:water ratio and concluded that in vitro enzymatic digestibility of starch in the flour was determined mostly by particle size. Reduction of particle size increases the surface area that is exposed to enzymes, which increases the digestion rate and elicits a higher glycemic and insulin response, whereas larger particles of whole and broken grains result in lower digestion rates and postprandial responses compared to fine flour (Edwards et al., 2015a, b). Similarly, Edwards et al. (2018) observed that when different particle size fractions (<0.25 mm, >0.25 mm, >0.5 mm and >1 mm) of raw smooth and raw wrinkled peas (*Pisum sativum* L.) were subjected to amylolysis using pancreatic α-amylase, the extent of starch digestion drastically decreased with increasing particle size. Guo et al. (2018) and Li et al. (2014a), indicated that grain milling increases starch susceptibility to digestibility because of both disruption of the surrounding plant tissue and damage to starch granules.

Several molecular and physicochemical events occur within a food during cooking due to the presence of heat and moisture. For instance, proteins may denature and aggregate, solidified fats may melt, and starch granules may gelatinize. The water activity – or the availability of water – is an important factor that determines the extent of starch digestibility through enzymatic hydrolysis. The K_m (Michaelis–Menten constant) of starch hydrolysis by α-amylase increased by a factor 2.5–4 at low moisture content compared to starch suspended in excess water. It has also been observed that V_{max} (maximal velocity) increased with increasing moisture content of the hydrolysis reaction medium (Slaughter et al., 2002). Processing results in an increase in the degree of in vitro starch hydrolysis in wheat, barley and oats (Bustos et al., 2017).

Starches in tubers and legumes are particularly well protected from the aqueous environment of luminal fluids, and even cereals such as wheat, may not have been accessed by α-amylase in the intestinal lumen unless they have been physically altered, which brings about a significant increase in starch digestibility due to gelatinization (Anguita et al., 2006). Cooking increases the rate of hydrolysis by gelatinizing the starch and making it more easily available for the enzymatic attack. The principal process facilitating starch availability for water penetration, and consequently eased α-amylase action, are physical processing and cooking by heating to 100°C for several minutes.

Bhattarai et al. (2016) found that pulses retained both cellular integrity and most of its ordered structure of entrapped starch granules, either of which slow the starch digestion rate, after the cooking process. Xiong et al. (2018) hypothesized on the factors that influence the kinetics of digestion, stating that crystalline structure is retained inside the cellular structure of whole pulses (e.g., pinto bean, garbanzo bean, green split pea, black-eyed peas), as a result from the limited swelling and gelatinization of the entrapped starch granules. Therefore, after a certain period of digestion, depending upon the natural porosity or wall penetration induced by the cooking process, enzymes could easily access the hydrated and partially gelatinized starch granules, thereby greatly increasing the digestion rate.

In the same sense, the multiscale structural orders of starch in cooked wheat flours are largely influenced by the water content during heating rather than by particle size. However, as mentioned earlier, in vitro enzymatic digestibility of starch in the flour is determined mostly by particle size, with little influence from the water content during cooking (Guo et al., 2018). Similarly, the change in water content of rice grains during cooking has been examined, showing that with an increase in cooking time the water content of the grains also increases. During this process, water molecules diffuse into the outer layer of the rice grains, and then gradually move into the inner regions, but the water diffusion rate is lower in the crystalline regions than in the amorphous regions (Kasai et al., 2005; Toutounji et al., 2019). The content of tightly bound water and free water decreases after black rice cooking, while the content of weakly bound water increases. This may be because amylose and amylopectin leach out of the granules during cooking and interact with free water molecules (Chen et al., 2019).

The impact of cooking on waxy and nonwaxy porous millet flours has been studied (Yang et al., 2019). Cooking did not change the crystalline type, but it progressively reduced the fraction of crystalline regions remaining in the starch. The waxy flours gelatinized earlier than the nonwaxy ones, indicating that the latter were more resistant to cooking.

It has been shown that cooking of starchy foods increases their RDS content but decreases the RS and SDS content (Hsu et al., 2015). This effect is attributed to gelatinization of the starch granules, inactivation of amylase inhibitors, and denaturation of the proteins surrounding the starch granules, changes that increase the ability of amylase to access and hydrolyze the starch molecules.

On the other hand, some rice varieties with similar high-amylose contents cooked for their minimum cooking time do not differ in starch digestibility. Thus, physicochemical properties of rice seem to be influenced by amylose content and amylose leaching during gelatinization, and may be a greater predictor of the starch-digestion rate than amylose content alone (Panlasigui et al., 1991; Wang et al., 2021).

During cooling or storage of cooked starchy foods, retrogradation of starch typically occurs. Starch retrogradation is the process by which disordered and disaggregated amylose and amylopectin chains (produced by gelatinization) partially reassociate to form more ordered assemblies (Capuano & Janssen, 2021). This results in the formation of starch fractions resistant to digestion (RS). Retrogradation can be divided into two stages: short term and long term (Wang et al., 2015; Wang & Copeland, 2013). The effect of food components (e.g., salts, dietary fiber, amino acids) on starch retrogradation depends on competition for water (Wang et al., 2015). Different factors affect starch retrogradation, including amylose content, the composition and structure of the starch molecules, and temperature (Singh et al., 2010).

Retrogradation by cold storage, and subsequent reordering of the starch molecules, has been demonstrated to lower the starch digestion rate (Li et al., 2014b). It was shown that the retrogradation process can reduce the starch digestibility of cooked rice, especially those cultivars with high levels of long linear amylopectin chains. In this case, the intrinsic biochemical properties of rice grains (i.e., starch supramolecular structure) had a significant impact on extrinsic processing (i.e., retrogradation), leading to reduced starch digestibility (Borah et al., 2017).

Retrograded amylose from peas, wheat and potatoes is highly resistant toward enzymatic hydrolysis. Storage of cooked food under refrigeration may lead to a reduction in their digestibility and in vitro estimated glycemic index (Frei et al., 2003). Potatoes were cooked and then cooled at refrigeration temperatures for up to 2 days before measuring their starch digestibility. Results indicated that the content of RDS decreased from 95% in the freshly cooked tubers to 45% in the refrigerated potatoes, which could be attributed to the retrogradation of starch and consequent RS starch formation. The dispersed polymers of the gelatinized starch during refrigerated storage have been observed to undergo retrogradation, which leads to the formation of semicrystalline structures that resist digestion by amylases (type 3 RS).

According to Guraya et al. (2001) when nonwaxy and waxy starch suspensions are debranched with pullulanase, followed by heating and cooling to allow crystallization or gelling, and then stirred, the digestibility decreases after cooling. This is probably due to the prevention or slowing of the formation of crystalline structures or double helices. Short-term retrogradation occurs as a result of the gelation and crystallization of the amylose fraction, leading to maximum SDS formation; in contrast, long-term retrogradation due to the amylopectin fraction occurs during storage of starch gels (Miao et al., 2010).

Processing techniques offer the possibility of improving starch nutritional functionality in processed food and foodstuffs through different combinations of moisture, temperature and time. Among these processing strategies, extrusion, autoclaving, baking, pressure-cooking, flaking and parboiling are known to influence starch availability and related physiological effects, through controlling gelatinization and food matrix interactions between starch and other components (Guo et al., 2018; Toutounji et al., 2019; Wang et al., 2021).

11.3.1.2 Cell Wall

One of the factors affecting starch and other macronutrient digestion is represented by structural barriers that prevent or delay the encounter between macromolecules and digestive enzymes. In plant-based foods, this structural barrier comprises the continuous network of cell walls. The cell wall provides mechanical resistance, support and protection to the plant. Since cell walls are made of nondigestible polysaccharides, they remain largely unaffected during digestion (Capuano & Pellegrini, 2019; Pallares et al., 2018). When encapsulated within intact cell walls, starch may be digested more slowly, as intact cells limit the access of α-amylase to starch (Gwala et al., 2019).

It has been reported that more than 15% of legume starch in legumes is not digested in the upper digestive tract due to physical inaccessibility, retrogradation or partial gelatinization; these starch fractions reach the colon and are fermented by the colonic microbiota (Tovar et al., 1992a; Paturi et al., 2021). The same may apply to cereals, but the amount escaping digestion in the upper tract will vary widely between different cereals, depending on the restriction to water and heat transfer imposed by the cell wall, which in turn affects starch gelatinization (Edwards et al., 2015b; Guo et al., 2018).

In cereals, starch is mainly located in the endosperm cellular compartment, which is embedded in a matrix formed by proteins and cell wall materials. Depending on the type of endosperm, dense packing of starch granules in the protein matrix of the vitreous endosperm significantly limits the rate of starch digestion (Bhattarai et al., 2018). Edwards et al. (2015a) investigated the effects of cell-wall encapsulation of wheat endosperm *on in vitro* starch bioaccessibility and digestion, as well as on glycemia, insulinemia and gut hormones in ileostomy participants. Their experiments clearly demonstrated that starch entrapped within the cells of the coarse wheat endosperm particles was hydrolyzed (by α-amylase) at a considerably lower rate than the exposed starch in the fine flour particles. Hence, the structural integrity of dietary fiber, as intact cell walls, can strongly influence the rate at which entrapped starch is made bioaccessible and thereby affect the time course of glucose absorption into the portal blood. The structural integrity of wheat endosperm is largely retained during gastroileal digestion and has a primary role in influencing the rate of starch amylolysis and, consequently, postprandial metabolism. In like manner, according to Dhital et al. (2017), the presence of an intact cell wall contributes to limiting starch digestion and also keeps a relatively tightly packed intracellular matrix in place, which represents an additional barrier for digestive enzymes.

Bhattarai et al. (2018) isolated individual intact cellular structures from cereals, wheat and sorghum, in order to elucidate the effect of intactness of cell walls on enzymatic hydrolysis of entrapped starch and showed that wheat and sorghum cell walls are effective barriers for the access of amylase, and that both an extensive protein matrix (particularly in sorghum) and noncatalytic binding of amylase to cell-wall surfaces can limit the amylolysis of starch within intact cells. Furthermore, the presence of incompletely gelatinized starch inside cooked intact cells suggests a limited swelling of cell-entrapped granules.

As dicots in general, pulses consist of three parts at the macrostructural level: seed coat, cotyledon and embryo. The starch and other macronutrients present in pulses are localized in plastids within the cotyledons, which in turn are surrounded by primary cell walls and a middle lamella. This complex, native structural organization plays a fundamental role during processing and subsequent digestion (Verkempinck et al., 2020).

Recent work by Xiong et al. (2018) showed that starch digestion in individual pulse cells is controlled by enzymatic accessibility (limited by the rigid cell wall and/or protein matrix) rather than by structural features of entrapped starch granules such as crystallinity and thermal parameters. Studies where the cell-wall barrier was disrupted using mechanical or enzymatic techniques showed that starch digestion rates are enhanced compared to the patterns observed when this digestive barrier is not disrupted or modified (Capuano & Pellegrini, 2019; Rovalino-Córdova et al., 2018; Verkempinck et al., 2020). Such evidence is in line with early studies showing that the magnitude of the postprandial glycemic and insulinemic responses to cooked legume seeds is largely dependent on the degree of process-induced disruption of cell-wall barriers (Tovar et al., 1992b).

In addition, scanning electron microscopy and chemical characterization revealed that enzymatic treatment modified cell-wall thickness and porosity without altering the cytoplasmic matrix, whereas mechanical treatment completely disrupted cell structure. Decreasing cell intactness increased the rate but not the extent of starch digestion in vitro. It was concluded that the cell wall serves as a permeable barrier limiting the access of digestive enzymes. The cytoplasmic matrix, on the other hand, further reduced the accessibility of amylase to starch affecting its hydrolysis rate. Also, it was proven that cell structural changes, if any, occurring during digestion had no effect on starch hydrolysis (Bhattarai et al., 2016).

Based on the results shown obtained by Rovalino-Córdova et al. (2018) about barriers affecting starch digestibility in beans, the compact organization of the cytoplasmic matrix is thought to represent an additional barrier to starch hydrolysis by α-amylase. The barrier effect exerted by the intact cell wall and also the packed cytoplasmic matrix proved to be key factors in the delay of starch hydrolysis in bean cotyledon cells. Pallares et al. (2018) demonstrated that the barrier role of cell walls during in vitro simulated digestion of starch in common bean cotyledon cells can be modified through variation of the thermal processing intensity. On the same line of evidence, Zahir et al. (2018) showed that the more severe the cell-wall damage in soybeans, the more completely emptied the cell after digestion.

In processed cereal foods, such as bread, bakery products and pasta, besides the cell wall, another structural barrier limits the digestibility of starch: a continuous protein network composed of glutenin and gliadin fractions (Capuano et al., 2018).

Bhattarai et al. (2017) showed the rate and extent of hydrolysis of starch and protein in legumes were greatly increased when the cell wall physical barrier was removed by either mechanical or enzymic processes. Furthermore, evidence for nonspecific binding of α-amylase on cell walls, hindering the hydrolysis of starch, was reported. This study suggested that the preservation of intactness of the cell wall, such as from legumes, could be a viable approach to achieve the targeted delivery of resistant starch to the colon.

Several investigations have suggested that it is possible that mechanical disruption of cell walls followed by a thermal treatment leads to increase starch digestibility as a result from deterioration of cell wall integrity combined with the release of otherwise physically inaccessible starch (Gwala et al., 2019; Wang et al., 2021; Zhu et al., 2019). Nevertheless, the cell wall of soybean cotyledons, as in other legumes, maintains their intact structure when the legumes are cooked as whole seeds (Zahir et al., 2018).

11.3.2 Intrinsic Factors Influencing Starch Digestion

11.3.2.1 Interaction of Starch with Other Constituents of the Food

The digestibility of starch can also be modulated by the presence of certain food components in the same starchy food or in other foods within the meal. The interaction of starch with proteins, lipids, nonstarch polysaccharides (e.g., cellulose, hemicellulose, gums) and other constituents of the food matrix may strongly impact its digestibility by increasing or decreasing the access of digestive enzymes to starch, and may also modify the sensory characteristics of the products (Bello-Perez et al., 2020).

11.3.2.2 Protein

Interactions between protein and starch in a food matrix play an important role in the rate of starch digestibility, since starch and protein usually coexist in foods.

First, the role of the cytoplasmic matrix in the starch digestion should be considered. The cytoplasmic matrix is referred to as the microstructural organization of cellular constituents, where globular starch granules are embedded in a protein matrix. This is a tightly packed network that has consequences on starch hydrolysis by limiting the surface area of interaction between enzymes and substrate, hindering α-amylase diffusion and catalysis (Rovalino-Córdova et al., 2018; Svihus et al., 2005). The proportion of protein increases toward the surface of the starch granule, and a large amount of the surface proteins are in the size range 5–60 kDa, while proteins associated with the internal zones of the starch granule are mainly in the 60–150 KDa size range (Nakamura, 2015).

Källman et al. (2013) observed a dense layer of protein matrix surrounding starch granules in poor malting barleys, while in good malting barleys, this layer was less apparent. Although this may indicate that the nature of the protein matrix and the interactions between the protein matrix and starch may affect starch digestibility, it is important to take into consideration that protein digestion usually precedes starch digestion. Thus, protein layers should be significantly degraded before starch digestion takes place. The number of broken starch granules is influenced by the strength of the bond between the starch and the protein matrix surrounding the starch granules. In maize, the bond between the protein and starch is quite strong, resulting in a large number of broken starch granules and great surface area.

Ye et al. (2018) reported a significant increase in starch digestibility after removal of proteins by protease treatment of native long-grain indica rice flour. It was suggested that the accessibility to the digestive enzymes was restricted due to attachment of endogenous proteins to the starch granule surface and a reduction in granular swelling.

Cooked rice is a common food matrix where protein and starch are the main components. In bread, pasta and bakery products, starch granules are also embedded in a continuous protein network composed of glutenin (Li et al., 2014a). Pasta has a more compact structure, with the gluten network tightly embedding the starch granules and thus limiting the contact with amylase (Zou et al., 2015). Bhattarai et al. (2016) found that gluten may be able to act as an inhibitor of α-amylase activity not only by being a physical barrier, but also by binding amylase and making it unavailable for starch hydrolysis. In this sense, the interaction between starch and protein in wheat bread reduced the availability of starch compared to gluten-free bread, which induces higher glucose response (Olawoye et al., 2020). In legumes, such as common beans (*Phaseolus vulgaris* L.), the protein matrix surrounding the starch granules appears to display physical interaction inside the cotyledon, which limits the accessibility of starch to amylolytic enzymes and thus reducing the rate of α-amylolysis (Dhital et al., 2016). Bhattarai et al. (2018) observed that the hydrolysis of entrapped starch and protein in legumes is affected by three distinct mechanisms: (1) residual molecular order in starches, (2) intactness of cells and (3) binding of enzymes to cell-wall components. In addition, interaction of starch and protein inside legume cells may also inhibit enzyme action since cells retain a compact structure of intracellular starch–protein even after the cell wall is removed.

11.3.2.3 Lipids

Lipids are present in natural foods as globules of different sizes that are dispersed in the aqueous cellular or extracellular environment and generally stabilized by multiple layers of emulsifying compounds. The relevance and presence of lipids in food is divided into three aspects, which indicate certain properties granted by these: quality (texture, lubrication, color and food taste), nutrition and biological features (Wissam, 2020).

The interactions between starch and lipids lead to the production of novel structures. Saturated monoacyl lipids or free fatty acids can form helicoidal inclusion complexes with amylose, whereas di- or triglycerides do not form complexes because of steric hindrance. The complexation of amylose with free fatty acids and monoglycerides results in significant changes in the physicochemical and nutritional properties of starch, including the change in X-ray diffraction pattern, reduced solubility, increased gelatinization temperature, retarded retrogradation and lowered digestibility (Miao et al., 2015). This amylose helicoidal complex with fatty acids exists in two forms: amorphous amylose–lipid complex (form I) and crystalline complex (form II) (Gelders et al., 2005).

The cognate lipids are associated with starch granules, which can be found on the surface (surface starch lipids) or inside the granule (true starch lipids) (Ai et al. 2013). Diverse studies have shown that starch digestibility may be reduced due to the formation of this type of complex between amylose and lipid molecules, since the network produced during gelation (food matrix) can restrict the accessibility of the polymer to the enzymes responsible for the hydrolysis process. Also, the interactions between phosphate groups and amylopectin chains change the starch structure in such a way that it is not efficiently recognized by amylolytic enzymes, thus becoming a bad substrate (Bhattarai et al., 2016; Toutounji et al., 2019; Wang et al., 2021).

Lipids in cereals are mainly stored in the form of spherical organelles, which are distributed in the aleurone layer plus the seed coat (~56%), in the embryo (~22%) and in the endosperm (~22%) (Nakamura, 2015). The most common fatty acids found in the starch granule from cereals usually consist of free fatty acids (FFAs) and lysophospholipids, associated in addition to amylose, to palmitic and linoleic acid. The effects of FFAs (lauric, myristic, palmitic, stearic and oleic acids) on the hydrolysis of starch using α-amylase and amyloglucosidase have been reported in the literature by Crowe et al. (2000). The addition of lauric, myristic, palmitic and oleic acids reduced the enzymatic hydrolysis of amylose by 35%. However, neither stearic acid nor cholesterol showed an inhibition. Lauric acid had no effect on the enzymatic breakdown of amylopectin, whereas the breakdown of whole starch was slightly inhibited by this acid.

In this context, Ye et al. (2018) investigated the starch digestibility of rice flour with and without the removal of lipids. The starch digestion of cooked rice flour containing endogenous lipids was significantly lower than samples that had lipids removed by petroleum ether extraction. In addition, the starch digestion of rice flour without lipids was slightly lower than the samples without proteins, despite the lipid content being ten times lower than protein content. Reduced starch digestibility was attributed to endogenous lipids attaching to the starch granular surface, thus reducing accessibility to enzymes. Furthermore, upon gelatinization of rice flour, endogenous lipids were suggested to reduce starch granular swelling, and thus reducing the substrate surface area for enzymatic degradation. It has been also observed that long-chain saturated monoglycerides are more resistant to enzymatic hydrolysis compared to short-chain saturated and unsaturated monoglycerides when complexed with cooked rice starch (Anderson & Guraya, 2006).

Bhattarai, Dhital, and Gidley (2016) observed a higher digestion in wheat flour when it was pre-exposed to α-amylase and pepsin, suggesting that the natural cereal endosperm interactions between starch, protein and lipids act as a barrier to enzymatic degradation. Ternary complex formation in starch–protein–FFA systems has been reported to involve three different structural elements: starch–FFA complex, protein–FFA complex and disulfide bond–linked protein aggregates (Zhang et al. 2003).

Starch–lipid complexes can be developed as a new type of resistant starch, via processing high-amylose starch with FFA. Rice containing this type of RS was found to be more thermostable (Kumar et al., 2018).

Besides FFA, other components of the food matrix may form complexes with starch, rendering it resistant to enzymatic hydrolysis. All these types of indigestible starch complexes, showing a V-type X-ray diffraction pattern, are included under the RS5 concept (Gutierrez & Tovar, 2021). Several authors have documented that processing conditions may increase the content of amylose–lipid complexes and thus may reduce digestibility of starch (Guo et al., 2018; Verkempinck et al., 2020; Zhu et al., 2019). This has been clearly noticed with high-amylose maize starch, where starch content in the lower small intestine, cecum and feces of rats increased substantially when this source of starch was heat-moisture treated (Saito et al., 2001). In fact, Hasjim et al. (2010) detected that starch–lipid complex in cereals had high melting temperatures ($\leq$126.6°C), and, as it has already been mentioned, they were resistant to enzyme hydrolysis.

11.3.2.4 Dietary Fiber
Analogously, to what was discussed for lipids and protein, the presence of dietary fiber (DF) in foods can modulate starch digestion. DF can be further subdivided into water-soluble (such as inulin, fructooligosaccharides, gums and pectins; many of them within the ability to create viscous solutions) or water insoluble (cellulose, some hemicellulose, lignin, cutin, suberin, chitin, chitosan and some resistant starches). Each type of DF has been associated with different physiological and health-beneficial effects, including reduction of the glycemic and insulinemic response to food, cholesterol-lowering action, and protective effects against colorectal cancer by the capacity to produce important levels of short-chain fatty acids, after fermentation in the colon (Goff et al., 2018; Sáyago-Ayerdi et al., 2011).

Many in vitro studies have revealed that interaction of starch with DF of various origins can delay amylolysis. Wu et al. (2016) observed that digestion of starch by salivary amylase was delayed when pectin was present. In fact, several researchers showed that DF can reduce the activity of pancreatic digestive enzymes including α-amylase. Dhital et al. (2017) have reported detailed kinetics of amylase inhibition by cellulose or wheat bran, showing that the inhibition of α-amylase activity was positively correlated with cellulose concentration, and that α-amylase was nonspecifically bound on the cellulose surface attenuating starch hydrolysis. It was thus proposed that DF can affect the catalytic efficiency of α-amylase through (1) interaction with substrate such as creation of DF–starch complexes, coating starch granules with DF and therefore acting as a physical barrier between substrate and enzyme; (2) interaction with enzymes, such as adsorption of enzymes to DF leading to noncompetitive inhibition; (3) minimizing water availability for starch hydrolysis and modifying thermal transition behavior of starch during processing, thereby lowering amylose leaching and maintaining native granular structure; (4) slowing diffusion of both enzymes and substrate due to increased viscosity of the reaction mixture and therefore slowing the frequency of interaction between enzymes and substrate; or (5) slowing diffusion of starch hydrolysis products, thereby leading to faster concentration rise of these amylase-inhibiting molecules in the reaction mixture (Sasaki et al., 2015; Shelat et al., 2010).

Fabek and Goff (2015) examined the effect of adding viscous soluble fibers on starch digestibility during simulated intestinal digestion. Starch solutions were prepared with native tapioca starch and different soluble fibers: guar gum, xanthan gum, soy-soluble polysaccharide and soluble flaxseed gum. Results showed that xanthan gum and guar gum were able to increase the digesta viscosity. Attenuation in the hydrolysis of starch was only observed for fibers that are able to retain a measurable viscosity during the simulated digestion, where those less viscous (during digestion) were much less effective. It is conceivable then that increasing the viscosity of the surrounding medium leads to less effective amylolysis of starch granules through interfering with the diffusion of enzymes onto the surface of starch granules. The effects may also be due to the ability of some gums to allow starch granules to agglomerate, thereby reducing the relative surface area exposed to digestive enzymes.

Brennan and Tudorica (2008) also reported that the rate of starch hydrolysis decreases when the starch granules and surrounding bread matrix are coated with a layer of galactomannan mucilage, which acts as a physical barrier to amylase–starch interactions and subsequent release of hydrolyzed products. In addition to its effect on digesta viscosity, guar gum may significantly reduce the rise in postprandial glycemic response as a result from a reduction in the rate of gastric emptying, which is also viscosity-sensitive.

Hence, incorporation of soluble, high molecular weight, poorly branched, rodlike indigestible polysaccharides in a food or meal would increase the viscosity of digesta in an almost dose-dependent fashion. The resulting delay of diffusive or convective transport of starch, amylase, and starch hydrolysis products to, and through, the mucus layer would determine the observed effects in some metabolic responses.

Indeed, the encapsulation of starch granules in a fiber network, such as biomimicking of a plant cell or tissue, decreases the substrate accessibility to enzymes and modulates the digestibility. Whole-grain kernels also have a fiber rich multiple-layered bran that may decrease the accessibility of hydrolytic enzymes to some starch granules. The botanical structure of whole-grain kernels seems a natural way to produce a physically related type of SDS. The native-form β-glucans in oat grains also encapsulate protein and starch to decrease the enzyme accessibility (Zhang et al., 2017). On the other hand, an alginate-based encapsulation of corn starch resulted in a slowly digestible microsphere with digestion site shift, from proximal to distal regions of the intestinal tract (Venkatachalam et al., 2009). Therefore, the content of SDS can be modulated according to the biopolymer type and concentration, starch type, and microsphere size (Miao & Hamaker, 2021).

The amount of DF that solubilizes can vary upon food processing because of changes in DF structure or in the architecture of the cell wall. As discussed earlier, starch in cooked, whole legumes is retained within rigid cell walls of cotyledon cells. This, together with a high DF content, results in a rather low rate of starch digestion (Sáyago-Ayerdi et al., 2011).

Bustos et al. (2017) evaluated in vitro digestion kinetics and bioaccessibility of starch in various cereal food products (white bread, gluten-free bread, sheeted pasta, extruded whole-grain spaghetti, whole-wheat cookies and peach cookies) and concluded that soluble DF reduced starch digestibility by changing the microstructure of food or by limiting water availability as a consequence of the soluble nonstarch polysaccharide hydration, which will restrict starch gelatinization (Sozer et al., 2014). In this context, malting of cassava flour with legumes decreased in vitro starch digestibility of cassava flour biscuits enriched with wheat and rice bran (Jisha et al., 2010). Such effect was also observed by Sáyago-Ayerdi et al. (2011) with tortilla prepared by blending maize with beans (20%), which exhibited a notably higher RS content. The combination of starchy foods with high RS content and/or the addition of natural RS sources or DF to common food products could provide potential health benefits related to increased RS intake.

11.3.2.5 Phenolic Compounds

Starch digestibility depends on and can be modulated, as it has been discussed before, by interactions with the macronutrients present in the food matrix.

The presence of various types of molecules, such as phenolic compounds (PCs), in different concentrations and proportions has received special attention because of their potential dietary and physiological role, mainly due to their bioactive action as antioxidants. However, some direct interactions between PCs and digestive enzymes have been demonstrated.

The α-amylase and α-glucosidase mutually can contribute to modulate their kinetics in the presence of PCs that establish noncovalent interactions with starch (Takahama & Hirota, 2018). The inhibition of these enzymes by PCs can retard the increase of postprandial blood glucose.

The relevance of the inhibition of these enzymes has attracted attention to the different mechanisms through what PCs can operate. Monomeric PCs can inactivate these enzymes by blocking the catalytic sites, but the polymeric phenolics are able to precipitate starch, forming nondigestible complexes (Barrett et al., 2018). Even more, PCs present in high concentrations may be able to interact with other components of the food matrix and form larger complexes (Amoako & Awika, 2016).

The action of dietary PCs, especially flavonoids, on the activity of α-amylase and thus retarding starch digestion, has been reported (Sun & Miao, 2020). This inhibitory activity is highly related to the structure of the phenolic compound. In the molecular structure of flavonoids, the hydroxyl (-OH) groups in the 5, 6 and 7 positions of the A ring and 4′ position in the B ring are able to exert an inhibitory action in the active site of the α-amylase (Miao et al., 2014). Besides this, the conjugation of the 4-carbonyl with the 2,3 double bonds also plays a role, as it enhances the electron delocalization between the A and C rings, which facilitates the stability of π-π assembling in the flavonoid aromatic ring and the tryptophan from the active site of α-amylase (Lo Piparo et al., 2008). The galloyl groups in the C ring have also been reported to increase the association of tea catechins and gallotannins with α-amylase. The catalytic amino acid residue (i.e., Glu[233]) and hydrophobic π-π conjugation between the galloyl benzene ring and the active site tryptophan can establish strong noncovalent interactions (Cao et al., 2020) (see Figure 11.2).

The methods used to evaluate these interactions are, generally, inhibition analysis (IC_{50} value and inhibition kinetics), spectroscopy analysis (fluorescence quenching), circular dichroism, Fourier transform infrared spectroscopy (FTIR), thermodynamic analysis (isothermal titration calorimetry) and differential scanning calorimetry (Sun et al., 2019). More recently, molecular docking has become a method used frequently to simulate binding sites between the PC molecules and enzymes. Molecular docking predicts the optimal relative orientation of a molecule with the active site of the enzyme. This technique is used to characterize the ligand–protein interaction mode through the amino acid residue involved in the binding with the PC (Wu et al., 2018).

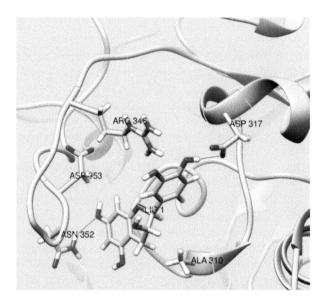

Figure 11.2   Docking of α-amylase with catechin interaction.

11.4 INFLUENCE OF THE FOOD MATRIX ON THE METABOLIC RESPONSE

The metabolic response to starchy foods depends on the food matrix. After the action of α-amylase and α-glucosidase during the digestion process, the released monosaccharides are able to cross the intestinal barrier. As mentioned before, the increase of postprandial glucose after starch digestion is the basis for its classification as a glycemic carbohydrate (Ming Miao & Hamaker, 2021). The concepts of glycemic index (GI) and glycemic load (GL) are based on the widely accepted, but nonetheless crude, measurement of the glycemic response expressed as the incremental area under the blood glucose curve (IAUC) for 2–3 h after a meal. These two measures express different aspects of glycemic potency (Monro & Shaw, 2008). The GI refers to the glycemic effect of available carbohydrate in food relative to the effect of an equal amount of glucose (or starch in white wheat bread) (Hardy et al., 2020). The quantity of food used in measuring GI is not necessarily a customary intake, but it delivers the same amount of carbohydrates (usually 50 g) as the reference product. Using the GI concept it is possible to classify different carbohydrate-rich foods with respect to their effect on postmeal glycemia. Using glucose as reference, foods can be classified into three main categories: low (<55), medium (55–69) and high (>70) GI (Foster-Powell et al., 2002). Foods with a high glycemic index are generally associated with obesity and noncommunicable diseases such as type 2 diabetes and cardiovascular diseases (Olawoye et al., 2020).

The GL of a typical serving of food is the product of the amount of available carbohydrate in that serving and the GI of the food (GL = GI × Available carbohydrate) (Singh et al., 2010). It is widely known that the continual ingestion of high GI and GL diets, which usually contain large amounts of high glycemic and refined carbohydrates and low fiber foods, over time are associated with increased risk of cardiometabolic diseases and mortality (Bhupathiraju et al., 2014; Hardy et al. 2020; Zhu et al., 2019).

Since the chronic diseases associated with hyperglycemia represent leading public health concerns worldwide, it is relevant to identify carbohydrate food materials that may contribute to daily blood glucose management. In addition to cereals and pulses, there are several starchy food materials in traditional diets; however, the postprandial glycemic response of them is rarely reported (Zhu et al., 2019). There is considerable interest in lowering the GI of highly digestible starch-rich products in order to promote reduced increases in blood glucose levels after meals (Gianluca et al., 2018). SDS, as a nutritional starch fraction that lies between RDS and RS, is digested slowly and possibly throughout the entire small intestine, thus providing a sustained glucose release corresponding to the defining characteristic of low GI foods (Ming Miao & Hamaker, 2021). The importance of this fraction (SDS) resides in that its ingestion results in a moderate postprandial blood glucose peak with a concomitant low insulinemic response, an important issue since the control of glycemic and insulinemic postprandial responses is of relevance not only for diabetics but also for healthy people (Breyton et al., 2021).

The physical structure of the food matrix is a characteristic that may contribute to a slow digestion of starch and may be controlled through the cooking degree. Pasta, whole grains and legumes, for instance, exhibit a structure that results in a slow starch digestion rate and slow release of glucose, which may be explained by the content of RS/SDS (Jenkins et al., 2002). However, there are reports indicating that some varieties of waxy whole grains could elicit a high glycemic response comparable to glucose, and domestic culinary methods such as prolonged boiling, baking, frying, and roasting have substantial impact on the glycemic response of reported low-GI starchy grains. In comparison to refined grains, they are rich in dietary fiber, RS, antioxidants, and other important micronutrients such as folic acid and other vitamins. These components of whole grain have relevant functional properties that can at least, in part, justify its health benefits (Della Pepa et al., 2018).

The relatively low GI of legumes is largely due to the presence of a thick and resistant cell wall limiting the contact of starch granules with amylase, as it was discussed in a previous section. Cell walls and intracellular proteins, however, are not the only structural components that can act as a barrier for amylase (Capuano & Janssen, 2021). In bread and pasta, on the other hand, starch granules are embedded in a gluten network. Pasta has a more compact structure, with the gluten network tightly embedding the starch granules, which would limit the contact with amylase. This is apparently one of the reasons for the relatively lower GI of pasta compared to bread. It has been suggested that a limited gelatinization of the starch granules located in the inner core of spaghetti may also contribute to the relatively low GI of pasta (Rovalino-Córdova et al., 2018; Zou et al., 2015).

Phenolic compounds, such as epigallocatechin gallate, tea phenolic extracts and chlorogenic acid, have demonstrated hypoglycemic effect in vivo, in part to their inhibitory activity on α-amylase (Li et al., 2019). The enzyme inhibition by PCs retards the increase in postprandial glucose and insulin levels after the ingestion of starchy foods. However, it is mainly the inhibition of the hydrolysis of the disaccharides by α- glucosidase that is responsible for the reduction in the postprandial response. Specifically, phlorizin has been suggested to retard the starch digestion (Li et al., 2019). PCs are able to inhibit the transport of glucose to the enterocytes by inhibiting the expression of the SGLT1 and GLUT2 transporters (Villa-Rodriguez et al., 2018). The reformulation of gluten-free bakery products is increasing the levels of starch in their formulations, using generally high-amylose starch (>60% amylose) in levels of 20% to 50% w/w. This replacement reduces the in vitro estimated GI (Giuberti et al., 2016). Similarly in a corn-based flatbread (arepa), an increased RS content was observed when high-amylose starch was used to substitute between 25% and 70% of regular corn, with marked impact on the glycemic response to the breads (Granfeldt et al., 1993). Besides, the ingestion of bread with palmitic acid–complexed RS5 resulted in substantially lower postprandial plasma–glucose and insulin responses in human subjects (Ai et al., 2013; Hasjim et al., 2010; Li et al., 2014a).

In addition to the impact of the interaction of starch with lipids present in the same food, further effects on starch digestibility and metabolic responses may be observed when different foods or ingredients are combined in meal, leading to another degree of food matrix complexity. This has been exemplified recently for oat-based drinks added with a conventional oil (rapeseed) or an oat oil preparation enriched in polar lipids (mainly galactolipids). When tested in a cohort of healthy volunteers, both oil-containing beverages promoted lower postprandial glucose and insulin responses than the reference drink (with no added oil), but the polar lipid–rich preparation exhibited a stronger effect (Hossain et al., 2021). Hence, the interactions of starch with cognate components of the food matrix and with other constituents in mixed meals can influence the metabolic response, a possibility that may be exploited for the formulation of commercial products targeting specific sectors of the population with particular nutritional needs.

11.5 EFFECT OF RESISTANT STARCH ON HUMAN GUT MICROBIOTA AND METABOLIC FUNCTIONS

Resistant starch (RS) is defined as the fraction of starch that escapes the small intestinal digestion and absorption (Asp, 1996). The increasing interest about RS as a fermentable nondigestible carbohydrate resides in its potential to modify the gut microbiota and different metabolic functions. The metabolites produced during the colonic fermentation of RS by gut microbiota, mainly short-chain fatty acids, have been shown capable to maintain colon health and their functional properties extended to extracolonic aspects of metabolism.

The already mentioned five classes of RS may coexist in the same food, and their overall action depends on the relative proportion in which they reach the human colon.

RS is an important substrate for gut bacteria, producing short-chain fatty acids, including butyrate, propionate, valerate, isovalerate and hexanoate (Upadhyaya et al., 2016). In the human colon, the Firmicutes (gram positive) phylum is mostly a butyrate producer; Bacteroidetes (gram negative) are mainly acetate and propionate producers. The utilization of energy from RS depends on the presence of specialist primary degraders such as *Bifidobacterium adolescentis* and *Rumminococcus bromii* with proven utilization of RS2 and RS3 as sources of energy. *Eubacterium rectale* and *Bacteroides thetaiotamicrom* species have shown limited ability to utilize RS2 and RS3 (Ze et al., 2012). The dietary modulation of gut microbiota has been demonstrated in intervention studies, showing that some circulating biochemical and proinflammatory markers, as well as certain anthropometric parameters, improved in patients with metabolic syndrome after intervention with RS4, where the *Bacteroides* species of *Parabacteroides, Oscillospira, Blautia, Ruminococcus, Eubacterium* and *Christensenella* proliferated in the RS4 group (Upadhyaya et al., 2016).

Butyric acid is absorbed and metabolized in the intestinal epithelial cells of the colon and immune cells through induction of intracellular or extracellular processes. It can suppress cancer cell proliferation and induce apoptotic cell death (Parada Venegas et al., 2019); restore the intestinal barrier function in inflammatory conditions in vitro; and is the main source of energy for colonocytes, which utilize more than 70% of their oxygen consumption for butyrate oxidation (Geirnaert et al., 2017). Butyric acid exerts anti-inflammatory effects in the intestinal mucosa, through the inhibition of histone deacetylases and by activating G-protein–coupled receptors of intestinal epithelial cells and immune cells, and suppressing lipopolysaccharide (LPS)-induced NF-KB activation (Parada Venegas et al, 2019). The production of butyrate by fermentation of RS by gut microbiota enhances the physiologic relevance of this metabolite in inflammation, transcription in colonic mucosa and metabolic responses.

There is still much to learn about the impact of DF and RS on different metabolic aspects, and their association with the gut microbiota composition and activity. But, for instance, an important reduction in postprandial glucose responses after the intake of whole grain barley breads, rich in DF and RS, has been shown. However, the magnitude of this beneficial effect seems to be related to the relative abundance of gut *Prevotella* species and an increased *Prevotella:Bacteroides* ratio (Sandberg et al., 2019). These observations stress the importance of further research on the link between carbohydrate contents, the form in which they are presented in the diet and their utilization by the host and the gut microbiota.

11.6 CONCLUSIONS AND FUTURE PERSPECTIVES

The historical approach to the study of the nutritional value of foods, based on their composition in terms of macro- and micronutrients, has proven insufficient to explain different metabolic aspects linked to the impact of the intake of complex foods on physiological processes. In the particular case of starch, there are multiple factors that may affect the digestibility of the polymer and the metabolic responses triggered by its intake. The simultaneous presence of other components of the food matrix leads to a variety of interactions that may limit both the rate and efficiency of starch digestion, with concomitant influence on the ileal absorption of the digestion products and their metabolic handling. Physical barriers represented by intact botanical architecture, cell walls and protein networks have been shown to hamper starch digestion, although their significance diminishes with thermomechanical processing. Beyond that, starch molecules establish complexes with certain types of lipids, proteins, nonstarch polysaccharides and phenolic compounds, resulting in structures of variable degree of digestibility. Such complexes may exist naturally in the food, but they may also arise as consequence of processing. Overall, restricted starch digestibility and absorption are followed by moderate postprandial glycemic and insulinemic responses with important metabolic effects, features that may be exploited for development of dietary strategies aiming for reduction of cardiometabolic risk in both healthy individuals and diabetic patients.

Two main research topics appear of immediate interest in this context. One relates to the fate of indigestible starch fractions reaching the large intestine and their influence on the colonic microbiota in terms of composition, activity and production of fermentation metabolites, whose metabolic implications for the host are yet to be fully understood. And additionally, besides the effects of individual food items, the second topic is related to the influence of combining different food items in complex meals – such as the ones normally consumed – on the overall starch digestibility and metabolic responses, which deserves more investigation. Further efforts will surely be devoted to these research questions in the coming years.

REFERENCES

Aguilera, J. M. (2019). The food matrix: Implications in processing, nutrition and health. *Critical Reviews in Food Science and Nutrition*, 59, 3612–3629.

Ai, Y. F., Hasjim, J., & Jane, J. L. (2013). Effects of lipids on enzymatic hydrolysis and physical properties of starch. *Carbohydrate Polymers*, 92(1), 120–127. https://doi .org/10.1016/j.carbpol.2012.08.092.

Amoako, D., & Awika, J. M. (2016). Polyphenol interaction with food carbohydrates and consequences on availability of dietary glucose. *Current Opinion in Food Science*, 8, 14–18. https://doi.org/10.1016/J.COFS.2016.01.010.

Anderson, A. K., & Guraya, H. S. (2006). Effects of microwave heat-moisture treatment on properties of waxy and non-waxy rice starches. *Food Chemistry*, 97(2), 318–323. https://doi.org/10.1016/j.foodchem.2005.04.025.

Anguita, M., Gasa, J., Martín-Orúe, S. M., & Pérez, J. F. (2006). Study of the effect of technological processes on starch hydrolysis, non-starch polysaccharides solubilization and physicochemical properties of different ingredients using a two-step in vitro system. *Animal Feed Science and Technology*, 129(1), 99–115. https://doi.org /10.1016/j.anifeedsci.2005.12.004.

Asp, N.-G. (1996). Dietary carbohydrates: Classification by chemistry and physiology. *Food Chemistry*, 57(1), 9–14. https://doi.org/10.1016/0308-8146(96)00055-6.

Barrett, A. H., Farhadi, N. F., & Smith, T. J. (2018). Slowing starch digestion and inhibiting digestive enzyme activity using plant flavanols/tannins: A review of efficacy and mechanisms. *LWT - Food Science and Technology*, 87, 394–399. https://doi. org/10.1016/J.LWT.2017.09.002.

Bello-Perez, L. A., Flores-Silva, P. C., Agama-Acevedo, E., & Tovar, J. (2020). Starch digestibility: Past, present, and future. *Journal of the Science of Food and Agriculture*. 100, 5009–5016. https://doi.org/10.1002/jsfa.8955.

Bhattarai, R. R., Dhital, S., & Gidley, M. J. (2016). Interactions among macronutrients in wheat flour determine their enzymic susceptibility. *Food Hydrocolloids*, 61, 415–425. https://doi.org/10.1016/j.foodhyd.2016.05.026.

Bhattarai, R. R., Dhital, S., Wu, P., Chen, X. D., & Gidley, M. J. (2017). Digestion of isolated legume cells in a stomach-duodenum model: Three mechanisms limit starch and protein hydrolysis. *Food and Function*, 8(7), 2573–2582. https://doi.org/10.1039/ c7fo00086c.

Bhattarai, R. R., Dhital, S., Mense, A., Gidley, M. J., & Shi, Y. C. (2018). Intact cellular structure in cereal endosperm limits starch digestion in vitro. *Food Hydrocolloids*, 81, 139–148. https://doi.org/10.1016/j.foodhyd.2018.02.027.

Bhupathiraju, S. N., Tobias, D. K., Malik, V. S., Pan, A., Hruby, A., Manson, J. E., Willett, W. C. and Hu, F. B. (2014). Glycemic index, glycemic load, and risk of type 2 diabetes: Results from 3 large US cohorts and an updated meta-analysis. *American Journal of Clinical Nutrition*, 100(1), 218–232. https://doi.org/10.3945/ajcn.113.079533.

Borah, P. K., Deka, S., & Duary, R. K. (2017). Effect of repeated cycled crystallization on digestibility and molecular structure of glutinous bora rice starch. *Food Chemistry*, 223, 31–39. https://doi.org/10.1016/j.foodchem.2016.12.022.

Brennan, C. S., & Tudorica, C. M. (2008). Evaluation of potential mechanisms by which dietary fibre additions reduce the predicted glycaemic index of fresh pastas. *International Journal of Food Science and Technology*, 43(12), 2151–2162. https://doi.org/10.1111/j.1365-2621.2008.01831.x.

Breyton, A. E., Goux, A., Lambert-Porcheron, S., Meynier, A., Sothier, M., VanDenBerghe, L., Brack, O., et al. (2021). Starch digestibility modulation significantly improves glycemic variability in type 2 diabetic subjects: A pilot study. *Nutrition, Metabolism and Cardiovascular Diseases*, 31(1), 237–246. https://doi.org/10.1016/j.numecd.2020.08.010.

Bustos, M. C., Vignola, M. B., Pérez, G. T., & León, A. E. (2017). In vitro digestion kinetics and bioaccessibility of starch in cereal food products. *Journal of Cereal Science*, 77, 243–250. https://doi.org/10.1016/j.jcs.2017.08.018.

Cao, J., Zhang, Y., Han, L., Zhang, S., Duan, X., & Sun, L. (2020). Number of galloyl moieties and molecular flexibility are both important in alpha-amylase inhibition by galloyl-based polyphenols. *Food & Function*, 11, 3838–3850.

Capuano, E., & Janssen, A. E. M. (2021). Food matrix and macronutrient digestion. *Annual Review of Food Science and Technology*, 12(1), 1–20. https://doi.org/10.1146/annurev-food-032519-051646.

Capuano, E., & Pellegrini, N.. (2019). An integrated look at the effect of structure on nutrient bioavailability in plant foods. *Journal of the Science of Food and Agriculture*, 99(2), 493–498. https://doi.org/10.1002/jsfa.9298.

Capuano, E., Oliviero, T., Fogliano, V., & Pellegrini, N. (2018). Role of the food matrix and digestion on calculation of the actual energy content of food. *Nutrition Reviews*, 76(4), 274–289. https://doi.org/10.1093/NUTRIT/NUX072.

Chávez-Mendoza, C., & Sánchez, E. (2017). Bioactive compounds from Mexican varieties of the common bean (Phaseolus vulgaris): Implications for health. *Molecules*, 22(8). https://doi.org/10.3390/molecules22081360.

Chen, L., Zhang, H., McClements, D. J., Zhang, Z., Zhang, R., Jin, Z., & Tian, Y. (2019). Effect of dietary fibers on the structure and digestibility of fried potato starch: A comparison of pullulan and pectin. *Carbohydrate Polymers*, 215, 47–57. https://doi.org/10.1016/j.carbpol.2019.03.046.

Chen, M.-H., Bergman, C. J, McClung, A. M., Everette, J. D., & Tabien, R. E. (2017). Resistant starch: Variation among high amylose rice varieties and its relationship with apparent amylose content, pasting properties and cooking methods. *Food Chemistry*, 234, 180–189. https://doi.org/10.1016/j.foodchem.2017.04.170.

Crowe, T. C., Seligman, S. A., & Copeland, L. (2000). Inhibition of enzymic digestion of amylose by free fatty acids in vitro contributes to resistant starch formation. *The Journal of Nutrition*, 130(8). https://doi.org/10.1093/jn/130.8.2006.

Dhital, S., Bhattarai, R. R., Gorham, J., & Gidley, M. J. (2016). Intactness of cell wall structure controls the in vitro digestion of starch in legumes. *Food & function*, 7(3), 1367–1379. https://doi.org/10.1039/c5fo01104c

Dhital, S., Warren, F. J., Butterworth, P. J., Ellis, P. R., & Gidley, M. J. (2017). Mechanisms of starch digestion by α-amylase: Structural basis for kinetic properties. *Critical Reviews in Food Science and Nutrition*, 57(5), 875–892. https://doi.org/10.1080/10408398.2014.922043.

Edwards, C. H., Grundy, M. M. L., Grassby, T., Vasilopoulou, D., Frost, G. S., Butterworth, P. J., Berry, S. E. E., Sanderson, J., & Ellis, P. R. (2015a). Manipulation of starch bioaccessibility in wheat endosperm to regulate starch digestion, postprandial glycemia, insulinemia, and gut hormone responses: A randomized controlled trial in healthy ileostomy participants. *American Journal of Clinical Nutrition*, 102(4), 791–800. https://doi.org/10.3945/ajcn.114.106203.

Edwards, C. H., Warren, F. J., Campbell, G. M., Gaisford, S., Royall, P. G., Butterworth, P. J., & Ellis, P. R. (2015b). A study of starch gelatinisation behaviour in hydrothermally-processed plant food tissues and implications for in vitro digestibility. *Food and Function*, 6(12), 3634–3641. https://doi.org/10.1039/c5fo00754b.

Edwards, C. H., Maillot, M., Parker, R., & Warren, F. J. (2018). A comparison of the kinetics of in vitro starch digestion in smooth and wrinkled peas by porcine pancreatic alpha-amylase. *Food Chemistry*, 244, 386–393. https://doi.org/10.1016/j.foodchem.2017.10.042.

Englyst, H., Kingman, S., & Cummings, J. (1992). Classification and measurement of nutritionally important starch fractions. *European Journal of Clinical Nutrition*, 46 (Suppl 2, November), S33–S50.

Fabek, H., & Goff, H. D. (2015). Simulated intestinal hydrolysis of native tapioca starch: Understanding the effect of soluble fibre. *Bioactive Carbohydrates and Dietary Fibre*, 6(2), 83–98. https://doi.org/10.1016/j.bcdf.2015.09.008.

Foster-Powell, K., Holt, S. H. A., & Brand-Miller, J. C. (2002). International table of glycemic index and glycemic load values: 2002. *American Journal of Clinical Nutrition*, no. January, 5–56.

Frei, M., Siddhuraju, P., & Becker, K. (2003). Studies on the in vitro starch digestibility and the glycemic index of six different indigenous rice cultivars from the Philippines. *Food Chemistry*, 395–402. https://doi.org/10.1016/S0308-8146(03)00101-8.

Gao, C., Rao, M., Huang, W., Wan, Q., Yan, P., Long, Y., Guo, M., Xu, Y., & Xu, Y. (2019). Resistant starch ameliorated insulin resistant in patients of type 2 diabetes with obesity: A systematic review and meta-analysis. *Lipids in Health and Disease*, 18(1), 205. https://doi.org/10.1186/s12944-019-1127-z.

Geirnaert, A., Calatayud M., Grootaert, C., Laukens, D., Devriese, S., Smagghe, G., De Vos, M., Boon, N., & Van de Wiele, T. (2017). Butyrate-producing bacteria supplemented in vitro to Crohn's disease patient microbiota increased butyrate production and enhanced intestinal epithelial barrier integrity. *Scientific Reports*, 7(1), 11450. https://doi.org/10.1038/s41598-017-11734-8.

Gelders, G. G., Duyck, J. P., Goesaert, H., & Delcour, J. A. (2005). Enzyme and acid resistance of amylose-lipid complexes differing in amylose chain length, lipid and complexation temperature. *Carbohydrate Polymers*, 60(3), 379–389. https://doi.org/10.1016/j.carbpol.2005.02.008.

Giuberti, G., & Gallo, A. (2018). Reducing the glycaemic index and increasing the slowly digestible starch content in gluten-free cereal-based foods: A review. *International Journal of Food Science and Technology*, 53(1), 50–60. https://doi.org/10.1111/ijfs.13552.

Giuberti, G., Gallo, A., Fiorentini, L., Fortunati, P., & Masoero, F. (2016). In vitro starch digestibility and quality attributes of gluten free 'tagliatelle' prepared with teff flour and increasing levels of a new developed bean cultivar. *Starch-Stärke*, 68, 374–378. https://doi.org/doi:10.1002/star.201500007.

Goff, H. D., Repin, N., Fabek, H., El Khoury, D., & Gidley, M. J. (2018). Dietary fibre for glycaemia control: Towards a mechanistic understanding. *Bioactive Carbohydrates and Dietary Fibre*, 14, 39–53. https://doi.org/10.1016/j.bcdf.2017.07.005.

Granfeldt, Y., A. Drews, & I. Bjorck. (1993). Starch bioavailability in arepas made from ordinary or high amylose corn: Concentration and gastrointestinal fate of resistant starch in rats. *Journal of Nutrition*, 123, 1676–1684.

Guo, P., Jinglin, Y., Wang, S., Wang, S., & Copeland, L. (2018). Effects of particle size and water content during cooking on the physicochemical properties and in vitro starch digestibility of milled durum wheat grains. *Food Hydrocolloids*, 77, 445–453. https://doi.org/10.1016/j.foodhyd.2017.10.021.

Guraya, H. S., James, C., & Champagne, E. T. (2001). Effect of enzyme concentration and storage temperature on the formation of slowly digestible starch from cooked debranched rice starch. *Starch–Stärke*, 53, 131–139.

Gutiérrez, T. J., & Tovar, J. (2021). Update of the concept of type 5 resistant starch (rs5): Self-assembled starch V-type complexes. *Trends in Food Science and Technology*, 109, 711–724. https://doi.org/10.1016/j.tifs.2021.01.078.

Gwala, S., Wainana, I., Pallares, A. P., Kyomugasho, C., Hendrickx, M., & Grauwet, T. (2019). Texture and interlinked post-process microstructures determine the in vitro starch digestibility of bambara groundnuts with distinct hard-to-cook levels. *Food Research International*, 120(February), 1–11. https://doi.org/10.1016/j.foodres.2019.02.022.

Hardy, D. S., Garvin, J. T., & Xu, H. (2020). Carbohydrate quality, glycemic index, glycemic load and cardiometabolic risks in the US, Europe and Asia: A dose–response meta-analysis. *Nutrition, Metabolism and Cardiovascular Diseases*, 30(6), 853–871. https://doi.org/10.1016/j.numecd.2019.12.050.

Hasek, L. Y., Phillips, R. J., Zhang, G., Kinzig, K. P., Kim, C. Y., Powley, T. L., & Hamaker, B. R. (2018). Dietary slowly digestible starch triggers the gut–brain axis in obese rats with accompanied reduced food intake. *Molecular Nutrition and Food Research*, 62(5), 1–22. https://doi.org/10.1002/mnfr.201700117

Hasjim, J., Lee, S. O., Hendrich, S., Setiawan, S., Ai, Y., & Jane, J. L. (2010). Characterization of a novel resistant-starch and its effects on postprandial plasma-glucose and insulin responses. *Cereal Chemistry*, 87(4), 257–262. https://doi.org/10.1094/CCHEM-87-4-0257.

Hossain, M. M., J. Tovar, L. Cloetens, M. Soria Florido, K. Petersson, O. Prothon, & A. Nilsson. (2021). Oat polar lipids improve cardiometabolic-related markers after breakfast and a subsequent standardized lunch: A randomized crossover study in healthy young adults. *Nutrients*, 13, 988. https://doi.org/10.3390/nu13030988.

Hsu, R. J.-C., Chen, H.-J., Lu, S., & Chiang, W. (2015). Effects of cooking, retrogradation and drying on starch digestibility in instant rice making. *Journal of Cereal Science*, 65, 154–161. https://doi.org/10.1016/j.jcs.2015.05.015.

Jenkins, D. J. A, Kendall, C. W. C., Augustin, L. S. A., Franceschi, A. L., Hamidi, M., & Marchie, A. (2002). Glycemic index: Overview of implications in health and disease. *American Journal of Clinical Nutrition*, 76, 266–273.

Jisha, S., Padmaja, G., & Sajeev, M. S. (2010). Nutritional and textural studies on dietary fiber-enriched muffins and biscuits from cassava-based composite flours. *Journal of Food Quality*, 33(Suppl. 1), 79–99. https://doi.org/10.1111/j.1745-4557.2010.00313.x.

Källman, A., Bertoft, E.,, Koch, K., Åman, P., & Andersson, R. (2013). On the interconnection of clusters and building blocks in barley amylopectin. *International Journal of Biological Macromolecules*, 55, 75–82. https://doi.org/10.1016/j.ijbiomac.2012.12.032.

Kasai, M., Lewis, A., Marica, F., Ayabe, S., Hatae, K., & Fyfe, C. A. (2005). NMR imaging investigation of rice cooking. *Food Research International*, 38(4), 403–410. https://doi.org/10.1016/j.foodres.2004.10.012.

Kellett, G. L., & Brot-Laroche, E. (2005). Apical GLUT2: A major pathway of intestinal sugar absorption. *Diabetes*, 54(October), 3056–3062.

Kumar, A., Sahoo, U., Baisakha, B., Okpani, O. A., Ngangkham, U., Parameswaran, C., Basak, N., Kumar, G., & Sharma, S. G.. (2018). Resistant starch could be decisive in determining the glycemic index of rice cultivars. *Journal of Cereal Science*, 79: 348–353. https://doi.org/10.1016/j.jcs.2017.11.013.

Lee, B. H., Rose, D. R., Lin, A. H. M., Quezada-Calvillo, R., Nichols, B. L., & Hamaker, B. R. (2016). Contribution of the individual small intestinal α-glucosidases to digestion of unusual α-linked glycemic disaccharides. *Journal of Agricultural and Food Chemistry*, 64(33), 6487–6494. https://doi.org/10.1021/acs.jafc.6b01816.

Li, D., Sun, L. J., Yang, Y. L., Wang, Z. C., Yang, X., & Zhao, T. (2019). Young apple polyphenols postpone starch digestion in vitro and in vivo. *Journal of Functional Foods*, 56, 127–135.

Li, E., Dhital, S., & Hasjim, J. (2014a). Effects of grain milling on starch structures and flour/starch properties. *Starch/Staerke*, 66(1–2), 15–27. https://doi.org/10.1002/star.201200224.

Li, J., Han, W., Xu, J., Xiong, S., & Zhao, S. (2014b). Comparison of morphological changes and in vitro starch digestibility of rice cooked by microwave and conductive heating. *Starch – Starke*, 66(5–6), 549–557. https://doi.org/10.1002/star.201300208.

Lin, A., Lee, B.-H, & Chang, W.-J. (2015). Small intestine mucosal α-glucosidase: A missing feature of in vitro starch digestibility. *Food Hydrocolloids*, 53.

Martinez, M. M. (2021). Starch nutritional quality: Beyond intraluminal digestion in response to current trends. *Current Opinion in Food Science*, 38, 112–121. https://doi.org/10.1016/j.cofs.2020.10.024.

Miao, M., & Hamaker, B. R.. (2021). Food matrix effects for modulating starch bioavailability. *Annual Review of Food Science and Technology*, 12(1), 1–23. https://doi.org/10.1146/annurev-food-070620-013937.

Miao, M., Zhang, T., Mu, W., & Jiang, B. (2010). Effect of controlled gelatinization in excess water on digestibility of waxy maize starch. *Food Chemistry*, 119(1), 41–48. https://doi.org/10.1016/j.foodchem.2009.05.035.

Miao, M., Jiang, B., Cui, S., Zhang, T., & Jin, Z., (2015) Slowly Digestible Starch—A Review, *Critical Reviews in Food Science and Nutrition*, 55(12), 1642–1657. doi: 10.1080/10408398.2012.704434

Miao, M., Jiang, H., Jiang, B., Zhang, T., Cui, S. W., & Jin, Z. (2014). Phytonutrients for controlling starch digestion: Evaluation of grape skin extract. *Food Chemistry*, 145, 205–211. https://doi.org/10.1016/j.foodchem.2013.08.056

Monk, J. M., Wu W., Lepp, D., Wellings, H. R., Hutchinson, A. L., Liddle, D. M., Graf, D., Peter Pauls, K., Robinson, L. E., & Power, K. A.. (2019). Navy bean supplemented high-fat diet improves intestinal health, epithelial barrier integrity and critical aspects of the obese inflammatory phenotype. *Journal of Nutritional Biochemistry*, 70, 91–104. https://doi.org/10.1016/j.jnutbio.2019.04.009.

Monro, J. A., & Shaw, M. (2008). Glycemic impact, glycemic glucose equivalents, glycemic index, and glycemic load: Definitions, distinctions, and implications.

American Journal of Clinical Nutrition, 87(1), 237–243. https://doi.org/10.1093/ajcn /87.1.237s.

Nakamura, Y.. (2015). Biosynthesis of reserve starch. In Nakamura, Yasunori (Ed.), *Starch: Metabolism and Structure* (pp. 425–441). Japan: Springer. https://doi.org /10.1007/978-4-431-55495-0_13.

Olawoye, B., Gbadamosi, S. O., Otemuyiwa, I. O., & Akanbi, C. T. (2020). Gluten-free cookies with low glycemic index and glycemic load: Optimization of the process variables via response surface methodology and artificial neural network. *Heliyon*, 6(10), e05117. https://doi.org/10.1016/j.heliyon.2020.e05117.

Pallares, A. P., Miranda, B. A., Truong, N. Q. A., Kyomugasho, C., Chigwedere, C. M., Hendrickx, M., & Grauwet, T. (2018). Process-induced cell wall permeability modulates the: In vitro starch digestion kinetics of common bean cotyledon cells. *Food and Function*, 9(12), 6544–6554. https://doi.org/10.1039/c8fo01619d.

Panlasigui, L. N., Thompson, L. U., Juliano, B. O., Perez, C. M., Yiu, S. H., & Greenberg, G. R. (1991). Rice varieties with similar amylose content differ in starch digestibility and glycemic response in humans. *American Journal of Clinical Nutrition*, 54(5), 871–877. https://doi.org/10.1093/ajcn/54.5.871.

Parada Venegas, D., De la Fuente, M. K., Landskron, G., González, M. J., Quera, R., Dijkstra, G., Harmsen, H. J. M., Faber, K. N., & Hermoso, M. A. (2019). Short chain fatty acids (SCFAS)-mediated gut epithelial and immune regulation and its relevance for inflammatory bowel diseases. *Frontiers in Immunology*. https://www .frontiersin.org/article/10.3389/fimmu.2019.00277.

Paturi, G., Mishra, S., Hedderley, D. I, & Monro, J. A. (2021). Gut microbiota responses to dietary fibre sources in rats fed starch-based or quasi-human background diets. *Journal of Functional Foods*, 83, 104565. https://doi.org/10.1016/j.jff.2021.104565.

Pepa, G. D., Vetrani, C., Vitale, M., & Riccardi, G. (2018). Wholegrain intake and risk of type 2 diabetes: Evidence from epidemiological and intervention studies. *Nutrients*, 10(9). https://doi.org/10.3390/nu10091288.

Piparo, E. Lo, H. Scheib, N. Frei, G. Williamson, M. G., & Chou, C. J.(2008). Flavonoids for controlling starch digestion: Structural requirements for inhibiting human alpha-amylase. *Journal of Medicinal Chemistry*, 51(12), 3555–3561. https://doi.org /10.1021/jm800115x.

Pletsch, E. A., & Hamaker, B. R. (2018). Brown rice compared to white rice slows gastric emptying in humans. *European Journal of Clinical Nutrition*, 72(3), 367–373. https://doi.org/10.1038/s41430-017-0003-z.

Rovalino-Córdova, A. M., Fogliano, V., & Capuano, E. (2018). A closer look to cell structural barriers affecting starch digestibility in beans. *Carbohydrate Polymers*, 181, 994–1002. https://doi.org/10.1016/j.carbpol.2017.11.050.

Saito, K., Ito, T., Kuribayashi, T., Mochida, K., Nakakuki, T., Shibata, M., & Sugawara, M. (2001). Effect of raw and heat-moisture treated high-amylose corn starch on fermentation by the rat cecal bacteria. *Starch/Staerke*, 53(9), 424–430. https://doi .org/10.1002/1521-379X(200109)53:9<424::AID-STAR424>3.0.CO;2-J.

Sandberg, J., Kovatcheva-Datchary, P., Björck, I., Bäckhed, F., & Nilsson, A. (2019). Abundance of gut prevotella at baseline and metabolic response to barley prebiotics. *European Journal of Nutrition*, 58, 2365–2376. https://doi.org/10.1007/ s00394-018-1788-9.

Sasaki, T., Sotome, I., & Okadome, H. (2015). In vitro starch digestibility and in vivo glucose response of gelatinized potato starch in the presence of non-starch polysaccharides. *Starch/Staerke*, 67(5–6), 415–423. https://doi.org/10.1002/star .201400214.

Sáyago-Ayerdi, S. G., Tovar, J., Blancas-Benítez, F. J., & Bello-Pérez, L. A. (2011). Resistant starch in common starchy foods as an alternative to increase dietary fibre intake. *Journal of Food and Nutrition Research*, 50(1), 1–12.

Shelat, K. J., Vilaplana, F., Nicholson, T. M., Wong K. H., Michael, J. G., & Gilbert, R. G. (2010). Diffusion and viscosity in arabinoxylan solutions: Implications for nutrition. *Carbohydrate Polymers*, 82(1), 46–53. https://doi.org/10.1016/j.carbpol.2010 .04.019.

Singh, J., Dartois, A., & Kaur, L. (2010). Starch digestibility in food matrix: A review. *Trends in Food Science and Technology*, 21(4), 168–180. https://doi.org/10.1016/j .tifs.2009.12.001.

Slaughter, S. L., Ellis, P. R., Jackson, E. C., & Butterworth, P. J. (2002). The effect of guar galactomannan and water availability during hydrothermal processing on the hydrolysis of starch catalysed by pancreatic alpha-amylase. *Biochimica et Biophysica Acta*, 1571(1), 55–63. https://doi.org/10.1016/s0304-4165(02)00209-x.

Sozer, N., Cicerelli, L., Heiniö, R. L., & Poutanen, K. (2014). Effect of wheat bran addition on invitro starch digestibility, physico-mechanical and sensory properties of biscuits. *Journal of Cereal Science*, 60(1), 105–113. https://doi.org/10.1016/j.jcs .2014.01.022.

Sun, L. J., Warren, F. J., & Gidley, M. J. (2019). Natural products for glycaemic control: Polyphenols as inhibitors of alpha-amylase. *Trends in Food Science & Technology*, 91, 262–273. https://doi.org/10.1016/j.tifs.2019.07.009

Sun, L., & Miao, M. (2020). Dietary polyphenols modulate starch digestion and glycaemic level: A review. *Critical Reviews in Food Science and Nutrition*, 60(4), 541–555. https://doi.org/10.1080/10408398.2018.1544883.

Svihus, B., Uhlen, A. K., & Harstad, O. M. (2005). Effect of starch granule structure, associated components and processing on nutritive value of cereal starch: A review. *Animal Feed Science and Technology*, 122(3–4), 303–320. https://doi.org/10.1016/j .anifeedsci.2005.02.025.

Takahama, U., & Hirota, S. (2018). Interactions of flavonoids with α-amylase and starch slowing down its digestion. *Food & Function*, 9(2), 677–687. https://doi.org/10. 1039/C7FO01539A.

Toutounji, M. R., Farahnaky, A., Santhakumar, A. B., Oli, P., Butardo, V. M., & Blanchard, C. L.. (2019). Intrinsic and extrinsic factors affecting rice starch digestibility. *Trends in Food Science and Technology*, 88(December 2018), 10–22. https://doi.org /10.1016/j.tifs.2019.02.012.

Tovar, J., Björck, I., & Asp, N.-G. (1992a). Effect of processing on blood glucose and insulin responses to starch in legumes. *Journal of Agricultural and Food Chemistry*, 40, 1846–1851. https://doi.org/10.1021/jf00022a024.

Tovar, J., Björck, I., & Asp, N.-G. (1992b). Incomplete digestion of legume starches in rats: A study of precooked flours containing retrograded and physically inaccessible starch fractions. *The Journal of Nutrition*, 122(7), 1500–1507. https://doi.org /10.1093/jn/122.7.1500.

Upadhyaya, B., McCormack, L., Fardin-Kia, A. R., Juenemann, R., Nichenametla, S., Clapper, J., Specker, B., & Dey, M. (2016). Impact of dietary resistant starch type 4 on human gut microbiota and immunometabolic functions. *Scientific Reports*, 6(June), 28797. https://doi.org/10.1038/srep28797.

Venkatachalam, M., Kushnick, M. R., Zhang, G., & Hamaker, B. R. (2009). Starch entrapped biopolymer microspheres as a novel approach to vary blood glucose profiles. *Journal of the American College of Nutrition*, 28(5), 583–590. https://doi.org/https://doi.org/10.1080/07315724.2009.10719790

Verkempinck, S., Pallares, A. P., Hendrickx, M., & Grauwet, T. (2020). Processing as a tool to manage digestive barriers in plant-based foods: Recent advances. *Current Opinion in Food Science*, 35, 1–9. https://doi.org/10.1016/j.cofs.2019.11.007.

Villa-Rodriguez, J. A., Kerimi, A., Abranko, L., Tumova, S., Ford, L., Blackburn, R. S., Rayner, C., & Williamson, G. (2018). Acute metabolic actions of the major polyphenols in chamomile: An in vitro mechanistic study on their potential to attenuate postprandial hyperglycaemia. *Scientific Reports*, 8, 5471. https://doi.org/10.1038/s41598-018-23736-1.

Wang, S., & Copeland, L. (2013). Molecular disassembly of starch granules during gelatinization and its effect on starch digestibility: A review. *Food and Function*, 4(11), 1564–1580. https://doi.org/10.1039/c3fo60258c.

Wang, S., Li, C., Copeland, L.,, Niu, Q., & Wang, S. (2015). Starch retrogradation: A comprehensive review. *Comprehensive Reviews in Food Science and Food Safety*, 14(5), 568–585. https://doi.org/10.1111/1541-4337.12143.

Wang, Y., Chen, L., Yang, T., Ma, Y., McClements, D. J., Ren, F., Tian, Y., & Jin, Z. (2021). A review of structural transformations and properties changes in starch during thermal processing of foods. *Food Hydrocolloids*, 113(July 2020), 106543. https://doi.org/10.1016/j.foodhyd.2020.106543.

Wissam, Z. (2020). Structured lipids: Synthesis, health effects, and nutraceutical applications. In Galanakis, C. M. (Ed.), *Lipids and Edible Oils* (pp. 289–327). Academic Press. https://doi.org/10.1016/B978-0-12-817105-9.00008-2.

Wu, P., Dhital, S., Williams B. A., Chen, X. D., & Gidley, M. J.. (2016). Rheological and microstructural properties of porcine gastric digesta and diets containing pectin or mango powder. *Carbohydrate Polymers*. 148, 216–226. https://doi.org/10.1016/j.carbpol.2016.04.037.

Wu, Q., Peng, Z., Zhang, Y., & Yang, J.. (2018). COACH-D: Improved protein–ligand binding sites prediction with refined ligand-binding poses through molecular docking. *Nucleic Acids Research*, 46, W438–W442.

Xiong, W., Zhang, B., Huang, Q., Li, C., Pletsch, E. A., & Fu, X. (2018). Variation in the rate and extent of starch digestion is not determined by the starch structural features of cooked whole pulses. *Food Hydrocolloids*, 83(January), 340–347. https://doi.org/10.1016/j.foodhyd.2018.05.022.

Xu, X., Chen, Y., Luo, Z., & Lu, X. (2019). Different variations in structures of A- and B-type starches subjected to microwave treatment and their relationships with digestibility. *LWT*, 99, 179–187. https://doi.org/10.1016/j.lwt.2018.09.072.

Yang, Q., Liu, L., Zhang, W., Li, J., Gao, X., & Feng, B. (2019). Changes in morphological and physicochemical properties of waxy and non-waxy proso millets during cooking process. *Foods*, 8(11). https://doi.org/10.3390/foods8110583.

Ye, J., Hu, X., Luo, S., McClements, D. J., Liang, L., & Liu, C. (2018). Effect of endogenous proteins and lipids on starch digestibility in rice flour. *Food Research International*, 106, 404–409. https://doi.org/10.1016/j.foodres.2018.01.008.

Yu, J., Wang, S., Wang, J., Li, C., Xin, Q., Huang, W., Zhang, Y., He, Z., & Wang, S. (2015). Effect of laboratory milling on properties of starches isolated from different flour millstreams of hard and soft wheat. *Food Chemistry*, 172, 504–514. https://doi.org/10.1016/j.foodchem.2014.09.070

Zahir, M., Fogliano, V., & Capuano, E. (2018). Food matrix and processing modulate: In vitro protein digestibility in soybeans. *Food and Function*, 9(12), 6326–6336. https://doi.org/10.1039/c8fo01385c.

Ze, X., Duncan, S. H., Louis, P., & Flint, H. J. (2012). Ruminococcus Bromii is a keystone species for the degradation of resistant starch in the human colon. *ISME Journal*, 6(8), 1535–1543. https://doi.org/10.1038/ismej.2012.4.

Zhang, G, Maladen, M. D., & Hamaker, B. R. (2003). Detection of a novel three component complex consisting of starch, protein, and free fatty acids. *Journal of Agricultural and Food Chemistry*, 51(9), 2801–2805. https://doi.org/doi:10.1021/jf030035.

Zhang, J., Luo, K., & Zhang, G. (2017). Impact of native form oat β-glucan on starch digestion and postprandial glycemia. *Journal of Cereal Science*, 73, 84–90. https://doi.org/https://doi.org/10.1016/j.jcs.2016.11.013

Zhu, R., Fan, Z., Han, Y., Li, S., Li, G., Wang, L., Ye, T., & Zhao, W. (2019). Acute effects of three cooked non-cereal starchy foods on postprandial glycemic responses and in vitro carbohydrate digestion in comparison with whole grains: A randomized trial. *Nutrients*, 11(3), 634–648. https://doi.org/10.3390/nu11030634.

Zou, W., Sissons, M., Gidley, M. J., Gilbert, R. G., & Warren, F. J. (2015). Combined techniques for characterizing pasta structure reveals how the gluten network slows enzymic digestion rate. *Food Chemistry*, 188, 559–568. https://doi.org/10.1016/j.foodchem.2015.05.032.

Index

W

Water solubility index (WSI), 88–93,
    227, 263
Wet extraction, 28
Wet milling, 24, 27, 31
Wheat, anatomy of, 19
Whole grains, 22, 24, 199, 252, 277, 279,
    292, 295, 307, 310, 313
Whole grain starches, 204–205

X

Xanthosoma sagittifolium, 253
X-ray diffractometry (XRD), 5, 49–50

Y

Yam tubers, 2
Yuca, 2